Annals of Mathematics Studies
Number 207

Berkeley Lectures on p-adic Geometry

Peter Scholze and Jared Weinstein

PRINCETON UNIVERSITY PRESS

PRINCETON AND OXFORD

2020

Published by Princeton University Press
41 William Street, Princeton, New Jersey 08540
6 Oxford Street, Woodstock, Oxfordshire OX20 1TR

press.princeton.edu

ISBN 978-0-691-20209-9
ISBN (pbk.) 978-0-691-20208-2
ISBN (e-book) 978-0-691-20215-0

British Library Cataloging-in-Publication Data is available

Editorial: Susannah Shoemaker
Production Editorial: Brigitte Pelner
Production: Jacqueline Poirier
Publicity: Matthew Taylor (US) and Katie Lewis (UK)

This book has been composed in LaTeX

The publisher would like to acknowledge the authors of this volume for providing the print-ready files from which this book was printed.

10 9 8 7 6 5 4 3 2 1

Contents

Foreword

This is a revised version of the lecture notes for the course on p-adic geometry given by P. Scholze in Fall 2014 at UC Berkeley. At a few points, we have expanded slightly on the material, in particular so as to provide a full construction of local Shimura varieties and general moduli spaces of shtukas, along with some applications to Rapoport-Zink spaces, but otherwise we have tried to keep the informal style of the lectures.

Let us give an outline of the contents:

In the first half of the course (Lectures 1–10) we construct the category of *diamonds*, which are quotients of perfectoid spaces by so-called pro-étale equivalence relations. In brief, diamonds are to perfectoid spaces as algebraic spaces are to schemes.

- Lecture 1 is an introduction, explaining the motivation coming from the Langlands correspondence and moduli spaces of shtukas.

- In Lectures 2–5 we review the theory of adic spaces [Hub94].

- In Lectures 6–7 we review the theory of perfectoid spaces [Sch12].

- In Lectures 8–10 we review the theory of diamonds [Sch17].

In the second half of the course (Lectures 11–25), we define spaces of mixed-characteristic local shtukas, which live in the category of diamonds. This requires making sense of products like $\operatorname{Spa} \mathbf{Q}_p \times S$, where S is an adic space over $\mathbf{F}_p$.

- In Lecture 11 we give a geometric meaning to $\operatorname{Spa} \mathbf{Z}_p \times S$, where S is a perfectoid space in characteristic p, and we define the notion of a mixed-characteristic local shtuka.

- In Lectures 12–15, we study shtukas with one leg, and their connection to p-divisible groups and p-adic Hodge theory.

- In Lecture 16, we prove the analogue of Drinfeld's lemma for the product $\operatorname{Spa} \mathbf{Q}_p \times \operatorname{Spa} \mathbf{Q}_p$.

- In Lectures 17–23, we construct a moduli space of shtukas for any triple $(G, b, \{\mu_1, \ldots, \mu_m\})$, for any reductive group $G/\mathbf{Q}_p$, any σ-conjugacy class

b, and any collection of cocharacters μ_i. This moduli space is a diamond, which is fibered over the m-fold product of $\mathrm{Spa}\,\mathbf{Q}_p$. Proving this is somewhat technical; it requires the technology of v-sheaves developed in Lecture 17.

- In Lecture 24, we show that our moduli spaces of shtukas specialize (in the case of one leg) to local Shimura varieties, which in turn specialize to Rapoport-Zink spaces. For this we have to relate local shtukas to p-divisible groups.

- In Lecture 25, we address the question of defining integral models for local Shimura varieties.

Since 2014, some of the material of this course has found its way to other manuscripts which discuss it in more detail, in particular [Sch17], and we will often refer to these references. In particular, the proper foundations on diamonds can only be found in [Sch17]; here, we only survey the main ideas in the same way as in the original lectures. In this way, we hope that this manuscript can serve as an informal introduction to these ideas.

During the semester at Berkeley, Laurent Fargues formulated his conjecture on the geometrization of the local Langlands conjecture, [Far16], which is closely related to the contents of this course, but leads to a radical change of perspective. We have kept the original perspective of the lecture course in these notes, so that Fargues' conjecture does not make an explicit appearance.

Acknowledgments. We thank the University of California at Berkeley for the opportunity to give these lectures and for hosting us in Fall 2014. Moreover, we thank all the participants of the course for their feedback, and we would like to thank especially Brian Conrad and João Lourenço for very detailed comments and suggestions for improvements. Part of this work was done while the first author was a Clay Research Fellow.

June 2019   Peter Scholze, Jared Weinstein

Lecture 1

Introduction

1.1 MOTIVATION: DRINFELD, L. LAFFORGUE, AND V. LAFFORGUE

The starting point for this course is Drinfeld's work [Dri80] on the global Langlands correspondence over function fields. Fix $X/\mathbf{F}_p$ a smooth projective curve, with function field K. The Langlands correspondence for GL_n/K is a bijection $\pi \mapsto \sigma(\pi)$ between the following two sets (considered up to isomorphism):

- Cuspidal automorphic representations of $\mathrm{GL}_n(\mathbf{A}_K)$, where $\mathbf{A}_K$ is the ring of adèles of K, and
- Irreducible representations $\mathrm{Gal}(\overline{K}/K) \to \mathrm{GL}_n(\overline{\mathbf{Q}}_\ell)$.

Whereas the global Langlands correspondence is largely open in the case of number fields K, it is a theorem for function fields, due to Drinfeld ($n = 2$, [Dri80]) and L. Lafforgue (general n, [Laf02]). The key innovation in this case is Drinfeld's notion of an X-*shtuka* (or simply shtuka, if X is clear from context).

Definition 1.1.1. A *shtuka* of rank n over an $\mathbf{F}_p$-scheme S is a pair $(\mathcal{E}, \varphi_\mathcal{E})$, where $\mathcal{E}$ is a rank n vector bundle over $S \times_{\mathbf{F}_p} X$ and $\varphi_\mathcal{E} \colon \mathrm{Frob}_S^* \mathcal{E} \dashrightarrow \mathcal{E}$ is a meromorphic isomorphism which is defined on an open subset $U \subset S \times_{\mathbf{F}_p} X$ that is fiberwise dense in X. Here, $\mathrm{Frob}_S \colon S \times_{\mathbf{F}_p} X \to S \times_{\mathbf{F}_p} X$ refers to the product of the pth power Frobenius map on S with the identity on X.

The Langlands correspondence for X is obtained by studying moduli spaces of shtukas.

Suppose we are given a shtuka $(\mathcal{E}, \varphi_\mathcal{E})$ of rank n over $S = \mathrm{Spec}\, k$, where k is algebraically closed. We can attach to it the following data:

1. The collection of points $x_1, \ldots, x_m \in X(k)$ where $\varphi_\mathcal{E}$ is undefined. We call these points the *legs* of the shtuka.
2. For each $i = 1, \ldots, m$, a conjugacy class μ_i of cocharacters $\mathbf{G}_\mathrm{m} \to \mathrm{GL}_n$, encoding the behaviour of $\varphi_\mathcal{E}$ near x_i.

The second item deserves some explanation. Let $x \in X(k)$ be a leg of the shtuka, and let $t \in \mathcal{O}_{X,x}$ be a uniformizing parameter at x. We have the completed stalks $(\mathrm{Frob}_S^* \mathcal{E})_x^\wedge$ and $\mathcal{E}_x^\wedge$. These are free rank n modules over

$\mathcal{O}_{X,x}^{\wedge} \cong k[\![t]\!]$, whose generic fibers are identified using $\varphi_{\mathcal{E}}$. In other words, we have two $k[\![t]\!]$-lattices in the same n-dimensional $k(\!(t)\!)$-vector space.

By the theory of elementary divisors, there exists a basis $e_1, \ldots, e_n$ of $\mathcal{E}_x^{\wedge}$ such that $t^{k_1} e_1, \ldots, t^{k_n} e_n$ is a basis of $(\mathrm{Frob}_S^* \mathcal{E})_x^{\wedge}$, where $k_1 \geq \cdots \geq k_n$. These integers depend only on the shtuka. Another way to package this data is as a conjugacy class μ of cocharacters $\mathbf{G}_{\mathrm{m}} \to \mathrm{GL}_n$, via $\mu(t) = \mathrm{diag}(t^{k_1}, \ldots, t^{k_n})$. Either way, we have one such datum for each of the legs $x_1, \ldots, x_m$.

Thus there are some discrete data attached to a shtuka: the number of legs m, and the ordered collection of cocharacters $(\mu_1, \ldots, \mu_m)$. Fixing these, we can define a moduli space $\mathrm{Sht}_{\mathrm{GL}_n, \{\mu_1, \ldots, \mu_m\}}$ whose k-points classify the following data:

1. An m-tuple of points $(x_1, \ldots, x_m)$ of $X(k)$.
2. A shtuka $(\mathcal{E}, \varphi_{\mathcal{E}})$ of rank n with legs $x_1, \ldots, x_m$, for which the relative position of $\mathcal{E}_{x_i}^{\wedge}$ and $(\mathrm{Frob}_S^* \mathcal{E})_{x_i}^{\wedge}$ is bounded by the cocharacter μ_i for all $i = 1, \ldots, m$.

It is known that $\mathrm{Sht}_{\mathrm{GL}_n, \{\mu_1, \ldots, \mu_m\}}$ is a Deligne-Mumford stack. Let

$$f \colon \mathrm{Sht}_{\mathrm{GL}_n, \{\mu_1, \ldots, \mu_m\}} \to X^m$$

map a shtuka onto its m-tuple of legs. One can think of $\mathrm{Sht}_{\mathrm{GL}_n, \{\mu_1, \ldots, \mu_m\}}$ as an equal-characteristic analogue of Shimura varieties, which are fibered over $\mathrm{Spec}\, \mathbf{Z}$ (or more generally over $\mathrm{Spec}\, \mathcal{O}_E[1/N]$, where E is a number field). But of course Shimura varieties are not fibered over anything like "$\mathrm{Spec}\, \mathbf{Z} \times_{\mathbf{F}_1} \mathrm{Spec}\, \mathbf{Z}$." In this sense the function field theory is more complete.

One can add level structures to these spaces of shtukas, parametrized by finite closed subschemes $N \subset X$ (that is, effective divisors). A level N structure on $(\mathcal{E}, \varphi_{\mathcal{E}})$ is then a trivialization of the pullback of $\mathcal{E}$ to N in a way which is compatible with $\varphi_{\mathcal{E}}$.

As a result we get a family of stacks $\mathrm{Sht}_{\mathrm{GL}_n, \{\mu_1, \ldots, \mu_m\}, N}$ and morphisms

$$f_N \colon \mathrm{Sht}_{\mathrm{GL}_n, \{\mu_1, \ldots, \mu_m\}, N} \to (X \backslash N)^m.$$

The stack $\mathrm{Sht}_{\mathrm{GL}_n, \{\mu_1, \ldots, \mu_m\}, N}$ carries an action of $\mathrm{GL}_n(\mathcal{O}_N)$, by altering the trivialization of $\mathcal{E}$ on N. The inverse limit $\varprojlim_N \mathrm{Sht}_{\mathrm{GL}_n, \{\mu_1, \ldots, \mu_m\}, N}$ even admits an action of $\mathrm{GL}_n(\mathbf{A}_K)$, via Hecke correspondences.

Recall that our motivation was the Langlands correspondence, which connects cuspidal automorphic representations of $\mathrm{GL}_n(\mathbf{A}_K)$ with ℓ-adic representations of $\mathrm{Gal}(\overline{K}/K)$. To do this, we consider the middle cohomology of the $\mathrm{Sht}_{\mathrm{GL}_n, \{\mu_1, \ldots, \mu_m\}, N}$. Let d be the relative dimension of f, and consider the cohomology $R^d(f_N)_! \overline{\mathbf{Q}}_\ell$, an étale $\overline{\mathbf{Q}}_\ell$-sheaf on X^m.

Before carrying out our analysis of $R^d(f_N)_! \overline{\mathbf{Q}}_\ell$, let us consider a simpler sort of object, namely a $\overline{\mathbf{Q}}_\ell$-sheaf $\mathbb{L}$ on X^m, which becomes lisse when restricted to U^m for some dense open subset $U \subset X$. Then we can think of $\mathbb{L}$ as a representation of the étale fundamental group $\pi_1(U^m)$ on an $\overline{\mathbf{Q}}_\ell$-vector space.[1] Ultimately we want to relate this to $\pi_1(U)$, because this is a quotient of $\mathrm{Gal}(\overline{K}/K)$. There is a natural homomorphism

$$\pi_1(U^m) \to \pi_1(U) \times \cdots \times \pi_1(U) \ (m \text{ copies}).$$

This map isn't surjective, because the target has m different Frobenius elements, while the source only has one. After extending the base field from $\mathbf{F}_p$ to $\overline{\mathbf{F}}_p$, this map indeed becomes surjective. But (regardless of the base) the map is not injective.[2]

These problems can be addressed by introducing *partial Frobenii*. For $i = 1, \dots, m$, we have a partial Frobenius map $F_i \colon X^m \to X^m$, which is Frob_X on the ith factor, and the identity on each other factor. For an étale morphism $V \to X^m$, let us say that a system of partial Frobenii on V is a commuting collection of isomorphisms $F_i^* V \cong V$ over X^m (and whose product is the relative Frobenius of $V \to X^m$). Finite étale covers of U^m equipped with partial Frobenii form a Galois category, and thus (recall we have already chosen a base point) they are classified by continuous actions of a profinite group $\pi_1(U^m/\text{partial Frob.})$ on a finite set.

Lemma 1.1.2 ([Dri80, Theorem 2.1]). *The natural map*

$$\pi_1(U^m/\text{partial Frob.}) \to \pi_1(U) \times \cdots \times \pi_1(U) \ (m \text{ copies})$$

is an isomorphism.

The lemma shows that if $\mathbb{L}$ is a $\overline{\mathbf{Q}}_\ell$-local system on U^m, which comes equipped with commuting isomorphisms $F_i^* \mathbb{L} \cong \mathbb{L}$, then $\mathbb{L}$ determines a representation of $\pi_1(U)^m$ on a $\overline{\mathbf{Q}}_\ell$-vector space. Studying the geometry of the moduli space of shtukas, one finds that it (essentially) admits partial Frobenii morphisms lying over the F_i, and therefore so does its cohomology.

Remark 1.1.3. We cannot literally apply this lemma to our sheaf $R^d(f_N)_! \overline{\mathbf{Q}}_\ell$ as it is not constructible (f_N is not of finite type, or even quasi-compact) and not a priori lisse on any subset of the form U^m. Drinfeld considers a "bounded" variant of this sheaf and shows that it extends to a lisse sheaf on $(X \backslash N)^m$.

Passing to the limit over N, one gets a big representation of $\mathrm{GL}_n(\mathbf{A}_K) \times \mathrm{Gal}(\overline{K}/K) \times \cdots \times \mathrm{Gal}(\overline{K}/K)$ on $\varinjlim_N R^d(f_N)_! \overline{\mathbf{Q}}_\ell$. Roughly, the way one expects

---

[1] Here and elsewhere in this introduction, we ignore base points.

[2] The standard counterexample is the Artin-Schreier cover $x^p - x = st$ of the product $\mathrm{Spec}\, \mathbf{F}_p[s] \times_{\mathbf{F}_p} \mathrm{Spec}\, \mathbf{F}_p[t]$, which corresponds to an $\mathbf{F}_p$-valued character of $\pi_1(\mathbf{A}_{\mathbf{F}_p}^2)$ which does not factor through $\pi_1(\mathbf{A}_{\mathbf{F}_p}^1)^2$. Generally, if X and Y are connected varieties over an algebraically closed field, the Künneth formula $\pi_1(X \times Y) \cong \pi_1(X) \times \pi(Y)$ is valid in characteristic 0, and in characteristic p if the varieties are proper.

(the cuspidal part of) this space to decompose is as follows:

$$\varinjlim_{N} R^d(f_N)_! \overline{\mathbf{Q}}_\ell = \bigoplus_{\pi} \pi \otimes (r_1 \circ \sigma(\pi)) \otimes \cdots \otimes (r_m \circ \sigma(\pi)),$$

where

- π runs over cuspidal automorphic representations of $\mathrm{GL}_n(\mathbf{A}_K)$,
- $\sigma(\pi)\colon \mathrm{Gal}(\overline{K}/K) \to \mathrm{GL}_n(\overline{\mathbf{Q}}_\ell)$ is the corresponding L-parameter, and
- $r_i\colon \mathrm{GL}_n \to \mathrm{GL}_{n_i}$ is an algebraic representation corresponding to μ_i. (If the μ_i are not minuscule, one should replace $\overline{\mathbf{Q}}_\ell$ with the intersection complex, and then r_i would be the irreducible representation of GL_n with heighest weight μ_i.)

Drinfeld ($n = 2$, [Dri80]) and L. Lafforgue (general n, [Laf02]) considered the case of $m = 2$, with μ_1 and μ_2 corresponding to the n-tuples $(1, 0, \ldots, 0)$ and $(0, \ldots, 0, -1)$ respectively. These cocharacters correspond to the tautological representation $r_1\colon \mathrm{GL}_n \to \mathrm{GL}_n$ and its dual r_2. Then the authors were able to prove the claimed decomposition, and in doing so constructed the Langlands correspondence $\pi \mapsto \sigma(\pi)$.

V. Lafforgue [Laf18] considered general reductive groups G in place of GL_n. There is a definition of G-shtuka, which involves G-bundles in place of vector bundles. Using moduli of G-shtukas, [Laf18] produces a correspondence $\pi \mapsto \sigma(\pi)$ from cuspidal automorphic representations of G to L-parameters (though it doesn't prove the full Langlands conjecture for G). In his work, all of the moduli spaces $\mathrm{Sht}_{G, \{\mu_1, \ldots, \mu_m\}}$ (with arbitrarily many legs, and arbitrary cocharacters) are used simultaneously in a crucial way.

Evidently Frobenius plays an important role in this story. We remark that *geometric Langlands* studies the stack Bun_G of G-bundles on X, even in circumstances where X is a complex curve; there is no Frobenius in that theory. Our story concerns *arithmetic Langlands*; it can be reformulated as a study of Bun_G together with its Frobenius map.

1.2   THE POSSIBILITY OF SHTUKAS IN MIXED CHARACTERISTIC

It would be desirable to have moduli spaces of shtukas over number fields, but as we noted earlier, the first immediate problem is that such a space of shtukas would live over something like $\mathrm{Spec}\,\mathbf{Z} \times \mathrm{Spec}\,\mathbf{Z}$, where the product is over $\mathbf{F}_1$ somehow.

In this course we will give a rigorous definition of $\mathrm{Spec}\,\mathbf{Z}_p \times \mathrm{Spec}\,\mathbf{Z}_p$, the completion of $\mathrm{Spec}\,\mathbf{Z} \times \mathrm{Spec}\,\mathbf{Z}$ at (p, p). It lives in the world of nonarchimedean analytic geometry, so it should properly be called $\mathrm{Spa}\,\mathbf{Z}_p \times \mathrm{Spa}\,\mathbf{Z}_p$. (The notation Spa refers to the adic spectrum.) Whatever it is, it should contain $\mathrm{Spa}\,\mathbf{Q}_p \times$

$\mathrm{Spa}\,\mathbf{Q}_p$ as a dense open subset. As a preview of material to come, we now give an explicit description of $\mathrm{Spa}\,\mathbf{Q}_p \times \mathrm{Spa}\,\mathbf{Q}_p$.

It may help to first spell out the equal characteristic analogue of these objects, replacing $\mathbf{Q}_p$ with the Laurent series field $K = \mathbf{F}_p((t))$. The product $\mathrm{Spa}\,\mathcal{O}_K \times_{\mathrm{Spa}\,\mathbf{F}_p} \mathrm{Spa}\,\mathcal{O}_K$ exists as an adic space. Let us rename the second copy of $\mathcal{O}_K$ as $\mathbf{F}_p[\![u]\!]$; then $\mathrm{Spa}\,\mathcal{O}_K \times_{\mathrm{Spa}\,\mathbf{F}_p} \mathrm{Spa}\,\mathcal{O}_K \cong \mathrm{Spa}\,\mathbf{F}_p[\![t, u]\!]$, the formal open polydisc of dimension 2 over $\mathbf{F}_p$. Within this, $\mathrm{Spa}\,K \times_{\mathrm{Spa}\,\mathbf{F}_p} \mathrm{Spa}\,K$ is the open subset defined by $tu \neq 0$.

Projection onto the first factor presents $\mathrm{Spa}\,K \times_{\mathrm{Spa}\,\mathbf{F}_p} \mathrm{Spa}\,K$ as the (rigid-analytic) punctured disc $\mathbf{D}_K^*$ over K, with coordinate u satisfying $0 < |u| < 1$, and t being in the field of constants. But all the same, the projection onto the second factor presents this adic space as $\mathbf{D}_K^*$ in a different way, with the roles of the two variables reversed.

Returning to characteristic 0, we can now present a model for the product $\mathrm{Spa}\,\mathbf{Q}_p \times \mathrm{Spa}\,\mathbf{Q}_p$, specified by picking one of the factors: one copy of $\mathbf{Q}_p$ appears as the field of scalars, but the other copy appears geometrically. Consider the open unit disc $\mathbf{D}_{\mathbf{Q}_p} = \{x \mid |x| < 1\}$ as a subgroup of (the adic version of) $\mathbf{G}_m$, via $x \mapsto 1 + x$. Then $\mathbf{D}_{\mathbf{Q}_p}$ is in fact a $\mathbf{Z}_p$-module with multiplication by p given by $x \mapsto (1 + x)^p - 1$, and we consider

$$\widetilde{\mathbf{D}}_{\mathbf{Q}_p} = \varprojlim_{x \mapsto (1+x)^p - 1} \mathbf{D}_{\mathbf{Q}_p}.$$

After base extension to a perfectoid field, this is a perfectoid space, which carries the structure of a $\mathbf{Q}_p$-vector space. Thus its punctured version $\widetilde{\mathbf{D}}_{\mathbf{Q}_p}^* = \widetilde{\mathbf{D}}_{\mathbf{Q}_p} \setminus \{0\}$ has an action of $\mathbf{Q}_p^\times$, and we consider the quotient $\widetilde{\mathbf{D}}_{\mathbf{Q}_p}^* / \mathbf{Z}_p^\times$. Note that this quotient does not exist in the category of adic spaces!

Definition 1.2.1. Let $\mathrm{Spa}\,\mathbf{Q}_p \times \mathrm{Spa}\,\mathbf{Q}_p = \widetilde{\mathbf{D}}_{\mathbf{Q}_p}^* / \mathbf{Z}_p^\times$, the quotient being taken in a formal sense.

On $\widetilde{\mathbf{D}}_{\mathbf{Q}_p}^* / \mathbf{Z}_p^\times$, we have an operator φ, corresponding to $p \in \mathbf{Q}_p^\times$. Let $X = (\widetilde{\mathbf{D}}_{\mathbf{Q}_p}^* / \mathbf{Z}_p^\times)/\varphi^{\mathbf{Z}} = \widetilde{\mathbf{D}}_{\mathbf{Q}_p}^* / \mathbf{Q}_p^\times$. One can define a finite étale cover of X simply as a $\mathbf{Q}_p^\times$-equivariant finite étale cover of $\widetilde{\mathbf{D}}_{\mathbf{Q}_p}^*$. There is a corresponding profinite group $\pi_1(X)$ which classifies such covers. We have the following theorem, which is a local version of Drinfeld's lemma in the case $m = 2$.

Theorem 1.2.2. We have

$$\pi_1(X) \cong \mathrm{Gal}(\overline{\mathbf{Q}}_p/\mathbf{Q}_p) \times \mathrm{Gal}(\overline{\mathbf{Q}}_p/\mathbf{Q}_p).$$

Similarly, in the case of $K = \mathbf{F}_p((t))$, one can form the quotient $X = \mathbf{D}_K^*/\varphi^{\mathbf{Z}}$ as an adic space, and then

$$\pi_1(X) \cong \mathrm{Gal}(\overline{K}/K) \times \mathrm{Gal}(\overline{K}/K).$$

This theorem suggests that if one could define a moduli space of $\mathbf{Q}_p$-shtukas which is fibered over products such as $\mathrm{Spa}\,\mathbf{Q}_p \times \mathrm{Spa}\,\mathbf{Q}_p$, then its cohomology would produce representations of $\mathrm{Gal}(\overline{\mathbf{Q}}_p/\mathbf{Q}_p) \times \mathrm{Gal}(\overline{\mathbf{Q}}_p/\mathbf{Q}_p)$.

What would a $\mathbf{Q}_p$-shtuka over S look like? It should be a vector bundle $\mathcal{E}$ over $\mathrm{Spa}\,\mathbf{Q}_p \times S$, together with a meromorphic isomorphism $\mathrm{Frob}_S^* \mathcal{E} \dashrightarrow \mathcal{E}$. In order for this to make any sense, we would need to give a geometric meaning to $\mathrm{Spa}\,\mathbf{Q}_p \times S$ (and to Frob_S) just as we gave one to $\mathrm{Spa}\,\mathbf{Q}_p \times \mathrm{Spa}\,\mathbf{Q}_p$.

If we turn to the equal characteristic setting for inspiration, the way forward is much clearer. Let K be a local field of characteristic p, say $K = \mathbf{F}_p((t))$. For a (topologically finite type, let's say) adic space S over $\mathrm{Spa}\,\mathbf{F}_p$, the product $\mathrm{Spa}\,K \times_{\mathrm{Spa}\,\mathbf{F}_p} S$ is again an adic space; namely, it is the punctured open unit disc over S. One can make sense of G-bundles on this, and thereby one can define shtukas. Moduli spaces of K-shtukas were studied by Hartl, Pink, Viehmann, and others; cf. [HP04], [HV11].

Returning to the case $K = \mathbf{Q}_p$, we will give a similar meaning to $\mathrm{Spa}\,\mathbf{Q}_p \times S$ whenever S is a perfectoid space of characteristic p, which lets us define moduli spaces of p-adic shtukas. In general, these are not representable by perfectoid spaces or classical rigid spaces, but instead they are diamonds: That is, quotients of perfectoid spaces by pro-étale equivalence relations. A large part of this course is about the definition of perfectoid spaces and diamonds.

There is an important special case where these moduli spaces of shtukas are classical rigid-analytic spaces. This is the case of local Shimura varieties. Some examples of these are the Rapoport-Zink spaces, [RZ96], which are moduli spaces of p-divisible groups. It was recently suggested by Rapoport-Viehmann, [RV14], that there should exist more general local Shimura varieties which do not have an interpretation as moduli spaces of p-divisible groups. We will prove their existence in general, and the comparison to Rapoport-Zink spaces.

Lecture 2

Adic spaces

In this lecture and the one that follows, we review the theory of *adic spaces* as developed by Huber in [Hub93] and [Hub94].

2.1 MOTIVATION: FORMAL SCHEMES AND THEIR GENERIC FIBERS

To motivate these, let us recall two familiar categories of geometric objects which arise in nonarchimedean geometry: *formal schemes* and *rigid-analytic varieties*.

Formal schemes. An *adic ring* is a topological ring carrying the I-adic topology for an ideal $I \subset A$, called an *ideal of definition*. (Examples: $A = \mathbf{Z}_p$ and $I = p\mathbf{Z}_p$, or $A = \mathbf{Z}_p[\![T]\!]$ and $I = (p, T)$, or A an arbitrary discrete ring, $I = (0)$.) Note that the topology of A is part of the data, but the ideal of definition is not, and indeed there may be many ideals of definition. More precisely, for ideals I and J of A, the I-adic and J-adic topology agree if and only if for some integer n, one has $I^n \subset J$ and $J^n \subset I$.

For an adic ring A, Spf A is the set of open prime ideals of A; this agrees with Spec A/I for any ideal of definition I. Spf A is given a topology and a sheaf of topological rings much in the same way as is done for the usual spectrum of a ring. To wit, for any $f \in A$, one defines the nonvanishing locus $D(f) \subset$ Spf A as usual, and then one declares that the $D(f)$ generate the topology of Spf A. Furthermore, the structure sheaf $\mathcal{O}_{\mathrm{Spf}\, A}$ is defined by setting $\mathcal{O}_{\mathrm{Spf}\, A}(D(f))$ to be the I-adic completion of $A[f^{-1}]$. A *formal scheme* is a topologically ringed space which is locally of the form Spf A for an adic ring A. In this discussion, we will only consider formal schemes which locally have a finitely generated ideal of definition.[1]

---

[1] One reason that this is a useful condition is that the following, surprisingly subtle, lemma holds true.

Lemma 2.1.1 ([Sta, Tag 00M9]). Let A be a ring, let M be an A-module, and let $I \subset A$ be a finitely generated ideal. Then for

$$\widehat{M} = \varprojlim M/I^n M,$$

The category of formal schemes contains the category of schemes as a full subcategory, via the functor which carries Spec A to Spf A, where A is considered with its discrete topology.

In applications (especially deformation theory), one often confuses a formal scheme X with its functor of points $R \mapsto X(R)$, where R is an adic ring. A typical example of a formal scheme is $X = \text{Spf } \mathbf{Z}_p[\![T]\!]$, the formal open unit disc over $\mathbf{Z}_p$; for any adic $\mathbf{Z}_p$-algebra R, we have $X(R) = R^{\circ\circ}$, the ideal of topologically nilpotent elements of R. In particular if $K/\mathbf{Q}_p$ is an extension of nonarchimedean fields, and $K^\circ \subset K$ is its ring of integers, $X(K^\circ) = K^{\circ\circ}$ is the open unit disc in K.

Rigid-analytic spaces. (References: [Con08], [BGR84].) Let K be a *nonarchimedean field*; that is, a field complete with respect to a nontrivial nonarchimedean absolute value $|\ |$. For each $n \geq 0$ we have the *Tate K-algebra* $K\langle T_1, \ldots, T_n \rangle$. This is the completion of the polynomial ring $K[T_1, \ldots, T_n]$ under the Gauss norm. Equivalently, $K\langle T_1, \ldots, T_n \rangle$ is the ring of formal power series in $T_1, \ldots, T_n$ with coefficients in K tending to 0. A *K-affinoid algebra* is a topological K-algebra A which is isomorphic to a quotient of some $K\langle T_1, \cdots, T_n \rangle$.

Suppose A is a K-affinoid algebra. For a point $x \in \text{MaxSpec } A$, the residue field of x is a finite extension of K, which therefore carries a unique extension of the absolute value $|\ |$. For $f \in A$, let $f(x)$ denote the image of f in this residue field, and let $|f(x)|$ denote its absolute value under this extension. For elements $f_1, \ldots, f_n, g \in A$ which generate the unit ideal, let

$$U \left(\frac{f_1, \ldots, f_n}{g} \right) = \left\{ x \in \text{MaxSpec } A \ \middle| \ |f_i(x)| \leq |g(x)|, \ i = 1, \ldots, n \right\};$$

this is called a *rational domain*. MaxSpec A is given a topology (actually a G-topology) in which rational domains are open, and one defines a sheaf of K-algebras on MaxSpec A by specifying its sections over a rational domain. The resulting topologically ringed G-topologized space is a *K-affinoid space*; a rational domain in a K-affinoid space is again a K-affinoid space. Finally, a *rigid-analytic space* over K is a G-topologized space equipped with a sheaf of K-algebras, which is locally isomorphic to a K-affinoid space.

A typical example is $X = \text{MaxSpec } \mathbf{Q}_p\langle T \rangle$, the rigid closed unit disc over $\mathbf{Q}_p$. For an extension $K/\mathbf{Q}_p$, $X(K) = K^\circ$. Within X we have the open subset U defined by $|T| < 1$; thus U is the rigid open unit disc, and $U(K)$ is the open unit disc in K. The subset U is not a rational subset, nor is it even an affinoid space; rather it is the ascending union of rational domains $U(T^n/p)$ for $n = 1, 2, \ldots$.

---

the I-adic completion of M, one has $\widehat{M}/I\widehat{M} = M/IM$.

This implies that $\widehat{M}$ is I-adically complete. The lemma fails in general if I is not finitely generated.

The two categories are closely linked. There is a *generic fiber functor* $X \mapsto X_\eta$ from a certain class of formal schemes over Spf $\mathbf{Z}_p$ (locally formally of finite type) to rigid-analytic spaces over $\mathbf{Q}_p$; cf. [Ber]. This has the property that $X(K^\circ) = X_\eta(K)$ for extensions $K/\mathbf{Q}_p$. Recall that an adic $\mathbf{Z}_p$-algebra A is formally of finite type if it is a quotient of $\mathbf{Z}_p[X_1, \ldots, X_n]\langle Y_1, \ldots, Y_m\rangle$ for some n, m.

The image of the formal open disc Spf $\mathbf{Z}_p[T]$ over $\mathbf{Z}_p$ under this functor is the rigid open disc over $\mathbf{Q}_p$. Berthelot's construction is somewhat indirect. It isn't literally a generic fiber, since Spf $\mathbf{Z}_p$ has only one point (corresponding to $p\mathbf{Z}_p$), so there is no generic point η with residue field $\mathbf{Q}_p$. (One might be tempted to define the generic fiber of Spf $\mathbf{Z}_p[T]$ as MaxSpec $\mathbf{Z}_p[T][1/p]$, but the latter ring is not a Tate algebra, and in any case we do not expect the generic fiber to be an affinoid. See Section 4.2.)

Our goal is to construct a category of *adic spaces* which contains both formal schemes and rigid-analytic spaces as full subcategories. We will use the notation $X \mapsto X^{\mathrm{ad}}$ to denote the functor from formal schemes to adic spaces. Objects in the new category will once again be topologically ringed spaces. But whereas Spf $\mathbf{Z}_p$ contains only one point, $(\mathrm{Spf}\,\mathbf{Z}_p)^{\mathrm{ad}}$ has two points: a generic point η and a special point s. Thus as a topological space it is the same as Spec $\mathbf{Z}_p$.

If X is a formal scheme over Spf $\mathbf{Z}_p$, then X^{ad} is fibered over $(\mathrm{Spf}\,\mathbf{Z}_p)^{\mathrm{ad}}$, and we can define the *adic generic fiber* of X by

$$X_\eta^{\mathrm{ad}} = X^{\mathrm{ad}} \times_{(\mathrm{Spf}\,\mathbf{Z}_p)^{\mathrm{ad}}} \{\eta\}.$$

If X is locally formally of finite type, then X_η^{ad} agrees with the adic space attached to Berthelot's X_η.

Just as formal schemes are built out of affine formal schemes associated to adic rings, and rigid-analytic spaces are built out of affinoid spaces associated to affinoid algebras, adic spaces are built out of affinoid adic spaces, which are associated to *pairs* of topological rings (A, A^+) (where A^+ plays a secondary role). The affinoid adic space associated to such a pair is written $\mathrm{Spa}(A, A^+)$, the *adic spectrum*.

In this lecture we will define which pairs (A, A^+) are allowed, and define $\mathrm{Spa}(A, A^+)$ as a topological space.

2.2 HUBER RINGS

Definition 2.2.1. A topological ring A is *Huber*[2] if A admits an open subring $A_0 \subset A$ which is adic with respect to a finitely generated ideal of definition. That is, there exists a finitely generated ideal $I \subset A_0$ such that $\{I^n | n \geq 0\}$ forms a

---

[2] We propose to use the term *Huber ring* to replace Huber's terminology *f-adic ring*. The latter poses a threat of confusion when there is also a variable f.

basis of open neighborhoods of 0. Any such A_0 is called a *ring of definition* of A.

Note that A_0 is not assumed to be I-adically complete.[3] One can always take the completion $\widehat{A}$ of A: $\widehat{A}$ is Huber and has an open subring $\widehat{A_0} \subset \widehat{A}$ that is simultaneously the closure of A_0 and the I-adic completion of A_0, and $\widehat{A} = \widehat{A_0} \otimes_{A_0} A$; cf. [Hub93, Lemma 1.6].

Example 2.2.2. We give three examples to indicate that adic spaces encompass schemes, formal schemes, and rigid spaces, respectively.

1. (Schemes) Any discrete ring A is Huber, with any $A_0 \subset A$ allowed (take $I = 0$ as an ideal of definition).
2. (Formal schemes) An adic ring A is Huber if it has a finitely generated ideal of definition. In that case, $A_0 = A$ is a ring of definition.
3. (Rigid spaces) Let A_0 be any ring, let $g \in A_0$ be a nonzero-divisor, and let $A = A_0[g^{-1}]$, equipped with the topology making $\{g^n A_0\}$ a basis of open neighborhoods of 0. This is Huber, with ring of definition A_0 and ideal of definition gA_0. For example, if A is a Banach algebra over a nonarchimedean field K, we can take $A_0 \subset A$ to be the unit ball, and $g \in K$ any nonzero element with $|g| < 1$. Then A is a Huber ring of this type.

The Banach algebras relevant to rigid analytic geometry (over a nonarchimedean field K with ring of integers $\mathcal{O}_K$) arise as quotients of the Tate algebra $A = K\langle T_1, \ldots, T_n \rangle$, consisting of power series in $T_1, \ldots, T_n$ whose coefficients tend to 0. This is a Banach K-algebra with unit ball $A_0 = \mathcal{O}_K\langle T_1, \ldots, T_n \rangle$.

Definition 2.2.3. A subset S of a topological ring A is *bounded* if for all open neighborhoods U of 0 there exists an open neighborhood V of 0 such that $VS \subset U$.

In verifying this condition for subsets of Huber rings, one is allowed to shrink U, and without loss of generality one may assume that U is closed under addition, because after all $\{I^n\}$ forms a basis of open neighborhoods of 0.

Lemma 2.2.4. A subring A_0 of a Huber ring A is a ring of definition if and only if it is open and bounded.

Proof. If A_0 is a ring of definition, it is open (by definition). Let U be an open neighborhood of 0 in A. Without loss of generality $U = I^n$, with $n \gg 0$. But then of course $V = I^n$ suffices. For the converse, see [Hub93, Proposition 1]. $\square$

The following special class of Huber rings will be especially relevant later on.

Definition 2.2.5. A Huber ring A is *Tate* if it contains a topologically nilpotent unit $g \in A$. A *pseudo-uniformizer* in A is a topologically nilpotent unit.

---

[3]Throughout, by "complete" we mean "separated and complete."

The Tate rings are exactly the rings that arise by the construction of Example (3) above. For example $\mathbf{Q}_p$ and $\mathbf{Q}_p\langle T\rangle$ are Tate with pseudo-uniformizer p, but $\mathbf{Z}_p$ and $\mathbf{Z}_p[\![T]\!]$ are not Tate.

Proposition 2.2.6.

1. If $A = A_0[g^{-1}]$ is as in Example (3), then A is Tate.
2. If A is Tate with topologically nilpotent unit g, and $A_0 \subset A$ is any ring of definition, then there exists n large enough so that $g^n \in A_0$, and then A_0 is g^n-adic. Furthermore $A = A_0[(g^n)^{-1}]$.
3. Suppose A is Tate with g as above and A_0 a ring of definition. A subset $S \subset A$ is bounded if and only if $S \subset g^{-n}A_0$ for some n.

Proof. 1. Since $g \in A = A_0[g^{-1}]$ is a topologically nilpotent unit, A is Tate by definition.

2. Let $I \subset A_0$ be an ideal of definition. Since g is topologically nilpotent, we can replace g by g^n for n large enough to assume that $g \in I$. Since gA_0 is the preimage of A_0 under the continuous map $g^{-1}\colon A \to A$, we have that $gA_0 \subset A_0$ is open, and thus it contains I^m for some m. Thus we have $g^m A_0 \subset I^m \subset gA_0$, which shows that A_0 is g-adic.
   It remains to show that $A = A_0[g^{-1}]$. Clearly $A_0[g^{-1}] \hookrightarrow A$. If $x \in A$ then $g^n x \to 0$ as $n \to \infty$, since g is topologically nilpotent. Thus there exists n with $g^n x \in A_0$, and therefore $x \in A_0[g^{-1}]$.

3. Left as exercise. $\square$

We remark that if A is a complete Tate ring and $A_0 \subset A$ is a ring of definition, with $g \in A_0$ a topologically nilpotent unit in A, then one can define a norm $|\cdot|\colon A \to \mathbf{R}_{\geq 0}$ by

$$|a| = \inf_{\{n \in \mathbf{Z} \mid g^n a \in A_0\}} 2^n$$

Thus $|g| = 1/2$ and $|g^{-1}| = 2$. Note that this really is a norm: if $|a| = 0$, then $a \in g^n A_0$ for all $n \geq 0$, and thus $a = 0$. Under this norm, A is a Banach ring whose unit ball is A_0.

This construction gives an equivalence of categories between the category of complete Tate rings (with continuous homomorphisms), and the category of Banach rings A that admit an element $g \in A^\times$, $|g| < 1$ such that $|g||g^{-1}| = 1$ (with bounded homomorphisms).

Remark 2.2.7. A slight generalization of the "Tate" condition has recently been proposed by Kedlaya, [Ked19a]. A Huber ring A is *analytic* if the ideal generated by topologically nilpotent elements is the unit ideal. Any Tate ring is analytic, but the converse does not hold true; cf. [Ked19a, Example 1.5.7]. We will discuss the relation further after defining the corresponding adic spectra in Proposition 4.3.1; that discussion will show that the "analytic" condition is in fact more natural.

Definition 2.2.8. Let A be a Huber ring. An element $x \in A$ is *power-bounded* if

$\{x^n | n \geq 0\}$ is bounded. Let $A° \subset A$ be the subring of power-bounded elements.

Example 2.2.9. If $A = \mathbf{Q}_p\langle T \rangle$, then $A° = \mathbf{Z}_p\langle T \rangle$, which as we have seen is a ring of definition. However, if $A = \mathbf{Q}_p[T]/T^2$, with ring of definition $\mathbf{Z}_p[T]/T^2$ carrying the p-adic topology, then $A° = \mathbf{Z}_p \oplus \mathbf{Q}_p T$. Since $A°$ is not bounded, it cannot be a ring of definition.

Proposition 2.2.10. 1. Any ring of definition $A_0 \subset A$ is contained in $A°$.
2. The ring $A°$ is the filtered union of the rings of definition $A_0 \subset A$. (The word *filtered* here means that any two subrings of definition are contained in a third.)

Proof. For any $x \in A_0$, $\{x^n\} \subset A_0$ is bounded, so $x \in A°$, giving part (1). For part (2), we check first that the poset of rings of definition is filtered: if A_0, A_0' are rings of definition, let $A_0'' \subset A$ be the ring they generate. We show directly that A_0'' is bounded. Let $U \subset A$ be an open neighborhood of 0; we have to find V such that $VA_0'' \subset U$. Without loss of generality, U is closed under addition. Pick U_1 such that $U_1 A_0 \subset U$, and pick V such that $VA_0' \subset U_1$.

A typical element of A_0'' is $\sum_i x_i y_i$, with $x_i \in A_0$, $y_i \in A_0'$. We have

$$\left(\sum_i x_i y_i \right) V \subset \sum_i (x_i y_i V) \subset \sum_i (x_i U_1) \subset \sum_i U = U.$$

Thus $VA_0'' \subset U$ and A_0'' is bounded.

For the claim that $A°$ is the union of the rings of definition of A, take any $x \in A°$, and let A_0 be any ring of definition. Then $A_0[x]$ is still a ring of definition, since it is still bounded. $\square$

Definition 2.2.11. A Huber ring A is *uniform* if $A°$ is bounded, or equivalently $A°$ is a ring of definition.

We remark that if A is separated, Tate, and uniform, then A is reduced. Indeed, assume $x \in A$ is nilpotent and $g \in A$ is a topologically nilpotent unit. Then for all n, the element $g^{-n}x \in A$ is nilpotent, and thus powerbounded, so that $g^{-n}x \in A°$, i.e., $x \in g^n A°$ for all $n \geq 0$. If $A°$ is bounded, it carries the g-adic topology, so if A is separated, then $A°$ is g-adically separated, so this implies $x = 0$.

Definition 2.2.12. 1. Let A be a Huber ring. A subring $A^+ \subset A$ is a *ring of integral elements* if it is open and integrally closed in A, and $A^+ \subset A°$.
2. A *Huber pair* is a pair (A, A^+), where A is Huber and $A^+ \subset A$ is a ring of integral elements.

We remark that one often takes $A^+ = A°$, especially in cases corresponding to classical rigid geometry. We also note that the subset $A^{°°} \subset A$ of topologically nilpotent elements is always contained in A^+. Indeed, if $f \in A$ is topologically nilpotent, then $f^n \in A^+$ for some n as A^+ is open, but then also $f \in A^+$ as A^+

is integrally closed.

2.3 CONTINUOUS VALUATIONS

We now define the set of continuous valuations on a Huber ring, which constitute
the points of an adic space.

Definition 2.3.1. A *continuous valuation* on a topological ring A is a map

$$|\cdot| : A \to \Gamma \cup \{0\}$$

into a totally ordered abelian group Γ such that

1. $|ab| = |a|\,|b|$
2. $|a + b| \leq \max(|a|, |b|)$
3. $|1| = 1$
4. $|0| = 0$
5. (Continuity) For all $\gamma \in \Gamma$ lying in the image of $|\cdot|$, the set $\{a \in A \mid |a| < \gamma\}$ is open in A.

(Our convention is that ordered abelian groups Γ are written multiplicatively,
and $\Gamma \cup \{0\}$ means the ordered monoid with $\gamma > 0$ for all $\gamma \in \Gamma$. Of course,
$\gamma 0 = 0$.)

Two continuous valuations $|\cdot|$, $|\cdot|'$ valued in Γ resp. Γ' are *equivalent* when it
is the case that $|a| \geq |b|$ if and only if $|a|' \geq |b|'$. In that case, after replacing Γ
by the subgroup generated by the image of A, and similarly for Γ', there exists
an isomorphism $\Gamma \cong \Gamma'$ such that

$$
\begin{array}{ccc}
 & \xrightarrow{\;|\cdot|\;} & \Gamma \cup \{0\} \\
A & & \downarrow{\scriptstyle\cong} \\
 & \xrightarrow{\;|\cdot|'\;} & \Gamma' \cup \{0\}
\end{array}
$$

commutes; cf. [Hub93, Definition 2.1].

Note that the kernel of $|\cdot|$ is a prime ideal of A.

Continuous valuations are like the multiplicative seminorms that appear in
Berkovich's theory. At this point we must apologize that continuous valuations
are not called "continuous seminorms," since after all they are written multiplicatively. On the other hand, we want to consider value groups of higher rank
(and indeed this is the point of departure from Berkovich's theory), which makes
the use of the word "seminorm" somewhat awkward.

Definition 2.3.2. The adic spectrum $\mathrm{Spa}(A, A^+)$ is the set of equivalence
classes of continuous valuations $|\cdot|$ on A such that $|A^+| \leq 1$. For $x \in \mathrm{Spa}(A, A^+)$,

write $g \mapsto |g(x)|$ for a choice of corresponding valuation.

The topology on $\mathrm{Spa}(A, A^+)$ is generated by open subsets of the form

$$\left\{ x \mid |f(x)| \leq |g(x)| \neq 0 \right\},$$

with $f, g \in A$.

The shape of these open sets is dictated by the desired properties that both $\{x \mid |f(x)| \neq 0\}$ and $\{x \mid |f(x)| \leq 1\}$ be open. These desiderata combine features of classical algebraic geometry and rigid geometry, respectively.

Huber shows that the topological space $\mathrm{Spa}(A, A^+)$ is reasonable (at least from the point of view of an algebraic geometer).

Theorem 2.3.3 ([Hub93, Theorem 3.5 (i)]). *The topological space $\mathrm{Spa}(A, A^+)$ is spectral.*

Here, we recall the following definition.

Definition 2.3.4 ([Hoc69]). A topological space T is *spectral* if the following equivalent conditions are satisfied.

1. $T \cong \mathrm{Spec}\, R$ for some ring R.
2. $T \cong \varprojlim T_i$ where $\{T_i\}$ is an inverse system of finite T_0-spaces. (Recall that T_0 means that given any two distinct points, there exists an open set which contains exactly one of them.)
3. T is quasicompact, there exists a basis of quasi-compact opens of T which is stable under finite intersection, and T is *sober*, i.e. every irreducible closed subset has a unique generic point.

Example 2.3.5. Let T_i be the topological space consisting of the first i primes (taken to be closed), together with a generic point whose closure is all of T_i. Let $\mathrm{Spec}\, \mathbf{Z} \to T_i$ be the map which sends the first i primes to their counterparts in T_i, and sends everything else to the generic point. Then there is a homeomorphism $\mathrm{Spec}\, \mathbf{Z} \cong \varprojlim T_i$.

Example 2.3.6. Let R be a discrete ring. Then $\mathrm{Spa}(R, R)$ is the set of valuations on R bounded by 1. We list the points of $\mathrm{Spa}(\mathbf{Z}, \mathbf{Z})$:

1. A point η, which takes all nonzero integers to 1,
2. A special point s_p for each prime p, which is the composition $\mathbf{Z} \to \mathbf{F}_p \to \{0, 1\}$, where the second arrow sends all nonzero elements to 1,
3. A point η_p for each prime p, which is the composition $\mathbf{Z} \to \mathbf{Z}_p \to p^{\mathbf{Z}_{\leq 0}} \cup \{0\}$, where the second arrow is the usual p-adic absolute value.

Then $\{s_p\}$ is closed, whereas $\overline{\{\eta_p\}} = \{\eta_p, s_p\}$, and $\overline{\{\eta\}} = \mathrm{Spa}(\mathbf{Z}, \mathbf{Z})$.

In general, for a discrete ring R, we have a map $\mathrm{Spec}\, R \to \mathrm{Spa}(R, R)$, which sends $\mathfrak{p}$ to the valuation $R \to \mathrm{Frac}(R/\mathfrak{p}) \to \{0, 1\}$, where the second map is 0 on

0 and 1 everywhere else. There is also a map $\mathrm{Spa}(R, R) \to \mathrm{Spec}\, R$, which sends a valuation to its kernel. The composition of these two maps is the identity on $\mathrm{Spec}\, R$. Both maps are continuous.

Example 2.3.7. Let K be a nonarchimedean field, let $A = K\langle T \rangle$, and let $A^+ = \mathcal{O}_K \langle T \rangle$. Then $X = \mathrm{Spa}(A, A^+)$ is the adic closed unit disc over K. In general there are five classes of points in X; these are discussed in detail in [Wed14, Example 7.57]. If K is algebraically closed, is spherically complete, and has valuation group $\mathbf{R}_{>0}$, then there are only three classes of points:

- To each $x \in K$ with $|x| \leq 1$ there corresponds a classical point in X, whose valuation is described by $|f(x)|$. Let us simply call this point x.

- For each $x \in K$ with $|x| \leq 1$ and each r in the interval $(0, 1]$, there is the Gauss point x_r, whose valuation is described by

$$|f(x_r)| = \sup_{y \in D(x,r)} |f(y)|\,,$$

  where $D(x, r) = \{y \in K \mid |x - y| \leq r\}$. Then x_r only depends on $D(x, r)$.

- For each $x \in K$ with $|x| \leq 1$, each $r \in (0, 1]$, and each sign $\pm$, there is the rank 2 point $x_{r\pm}$, defined as follows. (We exclude the sign $+$ if $r = 1$.) We take as our group Γ the product $\mathbf{R}_{>0} \times \gamma^{\mathbf{Z}}$, where the order is lexicographic (with $\gamma > 1$). Then if we write $f \in A$ as $\sum_{n \geq 0} a_n (T - x)^n$ with $a_n \in K$, we set

$$|f(x_{r\pm})| = \max_n |a_n| \, r^n \gamma^{\pm n}.$$

Then x_{r+} depends only on $D(x, r)$, whereas x_{r-} depends only on the open ball $\{y \in K \mid |x - y| < r\}$. (The points we have listed here are referred to in [Wed14] and elsewhere as the points of type (1), (2), and (5), respectively. The points of type (3) and (4) appear when one no longer assumes that K has value group $\mathbf{R}_{>0}$ or is spherically complete. For the present discussion, those points are a red herring.) The classical and rank 2 points are closed, and the closure of a Gauss point x' consists of x together with all rank 2 points $x_{r\pm} \in X$ for which $x_r = x'$. Thus X has Krull dimension 1.

We also remark that X is connected. One might try to disconnect X by decomposing it into two opens:

$$\begin{aligned} U &= \{|T(x)| = 1\} \\ V &= \bigcup_{\varepsilon > 0} \{|T(x)| \leq 1 - \varepsilon\}. \end{aligned}$$

But neither subset contains the rank 2 point x_{1-}, where $x \in K$ is any element with $|x| < 1$. (Recall that in the formalism of rigid spaces, the analogues of U and V really do cover $\mathrm{MaxSpec}\, K\langle T \rangle$, but they do not constitute an admissible covering in the G-topology. In this sense the extra points appearing in the adic theory allow us to work with an honest topology, rather than a G-topology.)

Example 2.3.8. The adic spectrum reduces to the Zariski-Riemann space in case $A = K$ is a discrete field. Recall that if K is a discrete field, then a subring $R \subset K$ is a *valuation ring* if for all $f \in K^{\times}$, one of f and f^{-1} lies in R. If $A \subset K$ is any subring, the *Zariski-Riemann space* $\mathrm{Zar}(K, A)$ is the set of valuation rings $R \subset K$ containing A. Each valuation ring $R \in \mathrm{Zar}(K, A)$ induces a valuation $|\ |_R$ on K; here the value group $\Gamma_{|\ |_R}$ is given by $K^{\times}/R^{\times}$, and the valuation by the obvious projection $K = K^{\times} \cup \{0\} \to K^{\times}/R^{\times} \cup \{0\}$. Conversely, every valuation on K gives rise to a valuation ring $R = \{f \in K \mid |f(x)| \leq 1\}$; this induces a bijective correspondence between equivalence classes of valuations and valuation rings. This set is given a topology and sheaf of rings which makes $\mathrm{Zar}(K, A)$ a quasi-compact ringed space; cf. [Mat80, Theorem 10.5]. We get a homeomorphism $\mathrm{Zar}(K, A) \cong \mathrm{Spa}(K, A)$. In the simplest example, if K is a function field with field of constants k (meaning that K/k is a finitely generated field extension of transcendence degree 1), then $\mathrm{Zar}(K, k)$ is homeomorphic to the normal projective curve over k whose function field is K.

Moreover, we will need the following results about the adic spectrum.

Proposition 2.3.9 ([Hub93, Proposition 3.9]). *Let $(\widehat{A}, \widehat{A}^+)$ be the completion of a Huber pair (A, A^+). Then the natural map is a homeomorphism*

$$\mathrm{Spa}(\widehat{A}, \widehat{A}^+) \cong \mathrm{Spa}(A, A^+).$$

Usually, we will restrict attention to complete Huber pairs. The next proposition shows that the adic spectrum $\mathrm{Spa}(A, A^+)$ is "large enough":

Proposition 2.3.10. *Let (A, A^+) be a complete Huber pair.*

1. *If $A \neq 0$ then $\mathrm{Spa}(A, A^+)$ is nonempty.*
2. *One has $A^+ = \{f \in A \mid |f(x)| \leq 1, \text{ for all } x \in X\}$.*
3. *An element $f \in A$ is invertible if and only if for all $x \in X$, $|f(x)| \neq 0$.*

Proof. Part (1) is [Hub93, Proposition 3.6 (i)], and part (2) is [Hub93, Lemma 3.3 (i)]. For part (3), apply part (1) to the separated completion of A/f, noting that if the closure of the ideal generated by f contains 1, then there is some $g \in A$ such that $fg - 1$ is topologically nilpotent, in which case fg is invertible, so that f itself is invertible. $\square$

Lecture 3

Adic spaces II

Today we define adic spaces. The reference is [Hub94]. Recall from the previous lecture:

1. A Huber ring is a topological ring A that admits an open subring $A_0 \subset A$ which is adic with a finitely generated ideal of definition.
2. A Huber pair is a pair (A, A^+), where A is a Huber ring and $A^+ \subset A^\circ$ is an open and integrally closed subring.

We constructed a spectral topological space $X = \mathrm{Spa}(A, A^+)$ consisting of equivalence classes of continuous valuations $|\cdot|$ on A such that $|A^+| \leq 1$.

3.1   RATIONAL SUBSETS

Definition 3.1.1. Let $s \in A$ and $T \subset A$ be a finite subset such that $TA \subset A$ is open. We define the subset

$$ U\left(\tfrac{T}{s}\right) = \{x \in X \mid |t(x)| \leq |s(x)| \neq 0, \text{ for all } t \in T\} \ . $$

Subsets of this form are called *rational subsets*.

   Note that rational subsets are open because they are an intersection of a finite collection of open subsets $\{|t(x)| \leq |s(x)| \neq 0\}$, $t \in T$.

Proposition 3.1.2. The intersection of finitely many rational subsets is again rational.

Proof. Take two rational subsets

$$ U_1 = \{x \mid |t(x)| \leq |s(x)| \neq 0, t \in T\} \ , \quad U_2 = \{x \mid |t'(x)| \leq |s'(x)| \neq 0, t' \in T'\} \ . $$

Their intersection is

$$ \{x \mid |tt'(x)|, |ts'(x)|, |st'(x)| \leq |ss'(x)| \neq 0, t \in T, t' \in T'\} \ . $$

Now we just have to check that the tt' for $t \in T$, $t' \in T'$ generate an open ideal of A. By hypothesis there exists an ideal of definition I (of some auxiliary ring of definition $A_0 \subset A$) such that $I \subset TA$ and $I \subset T'A$. Then the ideal generated by the tt' contains I^2.   □

The following theorem shows that rational subsets are themselves adic spectra.

Theorem 3.1.3 ([Hub94, Proposition 1.3]). *Let $U \subset \mathrm{Spa}(A, A^+)$ be a rational subset. Then there exists a complete Huber pair $(A, A^+) \to (\mathcal{O}_X(U), \mathcal{O}_X^+(U))$ such that the map*

$$\mathrm{Spa}(\mathcal{O}_X(U), \mathcal{O}_X^+(U)) \to \mathrm{Spa}(A, A^+)$$

factors over U, and is universal for such maps. Moreover this map is a homeomorphism onto U. In particular, U is quasi-compact.

Proof. (Sketch.) Choose s and T such that $U = U(T/s)$. Choose $A_0 \subset A$ a ring of definition, $I \subset A_0$ a finitely generated ideal of definition. Now take any $(A, A^+) \to (B, B^+)$ such that $\mathrm{Spa}(B, B^+) \to \mathrm{Spa}(A, A^+)$ factors over U. Then

1. The element s becomes invertible in B by Proposition 2.3.10 (3), so that we get a map $A[1/s] \to B$.
2. All t/s for $t \in T$ are of $|\cdot| \leq 1$ everywhere on $\mathrm{Spa}(B, B^+)$, so that $t/s \in B^+ \subset B^\circ$ by Proposition 2.3.10 (2).
3. Since B° is the inductive limit of the rings of definition B_0, we can choose a B_0 which contains all t/s for $t \in T$. We get a map

$$A_0[t/s|t \in T] \to B_0.$$

Endow $A_0[t/s|t \in T]$ with the $IA_0[t/s|t \in T]$-adic topology.

Lemma 3.1.4. *This defines a ring topology on $A[1/s]$ making $A_0[t/s|t \in T]$ a ring of definition.*

The crucial point is to show that there exists n such that $\frac{1}{s}I^n \subset A_0[t/s|t \in T]$, so that multiplication by $1/s$ can be continuous. It is enough to show that $I^n \subset TA_0$, which follows from the following lemma.

Lemma 3.1.5. *If $T \subset A$ is a subset such that $TA \subset A$ is open, then TA_0 is open.*

Proof. After replacing I with some power we may assume that $I \subset TA$. Write $I = (f_1, \ldots, f_k)$. There exists a finite set R such that $f_1, \ldots, f_k \in TR$.

Since I is topologically nilpotent, there exists n such that $RI^n \subset A_0$. Then for all $i = 1, \ldots, k$, $f_iI^n \subset TRI^n \subset TA_0$. Sum this over all i and conclude that $I^{n+1} \subset TA_0$.   □

Back to the proof of the proposition. We have $A[1/s]$, a (non-complete) Huber ring. Let $A[1/s]^+$ be the integral closure of the image of $A^+[t/s|t \in T]$ in $A[1/s]$.

Let $(A\langle T/s \rangle, A\langle T/s \rangle^+)$ be its completion, a complete Huber pair. This has the desired universal property. For the claim that Spa of this pair is homeomorphic to U, use Proposition 2.3.9.   □

Definition 3.1.6. Define a presheaf $\mathcal{O}_X$ of topological rings on $\mathrm{Spa}(A, A^+)$: If $U \subset X$ is rational, $\mathcal{O}_X(U)$ is as in the theorem. On a general open $W \subset X$, we put
$$\mathcal{O}_X(W) = \varprojlim_{U \subset W \text{ rational}} \mathcal{O}_X(U).$$

One defines $\mathcal{O}_X^+$ similarly.

Proposition 3.1.7. For all $U \subset \mathrm{Spa}(A, A^+)$,
$$\mathcal{O}_X^+(U) = \{f \in \mathcal{O}_X(U)|\, |f(x)| \leq 1, \text{ all } x \in U\}.$$

In particular $\mathcal{O}_X^+$ is a sheaf if $\mathcal{O}_X$ is.

Proof. It suffices to check this if U is rational, in which case it follows from Proposition 2.3.10 (2).   □

Theorem 3.1.8 ([Hub94, Theorem 2.2]). Let (A, A^+) be a complete Huber pair. Then $\mathcal{O}_X$ is a sheaf of topological rings in the following situations.

1. (Schemes) The topological ring A is discrete.
2. (Formal schemes) The ring A is finitely generated over a noetherian ring of definition.
3. (Rigid spaces) The topological ring A is Tate and *strongly noetherian*, i.e. the rings
$$A\langle T_1, \ldots, T_n \rangle = \left\{ \sum_{\underline{i}=(i_1,\ldots,i_n)\geq 0} a_{\underline{i}} T^{\underline{i}} \,\middle|\, a_{\underline{i}} \in A,\ a_{\underline{i}} \to 0 \right\}$$
are noetherian for all $n \geq 0$.

Example 3.1.9. The case $A = \mathbf{C}_p$ relevant to rigid geometry is not covered by case 2, because $\mathcal{O}_{\mathbf{C}_p}$ is not noetherian. But $\mathbf{C}_p\langle T_1, \ldots, T_n \rangle$ is noetherian, so case 3 applies. The same goes for $A = \mathbf{C}_p\langle T_1, \ldots, T_n \rangle$.

Remark 3.1.10. There are examples due to Rost (see [Hub94], end of §1) where $\mathcal{O}_X$ is not a sheaf. See [BV18] and [Mih16] for further examples.

Definition 3.1.11. A Huber pair (A, A^+) is *sheafy* if $\mathcal{O}_X$ is a sheaf of topological rings. (This implies that $\mathcal{O}_X^+$ is a sheaf of topological rings as well.)

3.2   ADIC SPACES

Recall that a scheme is a ringed space which locally looks like the spectrum of a ring. An adic space will be something similar. First we have to define the adic version of "locally ringed space." Briefly, it is a topologically ringed topological space equipped with valuations.

Definition 3.2.1. We define a category (V) as follows. The objects are triples $(X, \mathcal{O}_X, (|\cdot(x)|)_{x \in X})$, where X is a topological space, $\mathcal{O}_X$ is a sheaf of topological rings, and for each $x \in X$, $|\cdot(x)|$ is an equivalence class of continuous valuations on $\mathcal{O}_{X,x}$. (Note that this data determines $\mathcal{O}_X^+$.) The morphisms are maps of topologically ringed topological spaces $f \colon X \to Y$ (so that the map $\mathcal{O}_Y(V) \to \mathcal{O}_X(f^{-1}(V))$ is continuous for each open $V \subset Y$) that make the following diagram commute up to equivalence for all $x \in X$:

$$
\begin{array}{ccc}
\mathcal{O}_{Y,f(x)} & \longrightarrow & \mathcal{O}_{X,x} \\
\downarrow & & \downarrow \\
\Gamma_{f(x)} \cup \{0\} & \longrightarrow & \Gamma_x \cup \{0\}
\end{array}
$$

An *adic space* is an object $(X, \mathcal{O}_X, (|\cdot(x)|)_{x \in X})$ of (V) that admits a covering by spaces U_i such that the triple $(U_i, \mathcal{O}_X|_{U_i}, (|\cdot(x)|)_{x \in U_i})$ is isomorphic to $\mathrm{Spa}(A_i, A_i^+)$ for a sheafy Huber pair (A_i, A_i^+).

For sheafy (A, A^+), the topological space $X = \mathrm{Spa}(A, A^+)$ together with its structure sheaf and continuous valuations is an *affinoid adic space*, which we continue to write as $\mathrm{Spa}(A, A^+)$.

Often we will write X for the entire triple $(X, \mathcal{O}_X, (|\cdot(x)|)_{x \in X})$. In that case we will use $|X|$ to refer to the underlying topological space of X.

Proposition 3.2.2 ([Hub94, Proposition 2.1]). *The functor*

$$
(A, A^+) \mapsto \mathrm{Spa}(A, A^+)
$$

from sheafy complete Huber pairs to adic spaces is fully faithful.

3.3   THE ROLE OF A^+

We reflect on the role of A^+ in the definition of adic spaces. The subring A^+ in a Huber pair (A, A^+) may seem unnecessary at first: why not just consider all continuous valuations on A? For a Huber ring A, let $\mathrm{Cont}(A)$ be the set of equivalence classes of continuous valuations on A, with topology generated by subsets of the form $\{|f(x)| \leq |g(x)| \neq 0\}$, with $f, g \in A$.

Proposition 3.3.1 ([Hub93, Corollary 3.2, Lemma 3.3]).

1. $\mathrm{Cont}(A)$ is a spectral space.
2. The following sets are in bijection:

   a) The set of subsets $F \subset \mathrm{Cont}(A)$ of the form $\bigcap_{f \in S} \{|f| \leq 1\}$, as S runs over arbitrary subsets of A°.

   b) The set of open and integrally closed subrings $A^+ \subset A^\circ$.

   The map is

   $$F \mapsto \{f \in A \,|\, |f(x)| \leq 1 \text{ for all } x \in F\}$$

   with inverse

   $$A^+ \mapsto \left\{x \in \mathrm{Cont}(A) \,|\, |f(x)| \leq 1 \text{ for all } f \in A^+\right\}.$$

Thus specifying A^+ keeps track of which inequalities have been enforced among the continuous valuations in $\mathrm{Cont}(A)$.

Finally, we can say why A^+ is necessary: If $U \subset \mathrm{Cont}(A)$ is a rational subset and $A \to B$ is the corresponding rational localization, then in general $\mathrm{Cont}(B)$ is not equal to U: Instead U is a strict open subset of it. One needs to specify B^+ in addition to keep track of this. We note that one can moreover not build a theory in which $A^+ = A^\circ$ holds always, as this condition is not in general stable under passage to rational subsets.

3.4 PRE-ADIC SPACES

What can be done about non-sheafy Huber pairs (A, A^+)? It really is a problem that the structure presheaf on $\mathrm{Spa}(A, A^+)$ isn't generally a sheaf. It ruins any hope of defining a general adic space as what one gets by gluing together spaces of the form $\mathrm{Spa}(A, A^+)$; indeed, without the sheaf property this gluing doesn't make any sense.

Here are some of our options for how to proceed:

1. Ignore them. Maybe non-sheafy Huber pairs just don't appear in nature, so to speak.
2. It is possible to redefine the structure sheaf on $X = \mathrm{Spa}(A, A^+)$ so that for a rational subset U, $\mathcal{O}_X(U)$ is a henselization rather than a completion. Then one can show that $\mathcal{O}_X$ is always a sheaf; cf., e.g., [GR16, Theorem 15.4.26]. However, proceeding this way diverges quite a bit from the classical theory of rigid spaces.
3. Construct a larger category of adic spaces using a "functor of points" approach. This is analogous to the theory of algebraic spaces, which are functors on the (opposite) category of rings which may not be representable.

We will essentially follow route (1), but we want to say something about (3).

This approach has been introduced in [SW13, Section 2.1].

Let CAff be the category of complete Huber pairs,[1] where morphisms are continuous homomorphisms. Let CAff^{op} be the opposite category. We turn this into a site, where for any rational cover $X = \mathrm{Spa}(A, A^+) = \bigcup_{i \in I} U_i$, the set of maps $(\mathcal{O}_X(U_i), \mathcal{O}_X^+(U_i))^{\mathrm{op}} \to (A, A^+)^{\mathrm{op}}$ in CAff^{op} is a cover, and these generate all covers. An object $X = (A, A^+)^{\mathrm{op}}$ of CAff^{op} induces a set-valued covariant functor on CAff^{op}, by $(B, B^+) \mapsto \mathrm{Hom}_{\mathrm{CAff}}((A, A^+), (B, B^+))$. Let $\mathrm{Spa}^Y(A, A^+)$ denote its sheafification; here Y stands for Yoneda.

Now we repeat [SW13, Definition 2.1.5].

Definition 3.4.1. Let $\mathcal{F}$ be a sheaf on CAff^{op}, and let (A, A^+) be a complete Huber pair with adic spectrum $X = \mathrm{Spa}(A, A^+)$. A map $\mathcal{F} \to \mathrm{Spa}^Y(A, A^+)$ is an open immersion if there is an open subset $U \subset X$ such that

$$\mathcal{F} = \varinjlim_{\substack{V \subset U \\ V \, \mathrm{rational}}} \mathrm{Spa}^Y(\mathcal{O}_X(V), \mathcal{O}_X^+(V)) \ .$$

If $f : \mathcal{F} \to \mathcal{G}$ is any map of sheaves on CAff^{op}, then f is an open immersion if for all complete Huber pairs (A, A^+) with a map $\mathrm{Spa}^Y(A, A^+) \to \mathcal{G}$, the fiber product $\mathcal{F} \times_{\mathcal{G}} \mathrm{Spa}^Y(A, A^+) \to \mathrm{Spa}^Y(A, A^+)$ is an open immersion. Note that an open immersion $\mathcal{F} \to \mathcal{G}$ is injective; in that case we will simply say that $\mathcal{F} \subset \mathcal{G}$ is open. Finally, a pre-adic space is a sheaf $\mathcal{F}$ on CAff^{op} such that

$$\mathcal{F} = \varinjlim_{\mathrm{Spa}^Y(A, A^+) \subset \mathcal{F}} \mathrm{Spa}^Y(A, A^+) \ .$$

In the appendix to this lecture, we explain that one can give an equivalent definition of pre-adic spaces that is closer in spirit to the definition of adic spaces, inasmuch as an adic space is something like a locally ringed topological space. In particular, our comparison shows that adic spaces are naturally a full subcategory of pre-adic spaces.

---

[1]This notation appears in [SW13], and anyway recall that Huber calls Huber pairs *affinoid algebras*.

Appendix to Lecture 3:
Pre-adic spaces

In this appendix, we give an alternative and slightly more concrete definition of pre-adic spaces. The reader is advised to skip this appendix.

Recall that the problem is that the structure presheaf on $X = \mathrm{Spa}(A, A^+)$ can fail to be a sheaf. This suggests simply keeping the topological space X and using the sheafification $\mathcal{O}_X^{\mathrm{sh}}$ of $\mathcal{O}_X$ to arrive at an object $\mathrm{Spa}^{\mathrm{sh}}(A, A^+)$ of the category (V), but then one runs into problems. For instance, the analogue of Proposition 3.2.2 fails, and it seems impossible to describe the maps between affinoids defined this way. Certainly one does not expect that any morphism $f\colon X = \mathrm{Spa}^{\mathrm{sh}}(A, A^+) \to Y = \mathrm{Spa}^{\mathrm{sh}}(B, B^+)$ in (V) arises from a map of complete Huber pairs $(B, B^+) \to (A, A^+)$. One could hope however that this is still true locally on X; namely, there might be a rational cover U_i of $\mathrm{Spa}(A, A^+)$ such that each $f|_{U_i}$ arises from a map $(B, B^+) \to (\mathcal{O}_X(U_i), \mathcal{O}_X(U_i)^+)$. But one cannot expect this either: in general one only has a map from B into $\mathcal{O}_X^{\mathrm{sh}}(X)$, which is a colimit of topological rings indexed by rational covers of X. This need not arise from a map from B into any of the rings appearing in the colimit.

Here, our idea is that instead of sheafifying $\mathcal{O}_X$ in the category of topological rings, one sheafifies in the category of ind-topological rings. This is the category whose objects are formal colimits of filtered direct systems A_i, $i \in I$, where each A_i is a topological ring. We will write "$\varinjlim_i$"A_i for the formal colimit of the A_i. Individual topological rings B become compact in the ind-category, so that $\mathrm{Hom}(B, \text{"}\varinjlim_i\text{"} A_i) = \varinjlim_i \mathrm{Hom}(B, A_i)$.

For an ind-topological ring $A = \text{"}\varinjlim_i\text{"} A_i$, a continuous valuation on A is defined to be a compatible system of continuous valuations on all A_i, or equivalently a continuous valuation on the topological ring $\varinjlim_i A_i$.

Definition 3.5.1. Define a category $(V)^{\mathrm{ind}}$ as follows. The objects are triples $(X, \mathcal{O}_X, (|\cdot(x)|)_{x \in X})$, where X is a topological space, $\mathcal{O}_X$ is a sheaf of ind-topological rings, and for each $x \in X$, $|\cdot(x)|$ is an equivalence class of continuous valuations on $\mathcal{O}_{X,x}$. The morphisms are maps of ind-topologically ringed topological spaces $f\colon X \to Y$, which make the following diagram commute up to equivalence:

$$\begin{array}{ccc} \mathcal{O}_{Y,f(x)} & \longrightarrow & \mathcal{O}_{X,x} \\ \downarrow & & \downarrow \\ \Gamma_{f(x)} \cup \{0\} & \longrightarrow & \Gamma_x \cup \{0\} \end{array}$$

For any Huber pair (A, A^+), we define $\mathrm{Spa}^{\mathrm{ind}}(A, A^+) \in (V)^{\mathrm{ind}}$ as the triple $(X, \mathcal{O}_X^{\mathrm{ind}}, (|\cdot(x)|)_{x \in X})$, where $X = \mathrm{Spa}(A, A^+)$ is the usual topological space, $\mathcal{O}_X^{\mathrm{ind}}$ is the sheafification of the presheaf $\mathcal{O}_X$ in the category of ind-topological rings, and the valuations stay the same.

Definition 3.5.2. A pre-adic space is an object of $(V)^{\mathrm{ind}}$ that is locally isomorphic to $\mathrm{Spa}^{\mathrm{ind}}(A, A^+)$ for some complete Huber pair (A, A^+).

Let us prove that this agrees with the "Yoneda-style" definition of pre-adic spaces of Definition 3.4.1; to distinguish the latter, we temporarily call them Yoneda-adic spaces.

Proposition 3.5.3. For any pre-adic space $X \in (V)^{\mathrm{ind}}$, the association

$$X^Y : (A, A^+)^{\mathrm{op}} \mapsto \mathrm{Hom}_{(V)^{\mathrm{ind}}}(\mathrm{Spa}^{\mathrm{ind}}(A, A^+), X)$$

defines a sheaf on $\mathrm{CAff}^{\mathrm{op}}$ that is a Yoneda-adic space. This defines an equivalence of categories between pre-adic spaces and Yoneda-adic spaces that takes $\mathrm{Spa}^{\mathrm{ind}}(B, B^+)$ to $\mathrm{Spa}^Y(B, B^+)$. Under this equivalence, an open immersion of pre-adic spaces corresponds to an open immersion of Yoneda-adic spaces.

Proof. For any rational cover of $\mathrm{Spa}(A, A^+)$, one can glue $\mathrm{Spa}^{\mathrm{ind}}(A, A^+)$ from the corresponding pieces on rational subsets; this implies that X^Y is indeed a sheaf on $\mathrm{CAff}^{\mathrm{op}}$, for any $X \in (V)^{\mathrm{ind}}$.

Next, we check that if $X = \mathrm{Spa}^{\mathrm{ind}}(B, B^+)$, then $X^Y = \mathrm{Spa}^Y(B, B^+)$. Equivalently, we have to see that

$$(A, A^+)^{\mathrm{op}} \mapsto \mathrm{Hom}_{(V)^{\mathrm{ind}}}(\mathrm{Spa}^{\mathrm{ind}}(A, A^+), \mathrm{Spa}^{\mathrm{ind}}(B, B^+))$$

is the sheafification of

$$(A, A^+)^{\mathrm{op}} \mapsto \mathrm{Hom}_{\mathrm{CAff}^{\mathrm{op}}}((A, A^+)^{\mathrm{op}}, (B, B^+)^{\mathrm{op}}) = \mathrm{Hom}_{\mathrm{CAff}}((B, B^+), (A, A^+)) .$$

First, assume that two maps $f, g : (B, B^+) \to (A, A^+)$ induce the same map $\mathrm{Spa}^{\mathrm{ind}}(A, A^+) \to \mathrm{Spa}^{\mathrm{ind}}(B, B^+)$. In particular, on global sections, we find that the two maps

$$B \to \underset{X = \bigcup_i U_i}{\text{"}\varinjlim\text{"}} \mathrm{eq}\left(\prod_i \mathcal{O}_X(U_i) \to \prod_{i,j} \mathcal{O}_X(U_i \cap U_j) \right)$$

induced by f and g agree, where $X = \mathrm{Spa}(A, A^+)$ and the index sets runs over rational covers of X. This means that for some rational cover $\{U_i\}$ of X, the maps $B \to \mathcal{O}_X(U_i)$ induced by f and g agree, and then the same is true for $B^+ \to \mathcal{O}_X^+(U_i)$. But this means that f and g induce the same element in the sheafification of $(A, A^+)^{\mathrm{op}} \mapsto \mathrm{Hom}_{\mathrm{CAff}}((B, B^+), (A, A^+))$.

It remains to see that any map $f : \mathrm{Spa}^{\mathrm{ind}}(A, A^+) \to \mathrm{Spa}^{\mathrm{ind}}(B, B^+)$ is induced from a map $(B, B^+) \to (A, A^+)$ locally on $X = \mathrm{Spa}(A, A^+)$. Arguing as above, one can ensure after passage to a rational cover that the map $B \to \mathcal{O}_{\mathrm{Spa}^{\mathrm{ind}}(A,A^+)}(X)$ factors over a map $B \to A$, in which case so does $B^+ \to A^+$, i.e., we get a map $g : (B, B^+) \to (A, A^+)$, which induces a map $f' : \mathrm{Spa}^{\mathrm{ind}}(A, A^+) \to \mathrm{Spa}^{\mathrm{ind}}(B, B^+)$. We have to show that $f = f'$. As points are determined by the valuations induced on A resp. B, one sees that the maps f and f' agree on topological spaces. It remains to see that the maps agree on sheaves of ind-topological rings. For any rational subset $V \subset Y = \mathrm{Spa}(B, B^+)$, the map $f_V^\sharp : \mathcal{O}_Y(V) \to \mathcal{O}_X^{\mathrm{ind}}(f^{-1}(V))$ is compatible with the map $g : B \to A$, which by the universal properties of rational subsets means that it is also compatible with the map $f_V'^\sharp : \mathcal{O}_Y(V) \to \mathcal{O}_X(f^{-1}(V)) \to \mathcal{O}_X^{\mathrm{ind}}(f^{-1}(V))$, i.e., $f_V^\sharp = f_V'^\sharp$. Passing to the sheafification $\mathcal{O}_Y^{\mathrm{ind}}$ of $\mathcal{O}_Y$ in ind-topological rings, we get the result.

Now if $X \in (V)^{\mathrm{ind}}$ is any object and $(A, A^+) \in \mathrm{CAff}$, one sees that the map

$$\mathrm{Hom}_{(V)^{\mathrm{ind}}}(\mathrm{Spa}^{\mathrm{ind}}(A, A^+), X) \to \mathrm{Hom}((\mathrm{Spa}^{\mathrm{ind}}(A, A^+))^Y, X^Y)$$
$$= \mathrm{Hom}(\mathrm{Spa}^Y(A, A^+), X^Y)$$

is a bijection, as $\mathrm{Hom}(\mathrm{Spa}^Y(A, A^+), X^Y)$ is by the Yoneda lemma the value of X^Y at $(A, A^+)^{\mathrm{op}}$, which was defined to be $\mathrm{Hom}_{(V)^{\mathrm{ind}}}(\mathrm{Spa}^{\mathrm{ind}}(A, A^+), X)$.

It is clear that open immersions into $\mathrm{Spa}^Y(A, A^+)$ correspond bijectively to open subsets of $\mathrm{Spa}(A, A^+)$, and then to open subsets of $\mathrm{Spa}^{\mathrm{ind}}(A, A^+)$. This implies in particular that if $f : X \to X'$ is any open immersion of objects of $(V)^{\mathrm{ind}}$, then the induced map $X^Y \to (X')^Y$ is an open immersion. Indeed, it suffices to check this after replacing X' by $\mathrm{Spa}^Y(A, A^+)$ (using the last displayed equation), and then it reduces to the previous assertion.

Thus, for a pre-adic space $X \in (V)^{\mathrm{ind}}$, any affinoid open subspace $\mathrm{Spa}^{\mathrm{ind}}(A, A^+) \subset X$ defines an open immersion $\mathrm{Spa}^Y(A, A^+) \subset X^Y$, and conversely any open immersion $\mathrm{Spa}^Y(A, A^+) \subset X^Y$ comes from an affinoid open subspace $\mathrm{Spa}^{\mathrm{ind}}(A, A^+) \subset X$. To see that X^Y is a Yoneda-adic space, it remains to see that

$$X^Y = \varinjlim_{\mathrm{Spa}^{\mathrm{ind}}(A, A^+) \subset X} \mathrm{Spa}^Y(A, A^+).$$

First, we check that the map

$$\varinjlim_{\mathrm{Spa}^{\mathrm{ind}}(A, A^+) \subset X} \mathrm{Spa}^Y(A, A^+) \to X^Y$$

is injective; note that here, the left-hand side is the sheafification of the functor

$$(B, B^+) \mapsto \varinjlim_{\mathrm{Spa}^{\mathrm{ind}}(A,A^+) \subset X} \mathrm{Hom}_{\mathrm{CAff}}((A, A^+), (B, B^+)) \, .$$

For checking injectivity, take any $(B, B^+) \in$ CAff, and assume that we have maps $(A, A^+) \to (B, B^+)$ and $(A', A'^+) \to (B, B^+)$ for affinoid open subspaces $\mathrm{Spa}^{\mathrm{ind}}(A, A^+) \subset X$, $\mathrm{Spa}^{\mathrm{ind}}(A', A'^+) \subset X$ such that the induced maps

$$\mathrm{Spa}^Y(B, B^+) \to \mathrm{Spa}^Y(A, A^+) \to X^Y \, , \; \mathrm{Spa}^Y(B, B^+) \to \mathrm{Spa}^Y(A', A'^+) \to X^Y$$

agree. It follows that the map $\mathrm{Spa}^Y(B, B^+) \to X^Y$ factors over the corresponding intersection of these two open subspaces, which we can cover by subsets $\mathrm{Spa}^Y(A'', A''^+) \subset X^Y$ that are simultaneously rational subsets of $\mathrm{Spa}^Y(A, A^+)$ and $\mathrm{Spa}^Y(A', A'^+)$. Thus, replacing (B, B^+) by a rational cover, we find that the two maps to $\varinjlim_{\mathrm{Spa}^{\mathrm{ind}}(A,A^+) \subset X} \mathrm{Spa}^Y(A, A^+)$ agree, as both factor over a common $\mathrm{Spa}^Y(A'', A''^+)$.

On the other hand, the map

$$\varinjlim_{\mathrm{Spa}^{\mathrm{ind}}(A,A^+) \subset X} \mathrm{Spa}^Y(A, A^+) \to X^Y$$

of sheaves is surjective, as locally any map to X^Y factors over $\mathrm{Spa}^Y(A, A^+)$ for some such open subset.

We have already checked that the functor $X \mapsto X^Y$ from pre-adic spaces to Yoneda-adic spaces satisfies

$$\mathrm{Hom}_{(V)^{\mathrm{ind}}}(X, X') = \mathrm{Hom}(X^Y, (X')^Y)$$

in case $X = \mathrm{Spa}^{\mathrm{ind}}(A, A^+)$ is affinoid. In general, arguing as above, we can write

$$X = \varinjlim_{\mathrm{Spa}^{\mathrm{ind}}(A,A^+) \subset X} \mathrm{Spa}^{\mathrm{ind}}(A, A^+)$$

in $(V)^{\mathrm{ind}}$, and this colimit is preserved by the functor $X \mapsto X^Y$ by what was proved above. This implies the full faithfulness in general. Preservation of open immersions is easy to see.

It remains to prove essential surjectivity, so assume $\mathcal{F}$ is a Yoneda-adic space. By Zorn's lemma, there is a maximal open immersion $\mathcal{F}' \subset \mathcal{F}$ such that $\mathcal{F}' = U^Y$ for some pre-adic space U. If $\mathcal{F}' \neq \mathcal{F}$, we can find an open immersion $V = \mathrm{Spa}^Y(A, A^+) \subset \mathcal{F}$ that does not factor over $\mathcal{F}'$. The intersection of V and $\mathcal{F}'$ defines an open subspace $W \subset U$. One gets an induced map $(U \sqcup_W \mathrm{Spa}^{\mathrm{ind}}(A, A^+))^Y \to \mathcal{F}$. This is still an open immersion, which contradicts maximality of $\mathcal{F}'$. $\qquad\qquad\square$

Lecture 4

Examples of adic spaces

Today, we will discuss various examples of adic spaces.

4.1 BASIC EXAMPLES

We gather here some examples of adic spaces. For brevity, we write $\mathrm{Spa}\,A :=$ $\mathrm{Spa}(A, A^\circ)$ from now on. Moreover, for any Huber pair (A, A^+), we use the notation $\mathrm{Spa}(A, A^+)$ to denote the associated pre-adic space.

- The final object is $\mathrm{Spa}\,\mathbf{Z} = \mathrm{Spa}(\mathbf{Z}, \mathbf{Z})$.

- (The adic closed unit disc.) The space $\mathrm{Spa}\,\mathbf{Z}[T] = \mathrm{Spa}(\mathbf{Z}[T], \mathbf{Z}[T])$ represents the functor $X \mapsto \mathcal{O}_X^+(X)$. Note that if K is a nonarchimedean field, then

$$\mathrm{Spa}\,\mathbf{Z}[T] \times \mathrm{Spa}\,K = \mathrm{Spa}\,K\langle T\rangle,$$

  which has been discussed in Example 2.3.7; cf. also [Sch12, Example 2.20].

- (The adic affine line.) The functor $X \mapsto \mathcal{O}_X(X)$ is also representable, by $\mathrm{Spa}(\mathbf{Z}[T], \mathbf{Z})$. If K is any nonarchimedean field, then

$$\mathrm{Spa}(\mathbf{Z}[T], \mathbf{Z}) \times \mathrm{Spa}\,K = \bigcup_{n \geq 1} \mathrm{Spa}\,K\langle \varpi^n T\rangle$$

  is an increasing union of closed discs $|T| \leq |\varpi|^{-n}$. Here $\varpi \in K$ is any *pseudo-uniformizer*, i.e., a topologically nilpotent unit. One can check this using the universal property. Indeed, $\mathrm{Spa}(\mathbf{Z}[T], \mathbf{Z}) \times \mathrm{Spa}\,K$ represents the functor which sends a Huber pair (R, R^+) over $(K, \mathcal{O}_K)$ to R, and then R is the union of the $\varpi^{-n} R^+$ for $n = 1, 2, \dots$ (since R^+ is open and ϖ is topologically nilpotent). Note that between this example and the previous one, it was the ring R^+ that made all the difference!

- (The closure of the adic closed unit disc in the adic affine line.) Let K be a nonarchimedean field, let $X = \mathrm{Spa}\,K\langle T\rangle$ be the adic closed unit disc over

K, and let $Y = \mathrm{Spa}(\mathbf{Z}[T], \mathbf{Z}) \times \mathrm{Spa}\, K$ be the adic affine line over K. Then $X \subset Y$ is an open immersion. The closure of X in Y is $\mathrm{Cont}(K\langle T \rangle)$: this is X together with a rank 2 "boundary point" x_1^+. Explicitly, let $\Gamma = \mathbf{R}_{>0} \times \gamma^{\mathbf{Z}}$, where $1 < \gamma < r$ for all real $r < 1$. Then x_1^+ is defined by

$$\sum_{n \geq 0} a_n T^n \mapsto \max_{n \geq 0} |a_n| \gamma^n.$$

Finally, note that $\mathrm{Cont}(K\langle T \rangle) = \mathrm{Spa}(A, A^+)$, where $A = K\langle T \rangle$ and $A^+ \subset A$ is the subring of power series $\sum_{n \geq 0} a_n T^n$ where $a_0 \in \mathcal{O}_K$ and $|a_n| < 1$ for all $n \geq 1$.

- (Fiber products do not exist in general.) In the sense of Hom-functors, the product
$$\mathrm{Spa}(\mathbf{Z}[T_1, T_2, \ldots], \mathbf{Z}) \times \mathrm{Spa}\, K$$
  equals
$$\varinjlim_{(n_i) \to \infty} \mathrm{Spa}\, K\langle \varpi^{n_1} T_1, \varpi^{n_2} T_2, \ldots \rangle \, .$$

But in this direct limit, the transition maps are not open immersions; they are given by infinitely many inequalities $|T_i| \leq |\varpi|^{-n_i}$. So this direct limit is not representable as an adic space. If one restricts the class of Huber pairs to those pairs (A, A^+) for which A is finitely generated over a ring of definition $A_0 \subset A^+$, then fiber products will always exist in the category of pre-adic spaces.

- (The open unit disc.) Let $\mathbf{D} = \mathrm{Spa}\, \mathbf{Z}[\![T]\!]$. Then
$$\begin{aligned} \mathbf{D}_K := \mathbf{D} \times \mathrm{Spa}\, K &= [\mathbf{D} \times \mathrm{Spa}\, \mathcal{O}_K] \times_{\mathrm{Spa}\, \mathcal{O}_K} \mathrm{Spa}\, K \\ &= \mathrm{Spa}\, \mathcal{O}_K[\![T]\!] \times_{\mathrm{Spa}\, \mathcal{O}_K} \mathrm{Spa}\, K \\ &= \bigcup_{n \geq 1} \mathrm{Spa}\, K\langle T, \frac{T^n}{\varpi} \rangle \end{aligned}$$

is the open unit disc over K. This is another adic space, even though we have not proved the intermediate space $\mathrm{Spa}\, \mathcal{O}_K[\![T]\!]$ to be one. This shows the importance of allowing pre-adic spaces—they may appear as auxiliary objects in some calculations, for example.

- (The punctured open unit disc.) Let $\mathbf{D}^* = \mathrm{Spa}\, \mathbf{Z}(\!(T)\!)$. Then
$$\mathbf{D}_K^* := \mathbf{D}^* \times \mathrm{Spa}\, K = \mathbf{D}_K \backslash \{T = 0\} \, .$$

- (The constant adic space associated to a profinite set.) Let S be a profinite set, and let $A = C^0(S, \mathbf{Z})$ be the ring of continuous (thus locally constant) $\mathbf{Z}$-valued functions on S, with its discrete topology. Then $\underline{S} = \mathrm{Spa}\, A$ represents the functor $X \mapsto \mathrm{Hom}(|X|, S)$, and furthermore $|\underline{S}| = S \times |\mathrm{Spa}\,\mathbf{Z}|$. If K is a nonarchimedean field then $\underline{S}_K = \underline{S} \times \mathrm{Spa}\,K$ is the constant adic space over K, and $|\underline{S}_K| = S$. This construction can be globalized to the case of locally profinite sets S as well.

4.2   EXAMPLE: THE ADIC OPEN UNIT DISC OVER $\mathbf{Z}_P$

Let us now discuss one example more in depth. The adic spectrum $\mathrm{Spa}\,\mathbf{Z}_p$ consists of two points, a special point and a generic point. The same is true for $\mathrm{Spa}\,\mathbf{F}_p[\![T]\!]$, and more generally for $\mathrm{Spa}\,A$ for any valuation ring A of rank 1.

But now consider $\mathbf{Z}_p[\![T]\!]$ with the (p, T)-adic topology; this is a complete regular local ring of dimension 2. Then $\mathrm{Spa}\,\mathbf{Z}_p[\![T]\!]$ falls under case (2) of Theorem 3.1.8. Let us try to describe $X = \mathrm{Spa}\,\mathbf{Z}_p[\![T]\!]$.

There is a unique point $x_{\mathbf{F}_p} \in X$ whose kernel is open. It is the composition $\mathbf{Z}_p[\![T]\!] \to \mathbf{F}_p \to \{0, 1\}$, where the second arrow is 1 on nonzero elements. Let $\mathcal{Y} = X \backslash \{x_{\mathbf{F}_p}\}$. All points in $\mathcal{Y}$ have a non-open kernel, i.e. they are *analytic*:

Definition 4.2.1. Let (A, A^+) be a Huber pair. A point $x \in \mathrm{Spa}(A, A^+)$ is *non-analytic* if the kernel of $|\cdot|_x$ is open. Otherwise x is *analytic*.

Let us discuss the structure of analytic points. Suppose $A_0 \subset A$ is a ring of definition, and $I \subset A_0$ is an ideal of definition. If $x \in \mathrm{Spa}(A, A^+)$ is analytic, then the kernel of $|\cdot|_x$, not being open, cannot contain I. Thus there exists $f \in I$ such that $|f(x)| \neq 0$. Let $\gamma = |f(x)| \in \Gamma = \Gamma_x$. Since $f^n \to 0$ as $n \to \infty$, we must have $|f(x)|^n \to 0$. This means that for all $\gamma' \in \Gamma$ there exists $n \gg 0$ such that $\gamma^n < \gamma'$.

Lemma 4.2.2. Let Γ be a totally ordered abelian group, and let $\gamma < 1$ in Γ. Suppose that for all $\gamma' \in \Gamma$ there exists $n \gg 0$ such that $\gamma^n < \gamma'$. Then there exists a unique order-preserving homomorphism $\Gamma \to \mathbf{R}_{>0}$ which sends γ to $1/2$. (The kernel of this map consists of elements which are "infinitesimally close to 1".)

Proof. Exercise.   □

As an example, if x has value group $\Gamma_x = \mathbf{R}_{>0} \times \delta^{\mathbf{Z}}$ where $r < \delta < 1$ for all $r \in \mathbf{R}$, $r < 1$, then the map $\Gamma_x \to \mathbf{R}_{>0}$ of the lemma is just the projection, up to scaling.

Thus, any analytic point x gives rise to a continuous valuation $\widetilde{x} \colon A \to \mathbf{R}_{\geq 0}$.

Definition 4.2.3. A *nonarchimedean field* is a complete nondiscrete topological field K whose topology is induced by a nonarchimedean norm $|\,| \colon K \to \mathbf{R}_{\geq 0}$.

For an analytic point $x \in X$, let $K(x)$ be the completion of $\mathrm{Frac}(A/\ker|\cdot|_x)$ with respect to $|\,|_x$. The lemma shows that if x is analytic, then $K(x)$ is a nonarchimedean field. At non-analytic points of x, we endow $K(x) = \mathrm{Frac}(A/\ker |\cdot|_x)$ with the discrete topology. (In the situation of the special point $x_{\mathbf{F}_p}$ of our example, $K(x_{\mathbf{F}_p}) = \mathbf{F}_p$.) Note that x endows $K(x)$ moreover with a continuous valuation, or equivalently with an open and bounded valuation subring $K(x)^+ \subset K(x)$.

Definition 4.2.4 ([Hub96, Definition 1.1.5]). An *affinoid field* is a Huber pair (K, K^+) where K is either a nonarchimedean field or a discrete field, and K^+ is an open and bounded valuation subring.

We note that this definition makes crucial use of the second component of a Huber pair. This leads to a different perspective on the adic spectrum, akin to regarding $\mathrm{Spec}\, A$ as equivalence classes of maps from A into fields.

Proposition 4.2.5 ([Hub96, pp. 40-41]). Let (A, A^+) be a Huber pair. Points of $\mathrm{Spa}(A, A^+)$ are in bijection with maps $(A, A^+) \to (K, K^+)$ to affinoid fields such that the subfield of K generated by the image of $A \to K$ is dense. If $x \in \mathrm{Spa}(A, A^+)$ is an analytic point corresponding to a map $(A, A^+) \to (K(x), K(x)^+)$, then generalizations y of x in $\mathrm{Spa}(A, A^+)$ correspond to maps $(A, A^+) \to (K(y), K(y)^+)$ with $K(y) = K(x)$ and $K(y)^+ \supset K(x)^+$. In particular, the set of generalizations of x forms a totally ordered chain of length given by the rank of the valuation x, and the maximal generalization corresponds to the rank-1-valuation $\widetilde{x}$ introduced above.

Let us return to our example $\mathcal{Y} = X\backslash\{x_{\mathbf{F}_p}\}$, with $X = \mathrm{Spa}\,\mathbf{Z}_p[\![T]\!]$. For $x \in \mathcal{Y}$, we have that $|T(x)|$ and $|p(x)|$ cannot both be zero. Both are elements of the value group which are topologically nilpotent. We can measure their relative position as an element of $[0, \infty]$.

Proposition 4.2.6. There is a unique continuous map

$$\kappa \colon |\mathcal{Y}| \to [0, \infty]$$

characterized by the following property: $\kappa(x) = r$ if and only for all rational numbers $m/n > r$, $|T(x)|^n \geq |p(x)|^m$, and for all $m/n < r$, $|T(x)|^n \leq |p(x)|^m$. The map κ is surjective.

Proof. (Sketch.) Any $x \in \mathcal{Y}$ is analytic, so there exists a maximal generalization $\widetilde{x}$ which is real-valued. We define

$$\kappa(x) = \frac{\log |T(\widetilde{x})|}{\log |p(\widetilde{x})|} \in [0, \infty].$$

The numerator and denominator both lie in $[-\infty, 0)$, with at most one being equal to $-\infty$, so the quotient is indeed well-defined in $[0, \infty]$. The continuity, uniqueness, and surjectivity of the map are left as an exercise. □

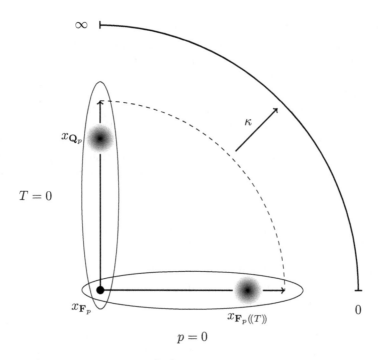

Figure 4.1: A depiction of $\operatorname{Spa}\mathbf{Z}_p[\![T]\!]$. The two closed subspaces $\operatorname{Spa}\mathbf{F}_p[\![T]\!]$ and $\operatorname{Spa}\mathbf{Z}_p$ appear as the x-axis and y-axis, respectively. Their intersection is the unique non-analytic point $x_{\mathbf{F}_p}$ of $\operatorname{Spa}A$. The complement of $x_{\mathbf{F}_p}$ in $\operatorname{Spa}\mathbf{Z}_p[\![T]\!]$ is the adic space $\mathcal{Y}$, on which the continuous map $\kappa\colon \mathcal{Y}\to[0,\infty]$ is defined.

We have $\kappa(x) = 0$ if and only if $|p(x)| = 0$, which is to say that $|\cdot|_x$ factors through $\mathbf{F}_p[\![T]\!]$. Similarly $\kappa(x) = \infty$ if and only if $|\cdot|_x$ factors through $\mathbf{Z}_p[\![T]\!] \to \mathbf{Z}_p$, $T \mapsto 0$. For an interval $I \subset [0,\infty]$, we define $\mathcal{Y}_I$ as the interior of $\kappa^{-1}(I)$.

An important subspace of $\mathcal{Y}$ is $\mathcal{Y}_{(0,\infty]}$, which is to say the locus $p \neq 0$. This is the generic fiber of $\operatorname{Spa}\mathbf{Z}_p[\![T]\!]$ over $\operatorname{Spa}\mathbf{Z}_p$. It is not quasi-compact (otherwise its image under κ would lie in a compact interval), and in particular it is not affinoid.

The failure of the generic fiber $\mathcal{Y}_{(0,\infty]}$ to be affinoid may be surprising. One might think that the fiber of $\operatorname{Spa}\mathbf{Z}_p[\![T]\!] \to \operatorname{Spa}\mathbf{Z}_p$ over $\operatorname{Spa}(\mathbf{Q}_p,\mathbf{Z}_p)$ should be something like $\operatorname{Spa}(\mathbf{Z}_p[\![T]\!][1/p],\mathbf{Z}_p[\![T]\!])$. The trouble arises when we consider what topology to put on $\mathbf{Z}_p[\![T]\!][1/p]$. If we give $\mathbf{Z}_p[\![T]\!][1/p]$ the topology induced from the (p,T)-adic topology of $\mathbf{Z}_p[\![T]\!]$, the result is not even a Huber ring! (The sequence $p^{-1}T^n$ approaches 0, but never enters $\mathbf{Z}_p[\![T]\!]$, and so $\mathbf{Z}_p[\![T]\!]$ is not an open subring.) Another explanation for this failure is that $\mathbf{Z}_p[\![T]\!][1/p]$ would have to be Tate (as p is a topologically nilpotent unit), but we have seen that any ring of definition of a Tate ring has ideal of definition generated by one element, in this case p.

We could also give $\mathbf{Z}_p[\![T]\!][1/p]$ the topology induced from the p-adic topology on $\mathbf{Z}_p[\![T]\!]$. The result is a Tate ring, but it does not receive a continuous ring homomorphism from $\mathbf{Z}_p[\![T]\!]$ (with the (p,T)-adic topology), since $T^n \to 0$ in the latter but not the former. Thus $\mathrm{Spa}(\mathbf{Z}_p[\![T]\!][1/p], \mathbf{Z}_p[\![T]\!])$ cannot be the desired generic fiber.

To review: the generic fiber of $\mathrm{Spa}\,\mathbf{Z}_p[\![T]\!] \to \mathrm{Spa}\,\mathbf{Z}_p$ is an adic space $\mathcal{Y}_{(0,\infty]}$ which is not quasi-compact. We can exhibit an explicit cover by rational subsets of $\mathrm{Spa}\,\mathbf{Z}_p[\![T]\!]$. Suppose $x \in \mathcal{Y}_{(0,\infty]}$. Since T is topologically nilpotent in $\mathbf{Z}_p[\![T]\!]$, we have $|T^n(x)| \to 0$. But also $|p(x)| \neq 0$, by definition of $\mathcal{Y}_{(0,\infty]}$. Therefore there exists $n \gg 0$ such that $|T^n(x)| \leq |p(x)|$, which is to say that x belongs to

$$\mathcal{Y}_{[1/n,\infty]} = \mathrm{Spa}\,\mathbf{Q}_p\langle T, T^n/p\rangle \ .$$

This is indeed a rational subset, because (T^n, p) is an open ideal. The $\mathcal{Y}_{[1/n,\infty]}$ exhaust $\mathcal{Y}_{(0,\infty]}$. Put another way, we have exhibited the open disc $\{|T| < 1\}$ over $\mathbf{Q}_p$ by an ascending union of affinoid subsets $\left\{|T| \leq |p|^{1/n}\right\}$.

To complete the picture of $\mathcal{Y}$, we discuss an affinoid neighborhood of the point $x_{\mathbf{F}_p(\!(T)\!)} \in \mathcal{Y}$, the unique point sent to 0 by κ. This is the point where $p = 0$ and $T \neq 0$. Let $U = \mathcal{Y}_{[0,1]}$ be the rational subset $\{|p(x)| \leq |T(x)| \neq 0\}$. Then $\mathcal{O}_X(U)$ is the completion of $\mathbf{Z}_p[\![T]\!][1/T]$ with respect to the T-adic topology on $\mathbf{Z}_p[\![T]\!][p/T]$. One might call this $\mathbf{Z}_p[\![T]\!]\langle p/T\rangle[1/T]$, an unfortunately complicated name. It is still a Tate ring, because T is topologically nilpotent. But it does not contain a nonarchimedean field! Thus one cannot make sense of it in the world of classical rigid spaces. Recently, adic spaces of this form have been given the name "pseudorigid spaces"—cf. [JN16], [Lou17]—and have found applications to the study of the boundary of the eigencurve and more general eigenvarieties, [AIP18], [JN16]. Lourenço's results [Lou17] will play an important role in Section 18.4.

As we progress in the course, we will encounter adic spaces similar to $\mathcal{Y}$ which are built out of much stranger rings, but for which the picture is essentially the same. Finally, we remark that the entire picture has a characteristic p analogue, in which $\mathbf{Z}_p$ is replaced with $\mathbf{F}_p[\![t]\!]$; one would begin with $X = \mathrm{Spa}\,\mathbf{F}_p[\![t,T]\!]$ and remove its sole non-analytic point to obtain an analytic adic space $\mathcal{Y}$. This object retains all of the features of its mixed characteristic counterpart, including the map κ, but has the additional feature that the roles played by t and T are completely symmetric.

4.3 ANALYTIC POINTS

The following proposition clarifies the relations between analytic rings and Tate rings.

Proposition 4.3.1. Let (A, A^+) be a complete Huber pair.

1. The Huber ring A is analytic—cf. Remark 2.2.7—if and only if all points of $\mathrm{Spa}(A, A^+)$ are analytic.
2. A point $x \in \mathrm{Spa}(A, A^+)$ is analytic if and only if there is a rational neighborhood $U \subset \mathrm{Spa}(A, A^+)$ of x such that $\mathcal{O}_X(U)$ is Tate.

Proof. Let $X = \mathrm{Spa}(A, A^+)$. Let $I \subset A$ be the ideal generated by the topologically nilpotent elements. Then a point $x \in X$ is analytic if and only if x does not lie in the vanishing locus of I. Indeed, if x is nonanalytic, there is some topologically nilpotent element which does not vanish at x; conversely, if all topologically nilpotent elements vanish on x, then x defines a continuous valuation of A/I, which is a discrete ring. Thus, X is analytic if and only if $I = A$, i.e., A is analytic. This is (1).

For (2), let $x \in X$ be an analytic point. We need to show that there exists a rational neighborhood U of x such that $\mathcal{O}_X(U)$ is Tate. Let $I \subset A_0$ be as usual. Take $f \in I$ to be such that $|f(x)| \neq 0$. Then $\{g \in A \mid |g(x)| \leq |f(x)|\}$ is open (by the continuity of the valuations). This means that there exists n so that this set contains I^n. Write $I^n = (g_1, \ldots, g_k)$. Then

$$ U = \left\{ y \in X \;\middle|\; |g_i(y)| \leq |f(y)| \neq 0 \right\} $$

is a rational subset. On U, the function f is a unit (because it is everywhere nonzero), but it must also be topologically nilpotent, because it is contained in I.

Conversely, suppose $x \in X$ has a rational neighborhood $U = U(T/s)$ such that $\mathcal{O}_X(U)$ is Tate; we claim that x is analytic. Assume otherwise; then $\ker(|\cdot|_x)$ contains $I \subset A_0$. Now suppose $f \in \mathcal{O}_X(U)$ is a topologically nilpotent unit. Then there exists $m \geq 1$ such that f^m lies in the closure of $I A_0[t/s \mid t \in T]$ in $\mathcal{O}_X(U)$. Since x lies in U, the valuation $|\cdot|_x$ extends to $\mathcal{O}_X(U)$, and since it is continuous, we must have $|f^m(x)| = 0$. This is a contradiction, since f is a unit in $\mathcal{O}_X(U)$. $\qquad\square$

In particular, this allows us to define the notion of analytic points for any pre-adic space.

Definition 4.3.2. Let X be a pre-adic space. A point $x \in X$ is analytic if there is some open affinoid neighborhood $U = \mathrm{Spa}(A, A^+) \subset X$ of x such that A is Tate. Moreover, X is analytic if all of its points are analytic.

Proposition 4.3.3. Let $f : Y \to X$ be a map of analytic pre-adic spaces. Then $|f| : |Y| \to |X|$ is generalizing. If f is quasicompact and surjective, then $|f|$ is a quotient mapping, i.e., a subset of $|X|$ is open if and only if its preimage in $|Y|$ is open.

Proof. For the first part, cf. [Hub96, Lemma 1.1.10], for the second [Sch17, Lemma 2.5]. $\qquad\square$

Analytic Huber rings are "as good as" Banach algebras over nonarchimedean fields, for example:

Proposition 4.3.4 ([Hub93, Lemma 2.4(i)]). *Analytic Huber rings satisfy Banach's open mapping theorem. That is, if A is an analytic Huber ring, and M and N are complete Banach A-modules, then any continuous surjective map $M \to N$ is also open.*

Lecture 5

Complements on adic spaces

Today's lecture is a collection of complements in the theory of adic spaces.

5.1 ADIC MORPHISMS

Definition 5.1.1. A morphism $f\colon A \to B$ of Huber rings is *adic* if for one (hence any) choice of rings of definition $A_0 \subset A$, $B_0 \subset B$ with $f(A_0) \subset B_0$, and $I \subset A_0$ an ideal of definition, $f(I)B_0$ is an ideal of definition.
  A morphism $(A, A^+) \to (B, B^+)$ of Huber pairs is adic if $A \to B$ is.

For example, the maps corresponding to rational subsets are adic.

Lemma 5.1.2. If A is Tate, then any $f\colon A \to B$ is adic.

Proof. If A contains a topologically nilpotent unit ϖ, then $f(\varpi) \in B$ is also a topologically nilpotent unit, and thus B is Tate as well. By Proposition 2.2.6(2), B contains a ring of definition B_0 admitting $f(\varpi)^n B_0$ as an ideal of definition for some $n \geq 1$. This shows that f is adic. $\qquad\square$

In fact, one has the following geometric characterization of adic maps.

Proposition 5.1.3 ([Hub93, Proposition 3.8]). A map $(A, A^+) \to (B, B^+)$ of complete Huber pairs is adic if and only if $\mathrm{Spa}(B, B^+) \to \mathrm{Spa}(A, A^+)$ carries analytic points to analytic points.

In particular, the previous lemma holds more generally when A is analytic.

Definition 5.1.4. A map $f\colon Y \to X$ of pre-adic spaces is *analytic* if it carries analytic points to analytic points.

The next proposition shows that pushouts exist in the category of (complete) Huber pairs when maps are adic.

Proposition 5.1.5. 1. If $(A, A^+) \to (B, B^+)$ is adic, then pullback along the associated map of topological spaces $\mathrm{Spa}(B, B^+) \to \mathrm{Spa}(A, A^+)$ preserves rational subsets.

2. Let $(B, B^+) \leftarrow (A, A^+) \rightarrow (C, C^+)$ be a diagram of Huber pairs where both morphisms are adic. Let A_0, B_0, C_0 be rings of definition compatible with the morphisms, and let $I \subset A_0$ be an ideal of definition. Let $D = B \otimes_A C$, and let D_0 be the image of $B_0 \otimes_{A_0} C_0$ in D. Make D into a Huber ring by declaring D_0 to be a ring of definition with ID_0 as its ideal of definition. Let D^+ be the integral closure of the image of $B^+ \otimes_{A^+} C^+$ in D. Then (D, D^+) is a Huber pair, and it is the pushout of the diagram in the category of Huber pairs.

Proof. In part (1), it is enough to observe that if TA is open for some finite subset $T \subset A$, then $I \subset TA$ for some ideal of definition $I \subset A_0$, in which case $IB_0 \subset B_0$ is also an ideal of definition, thus open, and so $TB \supset IB$ is open. Part (2) follows easily from the definitions. □

Remark 5.1.6. If the objects in the diagram were complete Huber pairs, then after completing (D, D^+), one would obtain the pushout of the diagram in the category of complete Huber pairs, and one also has $\mathrm{Spa}(D, D^+) = \mathrm{Spa}(B, B^+) \times_{\mathrm{Spa}(A, A^+)} \mathrm{Spa}(C, C^+)$.

Remark 5.1.7. An example of a non-adic morphism of Huber rings is $\mathbf{Z}_p \rightarrow \mathbf{Z}_p[\![T]\!]$. We claim that, as indicated in the last lecture, the diagram $(\mathbf{Z}_p[\![T]\!], \mathbf{Z}_p[\![T]\!]) \leftarrow (\mathbf{Z}_p, \mathbf{Z}_p) \rightarrow (\mathbf{Q}_p, \mathbf{Z}_p)$ has no pushout in the category of Huber pairs. Suppose it did, say (D, D^+); then we have a morphism

$$(D, D^+) \rightarrow (\mathbf{Q}_p \langle T, \tfrac{T^n}{p} \rangle, \mathbf{Z}_p \langle T, \tfrac{T^n}{p} \rangle)$$

for each $n \geq 1$. On the other hand, we have that $T \in D$ is topologically nilpotent, and $1/p \in D$. Therefore $T^n/p \rightarrow 0$ in D; now since $D^+ \subset D$ is open we have $T^n/p \in D^+$ for some n. But then D^+ cannot admit a morphism to $\mathbf{Z}_p \langle T, T^{n+1}/p \rangle$, a contradiction.

5.2   ANALYTIC ADIC SPACES

We discuss some recent results from [BV18], [Mih16], [KL15], and [Ked19a]. Let (A, A^+) be an analytic Huber pair, and let $X = \mathrm{Spa}(A, A^+)$. It turns out that there is a very general criterion for sheafyness. Recall that A is *uniform* if $A^\circ \subset A$ is bounded.

Theorem 5.2.1 (Berkovich, [Ber93]). For A uniform, the map

$$A \rightarrow \prod_{x \in \mathrm{Spa}(A, A^+)} K(x)$$

is a homeomorphism of A onto its image. Here $K(x)$ is the completed residue field. Moreover,

$$A^\circ = \left\{ f \in A \mid f \in \mathcal{O}_{K(x)}, \ \forall x \in X \right\}.$$

Remark 5.2.2. The theorem also follows from $A^+ = \{ f \in A \mid |f(x)| \leq 1, \ \forall x \in X \}$.

Corollary 5.2.3. Let $\widetilde{\mathcal{O}}_X$ be the sheafification of $\mathcal{O}_X$. If A is uniform, then $A \to H^0(X, \widetilde{\mathcal{O}}_X)$ is injective.

Proof. Indeed, the H^0 maps into $\prod_x K(x)$, into which A maps injectively. $\square$

In general, uniformity does not guarantee sheafyness, but a strengthening of the uniformity condition does.

Definition 5.2.4. A complete analytic Huber pair (A, A^+) is *stably uniform* if $\mathcal{O}_X(U)$ is uniform for all rational subsets $U \subset X = \mathrm{Spa}(A, A^+)$.

Theorem 5.2.5 ([BV18, Theorem 7], [KL15, Theorem 2.8.10], [Mih16], [Ked19a]). If the complete analytic Huber pair (A, A^+) is stably uniform, then it is sheafy.

Moreover, sheafyness, without any extra assumptions, implies other good properties.

Theorem 5.2.6 ([KL15, Theorem 2.4.23], [Ked19a]). If the complete analytic Huber pair (A, A^+) is sheafy, then $H^i(X, \mathcal{O}_X) = 0$ for $i > 0$.

The strategy of the proof is to use combinatorial arguments going back to Tate to reduce to checking everything for a simple Laurent covering $X = U \cup V$, where $U = \{|f| \leq 1\}$ and $V = \{|f| \geq 1\}$ (plus one other similar case if A is not Tate). Then $\mathcal{O}_X(U) = A\langle T \rangle / (T - f)$ and $\mathcal{O}_X(V) = A\langle S \rangle / (Sf - 1)$. We have $\mathcal{O}_X(U \cap V) = A\langle T, T^{-1} \rangle / (T - f)$. We need to check that the Čech complex for this covering is exact. It is

$$0 \to A \to A\langle T \rangle / \overline{(T - f)} \oplus A\langle S \rangle / \overline{(Sf - 1)} \xrightarrow{\alpha} A\langle T, T^{-1} \rangle / \overline{(T - f)} \to 0.$$

Lemma 5.2.7. 1. The map α is surjective.
2. If A is uniform, then the ideals $(T - f) \subset A\langle T \rangle$, $(Sf - 1) \subset A\langle S \rangle$ and $(T - f) \subset A\langle T, T^{-1} \rangle$ are closed, and the kernel of α is A.

Proof. 1) is clear. For 2), the hard part is to show that the ideals are closed. By Theorem 5.2.1, the norm on $A\langle T \rangle$ is the supremum norm, and then for all $f \in A\langle T \rangle$,

$$|f|_{A\langle T \rangle} = \sup_{x \in \mathrm{Spa}(A, A^+)} |f|_{K(x)\langle T \rangle}.$$

For each $x \in \mathrm{Spa}(A, A^+)$, the norm on $K(x)\langle T \rangle$ is the Gauss norm, and it is multiplicative.

We claim that for all $g \in A\langle T \rangle$, $|(T - f)g|_{A\langle T \rangle} \geq |g|_{A\langle T \rangle}$. By the above, it is enough to prove this when A is replaced by $K(x)$. But then the norm is multiplicative, and $|T - f|_{K(x)\langle T \rangle} = 1$, so we get the result. Using this observation, it is easy to see that the ideal generated by $T - f$ is closed. □

Moreover, there is a good theory of vector bundles.

Theorem 5.2.8 ([KL15, Theorem 2.7.7], [Ked19a]). *Let (A, A^+) be a sheafy analytic Huber pair and $X = \mathrm{Spa}(A, A^+)$. The functor from finite projective A-modules to locally finite free $\mathcal{O}_X$-modules is an equivalence.*

It is not immediately clear how to get a good theory of coherent sheaves on adic spaces, but in [KL16] and [Ked19a], Kedlaya-Liu define a category of $\mathcal{O}_X$-modules which are "pseudo-coherent," with good properties. We will not use these results.

The strategy of proof of the theorem is to reduce to simple Laurent coverings, and then imitate the proof of Beauville-Laszlo [BL95], who prove the following lemma.

Lemma 5.2.9. *Let R be a commutative ring, let $f \in R$ be a non-zero-divisor, and let $\widehat{R}$ be the f-adic completion of R. Then the category of R-modules M where f is not a zero-divisor is equivalent to the category of pairs $(M_{\widehat{R}}, M[f^{-1}], \beta)$, where $M_{\widehat{R}}$ is an $\widehat{R}$-module such that f is not a zero-divisor, $M_{R[f^{-1}]}$ is an $R[f^{-1}]$-module, and $\beta \colon M_{\widehat{R}}[f^{-1}] \to M_{R[f^{-1}]} \otimes_R \widehat{R}$ is an isomorphism. Under this equivalence, M is finite projective if and only if $M_{\widehat{R}}$ and $M_{R[f^{-1}]}$ are finite projective.*

This does not follow from fpqc descent because of two subtle points: $R \to \widehat{R}$ might not be flat if R is not noetherian, and also we have not included a descent datum on $\widehat{R} \otimes_R \widehat{R}$.

5.3   CARTIER DIVISORS

For later use, we include a discussion of Cartier divisors on adic spaces, specialized to the case of interest.

Definition 5.3.1. An adic space X is *uniform* if for all open affinoid $U = \mathrm{Spa}(R, R^+) \subset X$, the Huber ring R is uniform.

Note that if X is an analytic adic space, then X is uniform if and only if X is covered by open affinoid $U = \mathrm{Spa}(R, R^+)$ with R stably uniform.

Definition 5.3.2. Let X be a uniform analytic adic space. A *Cartier divisor* on X is an ideal sheaf $\mathcal{I} \subset \mathcal{O}_X$ that is locally free of rank 1. The support of a Cartier divisor is the support of $\mathcal{O}_X / \mathcal{I}$.

Proposition 5.3.3. Let (R, R^+) be a stably uniform analytic Huber pair, and let $X = \mathrm{Spa}(R, R^+)$. If $\mathcal{I} \subset \mathcal{O}_X$ is a Cartier divisor, then its support $Z \subset X$ is a nowhere dense closed subset of X. The map $I \mapsto \mathcal{I} = I \cdot \mathcal{O}_X$ induces a bijective correspondence between invertible ideals $I \subset R$ such that the vanishing locus of I in X is nowhere dense and Cartier divisors on X.

Proof. First, note that by Theorem 5.2.8, any Cartier divisor is of the form $I \otimes_R \mathcal{O}_X$ for an invertible ideal $I \subset R$. We need to see that the map $I \otimes_R \mathcal{O}_X \to \mathcal{O}_X$ is injective as a map of sheaves on X if and only if the vanishing locus of I is nowhere dense. By localization, we can assume that $I = (f)$ is principal. If the vanishing locus contains an open subset, then on this open subset, $f = 0$ as X is uniform, so the map $f : \mathcal{O}_X \to \mathcal{O}_X$ is not injective. Conversely, we have to see that f is a nonzerodivisor if its vanishing locus $Z \subset X$ is nowhere dense. Assume $fg = 0$ for some $g \in R$. The locus $S = \{g = 0\} \subset X$ is closed and contains $X \setminus Z$. In other words, $X \setminus S$ is an open subset of X that is contained in Z; as Z is nowhere dense, this implies that $S = X$. Therefore g vanishes at all points of X, which implies $g = 0$ by uniformity. $\square$

Proposition 5.3.4. Let X be a uniform analytic adic space and $\mathcal{I} \subset \mathcal{O}_X$ a Cartier divisor with support Z and $j: U = X \setminus Z \hookrightarrow X$. There are injective maps of sheaves

$$\mathcal{O}_X \hookrightarrow \varinjlim_n \mathcal{I}^{\otimes -n} \hookrightarrow j_* \mathcal{O}_U .$$

Proof. We may assume $X = \mathrm{Spa}(R, R^+)$ and $\mathcal{I} = f\mathcal{O}_X$ for some nonzerodivisor $f \in R$ whose vanishing locus Z is nowhere dense, and check on global sections. Then we get the maps

$$R \to R[f^{-1}] \to H^0(U, \mathcal{O}_U) .$$

It suffices to see that $R \to H^0(U, \mathcal{O}_U)$ is injective. But if $g \in R$ vanishes on U, then the vanishing locus of g is a closed subset containing U, which implies that it is all of X as Z is nowhere dense. By uniformity, $g = 0$. $\square$

Definition 5.3.5. In the situation of Proposition 5.3.4, a function $f \in H^0(U, \mathcal{O}_U)$ is meromorphic along the Cartier divisor $\mathcal{I} \subset \mathcal{O}_X$ if it lifts (necessarily uniquely) to $H^0(X, \varinjlim_n \mathcal{I}^{\otimes -n})$.

Remark 5.3.6. A Cartier divisor is not in general determined by its support, and also this meromorphy condition depends in general on the Cartier divisor and not only its support.

We note that if $\mathcal{I} \subset \mathcal{O}_X$ is a Cartier divisor, one can form the quotient $\mathcal{O}_X/\mathcal{I}$. There may or may not be an adic space $Z \subset X$ whose underlying space is the support of $\mathcal{O}_X/\mathcal{I}$ and with $\mathcal{O}_Z = \mathcal{O}_X/\mathcal{I}$.

Definition 5.3.7. Let X be a uniform analytic adic space. A Cartier divisor $\mathcal{I} \subset \mathcal{O}_X$ on X with support Z is *closed* if the triple $(Z, \mathcal{O}_X/\mathcal{I}, (|\cdot(x)|, x \in Z))$

is an adic space.

The term closed Cartier divisor is meant to evoke a closed immersion of adic spaces; moreover, the next proposition shows that it is equivalent to asking that $\mathcal{I} \hookrightarrow \mathcal{O}_X$ have closed image.

Proposition 5.3.8. Let X be a uniform analytic adic space. A Cartier divisor $\mathcal{I} \subset \mathcal{O}_X$ is closed if and only if the map $\mathcal{I}(U) \hookrightarrow \mathcal{O}_X(U)$ has closed image for all open affinoid $U \subset X$. In that case, for all open affinoid $U = \mathrm{Spa}(R, R^+) \subset X$, the intersection $U \cap Z = \mathrm{Spa}(S, S^+)$ is an affinoid adic space, where $S = R/I$ and S^+ is the integral closure of the image of R^+ in S.

Proof. The condition on closed image can be checked locally. If the Cartier divisor is closed, then in particular $(\mathcal{O}_X/\mathcal{I})(U)$ is separated on an open cover of X, which means that $\mathcal{I}(U) \hookrightarrow \mathcal{O}_X(U)$ has closed image.

Conversely, assume that $\mathcal{I}(U) \hookrightarrow \mathcal{O}_X(U)$ has closed image for all $U \subset X$; we want to see that the Cartier divisor is closed. We can assume $X = \mathrm{Spa}(R, R^+)$ is affinoid. The ring $S = R/I$ is then separated and complete, and is thus a complete Huber ring. Let S^+ be the integral closure of the image of R^+ in S. Then $Z = \mathrm{Spa}(S, S^+) \to X = \mathrm{Spa}(R, R^+)$ is a closed immersion on topological spaces (with image the support of the Cartier divisor), and we get a natural map of topological sheaves $\mathcal{O}_X/\mathcal{I} \to \mathcal{O}_Z$. It is now easy to see that this is an isomorphism on rational subsets of X by repeating the above argument.   □

Remark 5.3.9. We make the warning that if A is Tate and stably uniform, $X = \mathrm{Spa}(A, A^+)$, and one has a Cartier divisor corresponding to a regular element $f \in A$ for which the ideal $fA \subset A$ is closed, it does not follow that $\mathcal{I} = f\mathcal{O}_X \hookrightarrow \mathcal{O}_X$ is a closed Cartier divisor; cf. [Ked19a, Example 1.4.8]. The problem is that one has to check that $f\mathcal{O}_X(U) \hookrightarrow \mathcal{O}_X(U)$ has closed image also on rational subsets, and this may in general fail. For a specific example, consider the affinoid subset $X = \mathcal{X}^*_{\Gamma(p^\infty)}(\epsilon)_a$ of the modular curve at infinite level, and the function given by the Hodge-Tate period map, in the notation of [Sch15]. Looking at the Shilov boundary (which is contained in the supersingular locus), one checks that it is a nonzerodivisor generating a closed ideal; however, on the open subset $\mathcal{X}^*_{\Gamma(p^\infty)}(0)_a$, it vanishes on a whole connected component.

Lecture 6

Perfectoid rings

Today we begin discussing perfectoid spaces. Let p be a fixed prime throughout.

6.1 PERFECTOID RINGS

Recall that a Huber ring R is Tate if it contains a topologically nilpotent unit; such elements are called pseudo-uniformizers. The following definition is due to Fontaine [Fon13].

Definition 6.1.1. A complete Tate ring R is *perfectoid* if R is uniform and there exists a pseudo-uniformizer $\varpi \in R$ such that $\varpi^p | p$ holds in R°, and such that the pth power Frobenius map

$$\Phi \colon R^\circ/\varpi \to R^\circ/\varpi^p$$

is an isomorphism.

Remark 6.1.2. One can more generally define when an analytic Huber ring is perfectoid; see [Ked19a]. There are also notions of integral perfectoid rings which are not analytic, e.g., [BMS18, Section 3], [GR16]. In this course, our perfectoid rings are all Tate. It would have been possible to proceed with the more general definition of perfectoid ring as a kind of anaytic Huber ring (in particular, the resulting category of perfectoid spaces would be the same). Being analytic is, however, critical for our purposes.

Hereafter we use the following notational convention. If R is a ring, and $I, J \subset R$ are ideals containing p such that $I^p \subset J$, then $\Phi \colon R/I \to R/J$ will refer to the ring homomorphism $x \mapsto x^p$.

Remark 6.1.3. Let us explain why the isomorphism condition above is independent of ϖ. For any complete Tate ring R and pseudo-uniformizer ϖ satisfying $\varpi^p | p$ in R°, the Frobenius map $\Phi \colon R^\circ/\varpi \to R^\circ/\varpi^p$ is necessarily injective. Indeed, if $x \in R^\circ$ satisfies $x^p = \varpi^p y$ for some $y \in R^\circ$ then the element $x/\varpi \in R$ lies in R° since its pth power does. Thus, the isomorphism condition on Φ in Definition 6.1.1 is really a *surjectivity* condition. In fact, this surjectivity condition is equivalent to the surjectivity of the Frobenius map

$$R^\circ/p \to R^\circ/p \ .$$

Clearly, if the Frobenius on R°/p is surjective, then so is $\Phi\colon R^\circ/\varpi \to R^\circ/\varpi^p$. Conversely, if $\Phi\colon R^\circ/\varpi \to R^\circ/\varpi^p$ is surjective, then by successive approximation, any element $x \in R^\circ$ can be written in the form

$$x = x_0^p + \varpi^p x_1^p + \varpi^{2p} x_2^p + \ldots$$

with all $x_i \in R^\circ$, in which case

$$x - (x_0 + \varpi x_1 + \varpi^2 x_2 + \ldots)^p \in pR^\circ \ .$$

Remark 6.1.4. The nonarchimedean field $\mathbf{Q}_p$ is not perfectoid, even though the Frobenius map on $\mathbf{F}_p$ is an isomorphism. The problem is that there is no topologically nilpotent element $\varpi \in \mathbf{Z}_p$ whose pth power divides p. More generally, a discretely valued non-archimedean field K cannot be perfectoid. Indeed, if ϖ is a pseudo-uniformizer as in Definition 6.1.1, then ϖ is a non-zero element of the maximal ideal, so the quotients K°/ϖ and K°/ϖ^p are Artin local rings of different lengths and hence they cannot be isomorphic.

Example 6.1.5. The following are examples of perfectoid Tate rings.

1. The cyclotomic extension $\mathbf{Q}_p^{\mathrm{cycl}}$, the completion of $\mathbf{Q}_p(\mu_{p^\infty})$.
2. The completion of $\mathbf{F}_p((t))(t^{1/p^\infty})$, which we will write as $\mathbf{F}_p((t^{1/p^\infty}))$.
3. $\mathbf{Q}_p^{\mathrm{cycl}}\langle T^{1/p^\infty}\rangle$. This is defined as $A[1/p]$, where A is the p-adic completion of $\mathbf{Z}_p^{\mathrm{cycl}}[T^{1/p^\infty}]$.
4. (An example which does not live over a field.) Recall from our discussion in Section 4.2 the ring $\mathbf{Z}_p[\![T]\!]\langle p/T\rangle[1/T]$, which is Tate with pseudo-uniformizer T, but which does not contain a nonarchimedean field. One can also construct a perfectoid version of it,

$$R = \mathbf{Z}_p^{\mathrm{cycl}}[\![T^{1/p^\infty}]\!]\langle (p/T)^{1/p^\infty}\rangle[1/T].$$

Here we can take $\varpi = T^{1/p}$, because $\varpi^p = T$ divides p in R°.

Proposition 6.1.6. Let R be a complete Tate ring with $pR = 0$. The following are equivalent:

1. R is perfectoid.
2. R is perfect.

Here, perfect means that $\Phi\colon R \to R$ is an isomorphism of rings.

Proof. We have seen above that R is perfectoid if and only if it is uniform and $\Phi\colon R^\circ \to R^\circ$ is surjective (as now $R^\circ = R^\circ/p$). The latter is equivalent to

$\Phi \colon R \to R$ being surjective, and uniformity implies that R is reduced, so that $\Phi \colon R \to R$ is an isomorphism.

It remains to see that if R is a perfect complete Tate ring, then R is uniform. For this, pick any ring of definition $R_0 \subset R$. Then $\Phi(R_0) \subset R$ is open by Banach's open mapping theorem, so that $\varpi R_0 \subset \Phi(R_0)$ for some choice of pseudo-uniformizer ϖ. But then $\Phi^{-1}(R_0) \subset \varpi^{-1/p} R_0$, and so $\Phi^{-2}(R_0) \subset \varpi^{-1/p^2} \Phi^{-1}(R_0) \subset \varpi^{-1/p^2 - 1/p} R_0$, etc. This shows that the intersection $R'_0 = \bigcap_n \Phi^{-n}(R_0) \subset \varpi^{-1} R_0$ is bounded, and therefore is a ring of definition, on which Φ is an isomorphism. But then $R^{\circ\circ} \subset R'_0$, and so $R^\circ \subset \varpi^{-1} R^{\circ\circ} \subset \varpi^{-1} R'_0$ is bounded. $\qquad\square$

Definition 6.1.7. A *perfectoid field* is a perfectoid Tate ring R which is a nonarchimedean field.

Remark 6.1.8. By [Ked18], a perfectoid Tate ring R whose underlying ring is a field is a perfectoid field.

Proposition 6.1.9. Let K be a nonarchimedean field. Then K is a perfectoid field if and only if the following conditions hold:

1. K is not discretely valued,
2. $|p| < 1$, and
3. $\Phi \colon \mathcal{O}_K/p \to \mathcal{O}_K/p$ is surjective.

Proof. The conditions are clearly necessary. Conversely, if K is not discretely valued, one can find $\varpi \in K^\times$ with $|\varpi| < 1$ such that $\varpi^p | p$ holds in $\mathcal{O}_K$, by taking the valuation small enough. $\qquad\square$

Theorem 6.1.10 ([Sch12, Theorem 6.3], [KL15, Theorem 3.6.14]). Let (R, R^+) be a Huber pair such that R is perfectoid. Then for all rational subsets $U \subset X = \mathrm{Spa}(R, R^+)$, $\mathcal{O}_X(U)$ is again perfectoid. In particular, (R, R^+) is stably uniform, hence sheafy by Theorem 5.2.5.

The hard part of Theorem 6.1.10 is showing that $\mathcal{O}_X(U)$ is uniform. The proof of this fact makes essential use of the process of *tilting*.

6.2 TILTING

Definition 6.2.1. Let R be a perfectoid Tate ring. The *tilt* of R is

$$R^\flat = \varprojlim_{x \mapsto x^p} R,$$

given the inverse limit topology. A priori this is only a topological multiplicative monoid. We give it a ring structure where the addition law is

$$(x^{(0)}, x^{(1)}, \dots) + (y^{(0)}, y^{(1)}, \dots) = (z^{(0)}, z^{(1)}, \dots)$$

where

$$z^{(i)} = \lim_{n \to \infty} (x^{(i+n)} + y^{(i+n)})^{p^n} \in R.$$

Lemma 6.2.2. The limit $z^{(i)}$ above exists and defines a ring structure making $R^\flat$ a topological $\mathbf{F}_p$-algebra that is a perfect complete Tate ring. The subset $R^{\flat\circ}$ of power-bounded elements is given by the topological ring isomorphism

$$R^{\flat\circ} = \varprojlim_{x \mapsto x^p} R^\circ \cong \varprojlim_\Phi R^\circ/p \cong \varprojlim_\Phi R^\circ/\varpi,$$

where $\varpi \in R$ is a pseudo-uniformizer which divides p in R°. Furthermore there exists a pseudo-uniformizer $\varpi \in R$ with $\varpi^p | p$ in R° that admits a sequence of pth power roots ϖ^{1/p^n}, giving rise to an element $\varpi^\flat = (\varpi, \varpi^{1/p}, \dots) \in R^{\flat\circ}$, which is a pseudo-uniformizer of $R^\flat$. Then $R^\flat = R^{\flat\circ}[1/\varpi^\flat]$.

Proof. (Sketch.) Certainly the pth power map on $R^\flat$ is an isomorphism by design. Let ϖ_0 be a pseudo-uniformizer of R. Let us check that the maps

$$\varprojlim_\Phi R^\circ \to \varprojlim_\Phi R^\circ/p \to \varprojlim_\Phi R^\circ/\varpi_0$$

are bijective. The essential point is that any sequence $(\bar{x}_0, \bar{x}_1, \dots) \in \varprojlim_\Phi R^\circ/\varpi_0$ lifts uniquely to a sequence $(x_0, x_1, \dots) \in \varprojlim_\Phi R^\circ$. Here $x^{(i)} = \lim_{n \to \infty} x_{n+i}^{p^n}$, where $x_j \in R^\circ$ is any lift of $\bar{x}_j$. (For the convergence of that limit, note that if $x \equiv y \pmod{\varpi_0^n}$, then $x^p \equiv y^p \pmod{\varpi_0^{n+1}}$.) This shows that we get a well-defined ring $R^{\flat\circ}$.

Now assume that $\varpi_0^p | p$ holds in R°. Any preimage of ϖ_0 under $R^{\flat\circ} = \varprojlim_\Phi R^\circ/\varpi_0^p \to R^\circ/\varpi_0^p$ is an element $\varpi^\flat$ with the right properties. It is congruent to ϖ_0 modulo ϖ_0^p, and therefore it is also a pseudo-uniformizer. Then $\varpi = \varpi^{\flat\sharp}$ is the desired pseudo-uniformizer of R°. $\qquad\qquad\square$

Remark 6.2.3. In the special case that $R = K$ is a perfectoid field, the construction of $K^\flat$ is due to Fontaine, [Fon82], as an intermediate step towards his construction of p-adic period rings. Here, $K^\flat$ is a complete nonarchimedean field with absolute value defined by $f \mapsto |f^\sharp|$, where $| \ |$ is the absolute value on K.

Example 6.2.4. Let $\mathbf{Q}_p^{\mathrm{cycl}}$ be the completion of $\mathbf{Q}_p(\mu_{p^\infty})$. Then $\mathbf{Q}_p^{\mathrm{cycl}}$ is a perfectoid field. Let $\zeta_p, \zeta_{p^2}, \dots$ be a compatible system of pth power roots of 1. Then $t = (1, \zeta_p, \zeta_{p^2}, \dots) - 1$ is a pseudouniformizer of $(\mathbf{Q}_p^{\mathrm{cycl}})^\flat$. In fact

$(\mathbf{Q}_p^{\mathrm{cycl}})^\flat = \mathbf{F}_p((t^{1/p^\infty}))$, the t-adic completion of the perfect field $\mathbf{F}_p(t^{1/p^\infty})$. Note that $\mathbf{Z}_p^\times = \mathrm{Gal}(\mathbf{Q}_p(\mu_{p^\infty})/\mathbf{Q}_p)$ acts on $\mathbf{Q}_p^{\mathrm{cycl}}$ and therefore on $\mathbf{F}_p((t^{1/p^\infty}))$. Explicitly, the action of $\gamma \in \mathbf{Z}_p^\times$ carries t onto $(1+t)^\gamma - 1$.

We have a continuous and multiplicative (but not additive) map $R^\flat \to \varprojlim R \to R$ by projecting onto the zeroth coordinate; call this $f \mapsto f^\sharp$. This projection defines a ring isomorphism $R^{\flat\circ}/\varpi^\flat \cong R^\circ/\varpi$. By topological nilpotence conditions, the integrally closed open subrings of $R^{\flat\circ}$ and R° correspond exactly to the integrally closed subrings of their common quotients modulo $\varpi^\flat$ and ϖ. This defines an inclusion-preserving bijection between the sets of open integrally closed subrings of $R^{\flat\circ}$ and R°. This correspondence can be made more explicit:

Lemma 6.2.5. The set of rings of integral elements $R^+ \subset R^\circ$ is in bijection with the set of rings of integral elements $R^{\flat+} \subset R^{\flat\circ}$, via $R^{\flat+} = \varprojlim_{x \mapsto x^p} R^+$. Also, $R^{\flat+}/\varpi^\flat = R^+/\varpi$.

The following two theorems belong to a pattern of "tilting equivalence."

Theorem 6.2.6 ([KL15], [Sch12]). Let (R, R^+) be a perfectoid Huber pair, with tilt $(R^\flat, R^{\flat+})$. There is a homeomorphism $X = \mathrm{Spa}(R, R^+) \cong X^\flat = \mathrm{Spa}(R^\flat, R^{\flat+})$ sending x to $x^\flat$, where $|f(x^\flat)| = |f^\sharp(x)|$. This homeomorphism preserves rational subsets. For any rational subset $U \subset X$ with image $U^\flat \subset X^\flat$, the complete Tate ring $\mathcal{O}_X(U)$ is perfectoid with tilt $\mathcal{O}_{X^\flat}(U^\flat)$.

Note that this theorem implies stable uniformity, and thus sheafyness. Note that in characteristic p, it is clear that the perfectness condition is preserved under rational localization, and we have seen that this implies uniformity in Proposition 6.1.6. To transfer this information to the general case (and thus prove the final sentence of the last theorem), one uses the following theorem.

Theorem 6.2.7 ([KL15], [Sch12]). Let R be a perfectoid ring with tilt $R^\flat$. Then there is an equivalence of categories between perfectoid R-algebras and perfectoid $R^\flat$-algebras, via $S \mapsto S^\flat$.

Let us describe the inverse functor in Theorem 6.2.7, along the lines of Fontaine's Bourbaki talk, [Fon13]. In fact we will answer a more general question. Given a perfectoid algebra R in characteristic p, what are all the untilts $R^\sharp$ of R? Let us start with a pair (R, R^+).

Lemma 6.2.8. Let $(R^\sharp, R^{\sharp+})$ be an untilt of (R, R^+), i.e., a perfectoid Tate ring $R^\sharp$ together with an isomorphism $R^{\sharp\flat} \to R$, such that $R^{\sharp+}$ and R^+ are identified under Lemma 6.2.5.

1. There is a canonical surjective ring homomorphism

$$\theta: W(R^+) \quad \to \quad R^{\sharp +}$$

$$\sum_{n \geq 0} [r_n] p^n \quad \mapsto \quad \sum_{n \geq 0} r_n^{\sharp} p^n.$$

2. The kernel of θ is generated by a nonzerodivisor ξ of the form $\xi = p + [\varpi]\alpha$, where $\varpi \in R^+$ is a pseudo-uniformizer, and $\alpha \in W(R^+)$.

We remark that there is no assumption that an untilt of R should have characteristic 0. In particular R itself is an untilt of R, corresponding to $\xi = p$.

Definition 6.2.9. An ideal $I \subset W(R^+)$ is *primitive of degree 1* if I is generated by an element of the form $\xi = p + [\varpi]\alpha$, with $\varpi \in R^+$ a pseudo-uniformizer and $\alpha \in W(R^+)$.

Lemma 6.2.10. *Any element $\xi \in W(R^+)$ of the form $\xi = p + [\varpi]\alpha$, with $\varpi \in R^+$ a pseudo-uniformizer and $\alpha \in W(R^+)$, is a nonzerodivisor.*

Proof. Assume that $\xi \sum_{n \geq 0} [c_n] p^n = 0$. Modulo $[\varpi]$, this reads $\sum_{n \geq 0} [c_n] p^{n+1} \equiv 0 \pmod{[\varpi]}$, meaning that all $c_n \equiv 0 \pmod{\varpi}$. We can then divide all c_n by ϖ, and induct. $\qquad\square$

Proof. (of Lemma 6.2.8) Fix $\varpi \in R^+$ a pseudo-uniformizer such that $\varpi^{\sharp} \in R^{\sharp +}$ satisfies $(\varpi^{\sharp})^p | p$. For part (1), it is enough to check that θ is a ring map modulo $(\varpi^{\sharp})^m$ for any $m \geq 1$. For this, we use the fact that the mth ghost map

$$W(R^{\sharp +}) \to R^{\sharp +}/(\varpi^{\sharp})^m \quad : \quad (x_0, x_1, \ldots) \mapsto \sum_{n=0}^{m} x_n^{p^{m-n}} p^n$$

factors uniquely over $W(R^{\sharp +}/\varpi^{\sharp})$, by obvious congruences; the induced map $W(R^{\sharp +}/\varpi^{\sharp}) \to R^{\sharp +}/(\varpi^{\sharp})^m$ must be a ring homomorphism. Now the composite

$$W(R^+) \to W(R^+/\varpi) = W(R^{\sharp +}/\varpi^{\sharp}) \to R^{\sharp +}/(\varpi^{\sharp})^m ,$$

where the first map is given by the mth component map $R^+ = \varprojlim_{x \mapsto x^p} R^{\sharp +}/\varpi^{\sharp} \to R^{\sharp +}/\varpi^{\sharp}$, is a ring map, which we claim is equal to θ modulo $(\varpi^{\sharp})^m$. This is a direct verification from the definitions.

For surjectivity of θ, we know that $R^+ \to R^{\sharp +}/\varpi^{\sharp}$ is surjective, which shows that $\theta \bmod [\varpi]$ is surjective. As everything is $[\varpi]$-adically complete, this implies that θ is surjective.

For part (2), we claim that there exists $f \in \varpi R^+$ such that the congruence $f^{\sharp} \equiv p \pmod{p\varpi^{\sharp} R^{\sharp +}}$ holds. Indeed, consider $\alpha = p/\varpi^{\sharp} \in R^{\sharp +}$. There exists $\beta \in R^+$ such that $\beta^{\sharp} \equiv \alpha \pmod{pR^{\sharp +}}$. Then $(\varpi\beta)^{\sharp} = \varpi^{\sharp}\alpha \equiv p \pmod{p\varpi^{\sharp} R^{\sharp +}}$. Take $f = \varpi\beta$.

Thus we can write $p = f^{\sharp} + p\varpi^{\sharp} \sum_{n \geq 0} r_n^{\sharp} p^n$, with $r_n \in R^+$. We can now define $\xi = p - [f] - [\varpi]\sum_{n\geq 0}[r_n]p^{n+1}$, which is of the desired form, and which lies in the kernel of θ. Finally we need to show that ξ generates $\ker(\theta)$. For this, note that θ induces a surjective map $f : W(R^+)/\xi \to R^{\sharp+}$. It is enough to show that f is an isomorphism modulo $[\varpi]$. But

$$W(R^+)/(\xi,[\varpi]) = W(R^+)/(p,[\varpi]) = R^+/\varpi = R^{\sharp+}/\varpi^{\sharp} ,$$

as desired. □

From here, one gets the following theorem, which implies Theorem 6.2.7.

Theorem 6.2.11 ([KL15], [Fon13]). There is an equivalence of categories between:

1. Perfectoid Tate-Huber pairs (S, S^+)
2. Triples $(R, R^+, \mathcal{J})$, where (R, R^+) is a perfectoid Tate-Huber pair of characteristic p and $\mathcal{J} \subset W(R^+)$ is primitive of degree 1.

In one direction the map is $(S, S^+) \mapsto (S^\flat, S^{\flat+}, \ker\theta)$, and in the other, it is $(R, R^+, \mathcal{J}) \mapsto (W(R^+)[[\varpi]^{-1}]/\mathcal{J}, W(R^+)/\mathcal{J})$.

6.3 SOUSPERFECTOID RINGS

Theorem 6.1.10 states that if (R, R^+) is a Huber pair with R perfectoid, then (R, R^+) is stably uniform, and thus sheafy. It will be useful to extend this theorem to a slightly broader class of Huber pairs. The following definition and results are due to Hansen–Kedlaya.

Definition 6.3.1. Let R be a complete Tate-$\mathbf{Z}_p$-algebra. Then R is *sousperfectoid* if there exists a perfectoid Tate ring $\widetilde{R}$ with an injection $R \hookrightarrow \widetilde{R}$ that splits as topological R-modules.

Example 6.3.2.

1. Any perfectoid ring is sousperfectoid.
2. A Tate algebra $R = \mathbf{Q}_p\langle T\rangle$ is sousperfectoid, by taking $\widetilde{R} = \mathbf{Q}_p^{\mathrm{cycl}}\langle T^{1/p^\infty}\rangle$.

The class of sousperfectoid rings has good stability properties.

Proposition 6.3.3. Let R be a complete Tate-$\mathbf{Z}_p$-algebra with a ring of integral elements $R^+ \subset R$, and assume that R is sousperfectoid.

1. If $U \subset X = \mathrm{Spa}(R, R^+)$ is a rational subset, then $\mathcal{O}_X(U)$ is sousperfectoid.
2. If S is a finite étale R-algebra, then S is sousperfectoid.
3. For all $n \geq 0$, the ring $R\langle T_1,\ldots,T_n\rangle$ is sousperfectoid.

Proof. Choose a perfectoid Tate ring $\widetilde{R}$ with an injection $R \hookrightarrow \widetilde{R}$ that splits as topological R-modules, and let $\widetilde{R}^+ \subset \widetilde{R}$ be a ring of integral elements containing R^+. Let $\widetilde{X} = \mathrm{Spa}(\widetilde{R}, \widetilde{R}^+) \to X = \mathrm{Spa}(R, R^+)$, and let $\widetilde{U} \subset \widetilde{X}$ be the preimage of U. Then $\widetilde{R} \widehat{\otimes}_R \mathcal{O}_X(U) = \mathcal{O}_{\widetilde{X}}(\widetilde{U})$ is a perfectoid Tate ring, of which $\mathcal{O}_X(U)$ is a topological $\mathcal{O}_X(U)$-module direct summand, so $\mathcal{O}_X(U)$ is sousperfectoid.

Similarly, if S is a finite étale R-algebra, then $\widetilde{S} = \widetilde{R} \otimes_R S = \widetilde{R} \widehat{\otimes}_R S$ is a perfectoid Tate ring by Theorem 7.4.5 (1) below, and again S is a topological S-module direct summand of $\widetilde{S}$, so S is sousperfectoid. Finally, the same argument applies to $R\langle T_1, \ldots, T_n \rangle \hookrightarrow \widetilde{R}\langle T_1^{1/p^\infty}, \ldots, T_n^{1/p^\infty} \rangle$. $\qquad\square$

In particular, we get the following proposition.

Proposition 6.3.4. Let (R, R^+) be a Tate-Huber pair such that R is sousperfectoid. Then (R, R^+) is stably uniform, and thus sheafy.

Proof. As being sousperfectoid is stable under rational localization by the previous proposition, it is enough to see that R is uniform. If $\widetilde{R}$ is a perfectoid ring as in the definition, it is enough to see that $R^\circ \hookrightarrow \widetilde{R}^\circ$, as $\widetilde{R}^\circ$ is bounded and the map $R \to \widetilde{R}$ is strict (as it splits as topological R-modules). But an element is powerbounded if and only if for all maps $R \to K$ to a nonarchimedean field, the image is powerbounded. Now the map $K \to \widetilde{R} \widehat{\otimes}_R K$ is still injective, as it splits as topological K-vector spaces. This implies that $\widetilde{R} \widehat{\otimes}_R K \neq 0$, and so it admits a map to a nonarchimedean field L. If $f \in R$ maps into $\widetilde{R}^\circ$, it maps into $\mathcal{O}_L$, and thus into $\mathcal{O}_L \cap K = \mathcal{O}_K$, as desired. $\qquad\square$

Lecture 7

Perfectoid spaces

This is the second lecture on perfectoid spaces. Recall that a *perfectoid Tate ring* R is a complete, uniform Tate ring containing a pseudo-uniformizer ϖ such that $\varpi^p | p$ in R° and such that $\Phi \colon R^\circ/\varpi \to R^\circ/\varpi^p$ is an isomorphism.

7.1 PERFECTOID SPACES: DEFINITION AND TILTING EQUIVALENCE

We also talked about tilting. Suppose (R, R^+) is a Huber pair, with R perfectoid. Let $R^\flat = \varprojlim_{x \mapsto x^p} R$, a perfectoid ring of characteristic p, together with a map $R^\flat \to R$ of multiplicative monoids, $f \mapsto f^\sharp$. In the last lecture, we said a few words about the proof of the following theorem.

Theorem 7.1.1. A Huber pair (R, R^+) with R perfectoid is sheafy. Let $X = \mathrm{Spa}(R, R^+)$, $X^\flat = \mathrm{Spa}(R^\flat, R^{\flat+})$; then there is a homeomorphism $X \to X^\flat$, $x \mapsto x^\flat$, which preserves rational subsets. It is characterized by $|f(x^\flat)| = |f^\sharp(x)|$. Moreover for a rational subset $U \subset X$ with image $U^\flat \subset X^\flat$, the complete Tate ring $\mathcal{O}_X(U)$ is perfectoid with tilt $\mathcal{O}_{X^\flat}(U^\flat)$.

Definition 7.1.2. A *perfectoid space* is an adic space covered by affinoid adic spaces $\mathrm{Spa}(R, R^+)$ with R perfectoid.

Remark 7.1.3. If (R, R^+) is some sheafy Tate-Huber pair for which $\mathrm{Spa}(R, R^+)$ is a perfectoid space, it is not clear whether R has to be perfectoid (although it is fine if we are in characteristic p). See [BV18] for a discussion. This means that the term "affinoid perfectoid space" is ambiguous; we will always mean a space of the form $\mathrm{Spa}(R, R^+)$, where R is perfectoid.

The tilting process glues to give a functor $X \mapsto X^\flat$ from perfectoid spaces to perfectoid spaces of characteristic p. Moreover, the tilting equivalence geometrizes.

Theorem 7.1.4. For any perfectoid space X with tilt $X^\flat$, the functor $Y \mapsto Y^\flat$ induces an equivalence between the categories of perfectoid spaces over X

resp. $X^\flat$.

7.2   WHY DO WE STUDY PERFECTOID SPACES?

Let us put forward a certain philosophy here, which indicates that perfectoid spaces may arise even when one is only interested in classical objects.

If X is a perfectoid space, all topological information (e.g. $|X|$, and even $X_{\text{ét}}$ as discussed below) can be recovered from $X^\flat$. However $X^\flat$ forgets the structure morphism $X \to \text{Spa}\,\mathbf{Z}_p$. The following will be made precise in the next two lectures: The category of perfectoid spaces over $\mathbf{Q}_p$ is equivalent to the category of perfectoid spaces X of characteristic p together with a "structure morphism $X \to \mathbf{Q}_p$".

Remarkably, we can still carry out this procedure if X is not perfectoid, but rather is an arbitrary analytic adic space over $\mathbf{Z}_p$. We will see in (8.2) that there is always a perfectoid space $\widetilde{X}$ and a *pro-étale covering* $\widetilde{X} \to X$. For example, if $X = \text{Spa}\,\mathbf{Q}_p\langle T^{\pm 1}\rangle$ is the "unit circle," this has a pro-étale covering by the perfectoid space $\widetilde{X} = \text{Spa}\,\mathbf{Q}_p^{\text{cycl}}\langle T^{\pm 1/p^\infty}\rangle$, this being the inverse limit of the finite étale covers $X_n = \text{Spa}\,\mathbf{Q}_p(\zeta_{p^n})\langle T^{\pm 1/p^n}\rangle$. Then we can tilt $\widetilde{X}$ to arrive at $\widetilde{X}^\flat$, a perfectoid space in characteristic p, which inherits a descent datum corresponding to the original cover $\widetilde{X} \to X$. In this way, one can "access" all analytic adic spaces over $\mathbf{Z}_p$ using pro-étale descent from perfectoid spaces in characteristic p. This leads us to the notion of a diamond, introduced in (8.1).

7.3   THE EQUIVALENCE OF ÉTALE SITES

The tilting equivalence extends to the étale site of a perfectoid space. That is, if X is a perfectoid space then there is an equivalence $X_{\text{ét}} \cong X^\flat_{\text{ét}}$. First we discuss the case where X is a single point.

Theorem 7.3.1 ([FW79], [KL15, Theorem 3.5.6], [Sch12]). Let K be a perfectoid field with tilt $K^\flat$.

1. If L/K is finite, then L is perfectoid.
2. The functor $L \mapsto L^\flat$ is an equivalence of categories between finite extensions of K and finite extensions of $K^\flat$ which preserves degrees. Thus, the absolute Galois groups of K and $K^\flat$ are isomorphic.

A related result is the following.

Theorem 7.3.2 ([Tat67], [GR03, Section 6.6]). Let K be a perfectoid field and L/K a finite extension. Then $\mathcal{O}_L/\mathcal{O}_K$ is *almost finite étale*.

For the precise meaning of almost finite étale, which is somewhat technical, we refer to [Sch12, Section 4]. What Tate actually proved (for certain perfectoid fields K) is that if $\mathrm{tr}\colon L \to K$ is the trace map, then $\mathrm{tr}(\mathcal{O}_L)$ contains $\mathfrak{m}_K$, the maximal ideal of $\mathcal{O}_K$.

Example 7.3.3. Say $K = \mathbf{Q}_p(p^{1/p^\infty})^\wedge$, a perfectoid field. Let $L = K(\sqrt{p})$ (and assume $p \neq 2$). Let $K_n = \mathbf{Q}_p(p^{1/p^n})$ and $L_n = K_n(\sqrt{p})$. Note that $p^{1/2p^n} \in L_n$, because $p^{1/2p^n} = (p^{1/p^n})^{(p^n+1)/2}p^{-1/2}$, and that

$$\mathcal{O}_{L_n} = \mathcal{O}_{K_n}[p^{1/2p^n}] = \mathcal{O}_{K_n}[x]/(x^2 - p^{1/p^n}).$$

Let $f(x) = x^2 - p^{1/p^n}$. The different ideal δ_{L_n/K_n} is the ideal of $\mathcal{O}_{L_n}$ generated by $f'(p^{1/2p^n})$, which is $p^{1/2p^n}$. The p-adic valuation of δ_{L_n/K_n} is $1/2p^n$, which tends to 0 as $n \to \infty$. Inasmuch as the different ideal measures ramification, this means that the extensions L_n/K_n are getting less ramified as $n \to \infty$.

In other words, one can almost get rid of ramification along the special fiber by passing to a tower whose limit is perfectoid. This is what Tate does (using the cyclotomic tower) to do computations in Galois cohomology, which is an essential part of p-adic Hodge theory.

In fact Theorem 7.3.2 implies Theorem 7.3.1, as observed in [Sch12]. The equivalence between finite étale algebras over K and $K^\flat$ goes according to the diagram, all of whose arrows are equivalences (and which uses the notations of [Sch12]). Here we use the notion of almost $\mathcal{O}_K$-algebras, written $\mathcal{O}_K^a$-algebras, which are algebra objects in the symmetric monoidal category of almost $\mathcal{O}_K$-modules introduced in the next section.

$$\{\text{fin. ét. } K\text{-algebras}\} \xrightarrow{\text{Theorem 7.3.2}} \{\text{fin. ét. } \mathcal{O}_K^a\text{-algebras}\}$$

$$\{\text{fin. ét. } (\mathcal{O}_K/\varpi)^a\text{-algebras}\}$$

$$\{\text{fin. ét. } (\mathcal{O}_{K^\flat}/\varpi^\flat)^a\text{-algebras}\}$$

$$\{\text{fin. ét. } K^\flat\text{-algebras}\} \xleftarrow{\text{Theorem 7.3.2}} \{\text{fin. ét. } \mathcal{O}_{K^\flat}^a\text{-algebras}\}$$

Philosophically, properties of K extend "almost integrally" to $\mathcal{O}_K$, which one can then pass to $\mathcal{O}_K/\varpi$ and tilt to the other side.

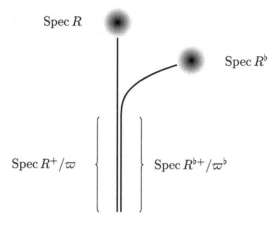

Spec R

Spec $R^\flat$

Spec R^+/ϖ

Spec $R^{\flat+}/\varpi^\flat$

Figure 7.1: A (not to be taken too seriously) depiction of the tilting process for a perfectoid ring R. The blue figure represents Spec R^+ and the red figure represents Spec $R^{\flat+}$. Objects associated with R can "almost" be extended to R^+ and then reduced modulo ϖ. But then $R^+/\varpi = R^{\flat+}/\varpi^\flat$, so one gets an object defined over $R^{\flat+}/\varpi^\flat$. The process can be reversed on the $R^\flat$ side, so that one gets a tilted object defined over $R^\flat$.

7.4 ALMOST MATHEMATICS, AFTER FALTINGS

Let R be a perfectoid Tate ring.

Definition 7.4.1. An R°-module M is *almost zero* if $\varpi M = 0$ for all pseudo-uniformizers ϖ. Equivalently, if ϖ is a fixed pseudo-uniformizer admitting pth power roots, then M is almost zero if $\varpi^{1/p^n} M = 0$ for all n. (Similar definitions apply to R^+-modules.)

Example 7.4.2. 1. If K is a perfectoid field, then the residue field $k = \mathcal{O}_K/\mathfrak{m}_K$ is almost zero. (A general almost zero module is a direct sum of such modules.)

2. If R is perfectoid and $R^+ \subset R^\circ$ is any ring of integral elements, then R°/R^+ is almost zero. Indeed, if ϖ is a pseudo-uniformizer, and $x \in R^\circ$, then ϖx is topologically nilpotent. Since R^+ is open, there exists n with $(\varpi x)^n \in R^+$, so that $\varpi x \in R^+$ by integral closedness.

Note that extensions of almost zero modules are almost zero. Thus the category of almost zero modules is a thick Serre subcategory of the category of all modules, and one can take the quotient.

Definition 7.4.3. The category of *almost R°-modules*, written $R^{\circ a}$-mod, is the quotient of the category of R°-modules by the subcategory of almost zero modules.

One can also define R^{+a}-mod, but the natural map R^{oa}-mod $\to R^{+a}$-mod is an equivalence.

Theorem 7.4.4. Let (R, R^+) be a perfectoid Tate-Huber pair, and let $X = \mathrm{Spa}(R, R^+)$. (Thus by Kedlaya-Liu, $H^i(X, \mathcal{O}_X) = 0$ for $i > 0$.) Then $H^i(X, \mathcal{O}_X^+)$ is almost zero for $i > 0$, and $H^0(X, \mathcal{O}_X^+) = R^+$.

Proof. By tilting one can reduce to the case of $pR = 0$. We will show that for any finite rational covering $X = \bigcup_i U_i$, all cohomology groups of the Cech complex

$$C^\bullet : \quad 0 \to R^+ \to \prod_i \mathcal{O}_X^+(U_i) \to \prod_{i,j} \mathcal{O}_X^+(U_i \cap U_j) \to \cdots$$

are almost zero. We know that $C^\bullet[1/\varpi]$ (replace $\mathcal{O}_X^+$ with $\mathcal{O}_X$ everywhere) is exact. Now we use Banach's Open Mapping Theorem: each cohomology group of $C^\bullet$ is killed by a power of ϖ. (From the exactness of $C^\bullet[1/\varpi]$ one deduces formally that each cohomology group is ϖ-power-torsion, but we assert that there is a single power of ϖ which kills everything. For the complete argument, see [Sch12, Proposition 6.10].)

Since R is perfect, Frobenius induces isomorphisms on all cohomology groups of $C^\bullet$. So if these are killed by ϖ^n, they are also killed by all the ϖ^{n/p^k}, so they are almost zero. □

This is a typical strategy: bound the problem up to a power of ϖ, and then use Frobenius to shrink the power to zero.

Theorem 7.4.5 ([Fal02a],[KL15],[Sch12]). Let R be perfectoid with tilt $R^\flat$.

1. For any finite étale R-algebra S, S is perfectoid.
2. Tilting induces an equivalence

$$\{\text{Finite étale } R\text{-algebras}\} \quad \to \quad \{\text{Finite étale } R^\flat\text{-algebras}\}$$
$$S \quad \mapsto \quad S^\flat$$

3. (Almost purity) For any finite étale R-algebras S, the algebra S° is almost finite étale over R°.

The line of argument is to prove (2) and deduce (1) and (3) (by proving them in characteristic p). Let us sketch the proof of (2). We reduce to the case of perfectoid fields via the following argument. Let $x \in X = \mathrm{Spa}(R, R^+)$ with residue field $K(x)$, and similarly define $K(x^\flat)$. Then we have $K(x)^\flat = K(x^\flat)$.

Lemma 7.4.6.

$$2\text{-}\varinjlim_{U \ni x} \{\text{finite étale } \mathcal{O}_X(U)\text{-algebras}\} \to \{\text{finite étale } K(x)\text{-algebras}\}$$

is an equivalence.

Remark 7.4.7. One has here a directed system of categories $\mathcal{C}_i$, indexed by a filtered category I, with functors $F_{ij} \colon \mathcal{C}_i \to \mathcal{C}_j$ for each morphism $i \to j$ in I. The *2-limit* $\mathcal{C} = 2\text{-}\varinjlim \mathcal{C}_i$ is a category whose objects are objects of any $\mathcal{C}_i$. If X_i and X_j belong to $\mathcal{C}_i$ and $\mathcal{C}_j$, respectively, then

$$\mathrm{Hom}_{\mathcal{C}}(X_i, X_j) = \varinjlim_{i,j \to k} \mathrm{Hom}_{\mathcal{C}_k}(F_{ik}(X_i), F_{jk}(X_j)),$$

the limit being taken over pairs of morphisms from i and j into a common third object k.

Admitting the lemma for the moment, we can complete this to a diagram

$$2\text{-}\varinjlim_{U \ni x} \{\text{finite étale } \mathcal{O}_X(U)\text{-algebras}\} \longrightarrow \{\text{finite étale } K(x)\text{-algebras}\}$$

$$\Big\downarrow \qquad\qquad\qquad\qquad\qquad\qquad\qquad \Big\downarrow$$

$$2\text{-}\varinjlim_{U \ni x} \{\text{finite étale } \mathcal{O}_{X^\flat}(U)\text{-algebras}\} \longrightarrow \{\text{finite étale } K(x)^\flat\text{-algebras}\}$$

Thus we get equivalences locally at every point, which we can glue together to deduce (2) (using Theorem 5.2.8 for example). It remains to address Lemma 7.4.6. This rests on the following theorem.

Theorem 7.4.8 ([Elk73], [GR03]). Let A be a Tate ring such that A is "topologically henselian." That is, for a ring of definition $A_0 \subset A$, and for $\varpi \in A_0$ a pseudo-uniformizer, the ring A_0 is henselian along ϖA_0. Then the functor

$$\{\text{finite étale } A\text{-algebras}\} \quad \to \quad \{\text{finite étale } \widehat{A}\text{-algebras}\}$$
$$B \quad \mapsto \quad \widehat{B} = B \otimes_A \widehat{A}$$

is an equivalence.

As a corollary, let A_i be a filtered directed system of complete Tate rings, $A_\infty = \widehat{\varinjlim A_i}$. Then

$$2\text{-}\varinjlim\{\text{finite étale } A_i\text{-algebras}\} \quad \cong \quad \{\text{finite étale } \varinjlim A_i\text{-algebras}\}$$
$$\cong \quad \{\text{finite étale } A_\infty\text{-algebras}\}.$$

To deduce Lemma 7.4.6 from Theorem 7.4.8, it remains to show that $K(x) = \widehat{\varinjlim \mathcal{O}_X(U)}$, for the topology making $\varinjlim \mathcal{O}_X^+(U)$ open and bounded. We have

$$0 \to I \to \varinjlim_{x \in U} \mathcal{O}_X^+(U) \to K(x)^+$$

where the image of the last arrow is dense. We claim that ϖ is invertible in

I. If $f \in \mathcal{O}_X^+(U)$ is such that $|f(x)| = 0$, then $V = \{|f(x)| \leq |\varpi|\}$ is an open neighborhood of x, and then $f \in \varpi\mathcal{O}_X^+(V)$, and so $f \in \varpi I$. Thus the ϖ-adic completion of $\varinjlim \mathcal{O}_X^+(U)$ is $K(x)^+$. $\square$

7.5   THE ÉTALE SITE

Finally, we define the étale site of perfectoid spaces.

Definition 7.5.1. 1. A morphism $f\colon X \to Y$ of perfectoid spaces is *finite étale* if for all $\mathrm{Spa}(B, B^+) \subset Y$ open, the pullback $X \times_Y \mathrm{Spa}(B, B^+)$ is $\mathrm{Spa}(A, A^+)$, where A is a finite étale B-algebra, and A^+ is the integral closure of the image of B^+ in A.

2. A morphism $f\colon X \to Y$ is *étale* if for all $x \in X$ there exists an open $U \ni x$ and $V \supset f(U)$ such that there is a diagram

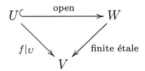

3. An *étale cover* is a jointly surjective family of étale maps.

The following proposition shows that this has the expected properties. We note that this characterization of étale maps does not hold true in the world of schemes, but it does in the case of analytic adic spaces, by [Hub96, Lemma 2.2.8].

Proposition 7.5.2. 1. Compositions of (finite) étale morphisms are (finite) étale.
2. Pullbacks of (finite) étale morphisms are (finite) étale.
3. If g and gf are (finite) étale, then so is f.
4. f is (finite) étale if and only if $f^\flat$ is (finite) étale.

In particular, we get the following corollary showing that the étale site is invariant under tilting.

Corollary 7.5.3. There exists an étale site $X_{\text{ét}}$, such that naturally $X_{\text{ét}} \cong X_{\text{ét}}^\flat$, and $H^i(X_{\text{ét}}, \mathcal{O}_X^+)$ is almost zero for $i > 0$ for affinoids X.

Lecture 8

Diamonds

8.1 DIAMONDS: MOTIVATION

Today we discuss the notion of a diamond. The idea is that there should be a functor

$$\{\text{analytic adic spaces over } \mathbf{Z}_p\} \;\to\; \{\text{diamonds}\}$$
$$X \;\mapsto\; X^{\diamond}$$

which "forgets the structure morphism to $\mathbf{Z}_p$." For a perfectoid space X, the functor $X \mapsto X^{\flat}$ has this property, so the desired functor should essentially be given by $X \mapsto X^{\flat}$ on such objects. In general, if $X/\mathbf{Z}_p$ is an analytic adic space, then X is pro-étale locally perfectoid:

$$X = \mathrm{Coeq}(\widetilde{X} \times_X \widetilde{X} \rightrightarrows \widetilde{X}),$$

where $\widetilde{X} \to X$ is a pro-étale (in the sense defined below) perfectoid cover; one can then show that the equivalence relation $R = \widetilde{X} \times_X \widetilde{X}$ is also perfectoid (as it is pro-étale over $\widetilde{X}$), at least after passing to a uniform completion. Then the functor should send X to $\mathrm{Coeq}(R^{\flat} \rightrightarrows \widetilde{X}^{\flat})$. The only question now is, what category does this live in? (There is also the question of whether this construction depends on the choices made.) Whatever this object is, it is pro-étale *under* a perfectoid space in characteristic p, namely $\widetilde{X}^{\flat}$.

Example 8.1.1. If $X = \mathrm{Spa}(\mathbf{Q}_p)$, then a pro-étale perfectoid cover of X is $\widetilde{X} = \mathrm{Spa}(\mathbf{Q}_p^{\mathrm{cycl}})$. Then $R = \widetilde{X} \times_X \widetilde{X}$ is essentially $\widetilde{X} \times \mathbf{Z}_p^{\times}$, which can again be considered as a perfectoid space, and so $X^{\diamond}$ should be the coequalizer of $\widetilde{X}^{\flat} \times \mathbf{Z}_p^{\times} \rightrightarrows \widetilde{X}^{\flat}$, which comes out to be the quotient $\mathrm{Spa}((\mathbf{Q}_p^{\mathrm{cycl}})^{\flat})/\mathbf{Z}_p^{\times}$, whatever this means.

8.2 PRO-ÉTALE MORPHISMS

The desired quotient in Example 8.1.1 exists in a category of sheaves on the site of perfectoid spaces with pro-étale covers. So our next task is to define pro-étale morphisms between perfectoid spaces.

Definition 8.2.1. A morphism $f\colon \mathrm{Spa}(B, B^+) \to \mathrm{Spa}(A, A^+)$ of affinoid perfectoid spaces is *affinoid pro-étale* if

$$(B, B^+) = \widehat{\varinjlim(A_i, A_i^+)}$$

is a completed filtered colimit of pairs (A_i, A_i^+) with A_i perfectoid, such that

$$\mathrm{Spa}(A_i, A_i^+) \to \mathrm{Spa}(A, A^+)$$

is étale. A morphism $f\colon X \to Y$ of perfectoid spaces is *pro-étale* if it is locally (on the source and target) affinoid pro-étale.

Example 8.2.2. An important example of pro-étale morphisms comes from profinite sets. If X is any perfectoid space and S is a profinite set, we can define a new perfectoid space $X \times \underline{S}$ as the inverse limit of $X \times S_i$, where $S = \varprojlim_i S_i$ is the inverse limit of finite sets S_i. Then $X \times \underline{S} \to X$ is pro-étale. This construction extends to the case that S is locally profinite.

Remark 8.2.3. Pro-étale morphisms are not necessarily open. For instance, we could have $Y = \underline{S}_K = \mathrm{Spa}(K, \mathcal{O}_K) \times \underline{S}$ for a perfectoid field K and profinite set S, and $f\colon X \to Y$ could be the inclusion of a point $X = \mathrm{Spa}(K, \mathcal{O}_K) \times \{x\}$, $x \in S$. Indeed, this is the completed inverse limit of morphisms $\mathrm{Spa}(K, \mathcal{O}_K) \times \underline{U_i} \to Y$, where $U_i \subset S$ is an open and closed subset and $\bigcap_i U_i = \{x\}$.

Lemma 8.2.4 ([Sch17, Proposition 7.10]). Let

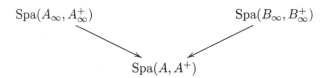

be a diagram of affinoid pro-étale morphisms of affinoid perfectoid spaces, where $(A_\infty, A_\infty^+) = \widehat{\varinjlim_{i \in I}(A_i, A_i^+)}$ is as in the definition, and similarly $(B_\infty, B_\infty^+) = \widehat{\varinjlim_{j \in J}(B_j, B_j^+)}$. Then

$$\mathrm{Hom}_{\mathrm{Spa}(A,A^+)}(\mathrm{Spa}(A_\infty, A_\infty^+), \mathrm{Spa}(B_\infty, B_\infty^+))$$
$$= \varprojlim_J \varinjlim_I \mathrm{Hom}_{\mathrm{Spa}(A,A^+)}\left(\mathrm{Spa}(A_i, A_i^+), \mathrm{Spa}(B_j, B_j^+)\right).$$

Proof. (Sketch.) Without loss of generality J is a singleton, and we can write $(B, B^+) = (B_\infty, B_\infty^+)$. Now we have to check that

$$\text{Hom}(\text{Spa}(A_\infty, A_\infty^+), \text{Spa}(B, B^+)) = \varinjlim_I \text{Hom}(\text{Spa}(A_i, A_i^+), \text{Spa}(B, B^+))$$

(where all Homs are over (A, A^+)). This can be checked locally on $\text{Spa}(B, B^+)$. An étale morphism is locally a composition of rational embeddings and finite étale morphisms. So without loss of generality $f \colon \text{Spa}(B, B^+) \to \text{Spa}(A, A^+)$ is one of these.

1. If f is a rational embedding: let $U = \text{Spa}(B, B^+) \hookrightarrow \text{Spa}(A, A^+)$, then the fact that $\text{Spa}(A_\infty, A_\infty^+) \to \text{Spa}(A, A^+)$ factors over U implies that there exists i such that $\text{Spa}(A_i, A_i^+) \to \text{Spa}(A, A^+)$ factors over U. Indeed, we can apply the following quasi-compactness argument, which applies whenever one wants to show that a "constructible" algebro-geometric property applies to a limit of spaces if and only if it applies to some stage of the limit.
   Topologically we have $\text{Spa}(A_\infty, A_\infty^+) = \varprojlim_i \text{Spa}(A_i, A_i^+)$, and thus

   $$\text{Spa}(A_\infty, A_\infty^+) \backslash \{\text{preimage of } U\} = \varprojlim \left(\text{Spa}(A_i, A_i^+) \backslash \{\text{preimage of } U\} \right)$$

   For each i, the space $\text{Spa}(A_i, A_i^+) \backslash \{\text{preimage of } U\}$ is closed in a spectral space, thus is spectral, and so it is compact and Hausdorff for the constructible topology. If the inverse limit is empty, one of the terms has to be empty: this is a version of Tychonoff's theorem; see [RZ10, Proposition 1.1.4]. We get that $\text{Spa}(A_i, A_i^+)$ equals the preimage of U for some i.

2. Suppose that f is finite étale. Recall from Theorem 7.4.8 that

   $$\{\text{finite étale } A_\infty\text{-algebras}\} = 2\text{-}\varinjlim \{\text{finite étale } A_i\text{-algebras}\}.$$

   This shows that

   $$\begin{aligned} \text{Hom}_A(B, A_\infty) &= \text{Hom}_{A_\infty}(B \otimes A_\infty, A_\infty) \\ &= \varinjlim_i \text{Hom}_{A_i}(B \otimes A_i, A_i) \\ &= \varinjlim_i \text{Hom}_A(B, A_i). \end{aligned}$$

   $\square$

Proposition 8.2.5 ([Sch17, Lemma 7.11]). 1. Compositions of pro-étale maps are pro-étale.

2. If

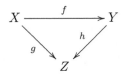

is a diagram of perfectoid spaces where g and h are pro-étale, then f is pro-étale.
3. Pullbacks of pro-étale morphisms are pro-étale. (Note: the category of affinoid perfectoid spaces has all connected limits. In particular fiber products of perfectoid spaces exist.)

Unfortunately, the property of being pro-étale cannot be checked pro-étale locally on the target. Still, we can define the pro-étale site.[1]

Definition 8.2.6 (The big pro-étale site). Consider the following categories:

- Perfd, the category of perfectoid spaces.
- Perf $\subset$ Perfd, the subcategory of perfectoid spaces in characteristic p.
- $X_{\text{proét}}$, the category of perfectoid spaces pro-étale over X, where X is any perfectoid space.

We endow each of these with the structure of a site by saying that a collection of morphisms $\{f_i \colon Y_i \to Y\}$ is a covering (a *pro-étale cover*) if the f_i are pro-étale, and if for all quasi-compact open $U \subset Y$, there exists a finite subset $I_U \subset I$, and quasicompact open subsets $U_i \subset Y_i$ for $i \in I_U$, such that $U = \cup_{i \in I_U} f_i(U_i)$.

Remark 8.2.7. It is not good enough to demand merely that the f_i be a topological cover. For instance, let X be a pro-finite set considered as a perfectoid space over some perfectoid field in characteristic p, and let $X_i \to X$ be a pro-étale morphism whose image is the point $i \in X$, as in Remark 8.2.3. The finiteness criterion in Definition 8.2.6 prevents the $X_i \to X$ from constituting a cover in Perfd. The same issue arises for the fpqc topology on the category of schemes: if X is an affine scheme, the collection of flat quasi-compact maps $\operatorname{Spec} \mathcal{O}_{X,x} \to X$ for $x \in X$ is generally not an fpqc cover.

Proposition 8.2.8 ([Sch17, Corollary 8.6]). Let X be a perfectoid space.

1. The presheaves $\mathcal{O}_X, \mathcal{O}_X^+$ defined on Perfd by $X \mapsto \mathcal{O}_X(X), \mathcal{O}_X^+(X)$ are sheaves. If X is affinoid, then $H^i(X_{\text{proét}}, \mathcal{O}_X) = 0$, and $H^i(X_{\text{proét}}, \mathcal{O}_X^+)$ is almost zero.
2. The presheaf on Perfd defined by $h_X(Y) = \operatorname{Hom}(Y, X)$ is a sheaf on the pro-étale site. (That is, all representable presheaves are sheaves, or equivalently the pro-étale site is subcanonical.)

---

[1]We ignore set-theoretic difficulties. They can be resolved by fixing a suitable cutoff cardinal; cf. [Sch17].

Proof. (1) Without loss of generality $X = \mathrm{Spa}(R, R^+)$ is a perfectoid affinoid. (This reduction proceeds because $\mathcal{O}_X$ and $\mathcal{O}_X^+$ are already sheaves for the analytic topology.) Furthermore, we need only check the sheaf property and the almost vanishing of Čech cohomology relative to affinoid pro-étale covers $Y \to X$, where $Y = \mathrm{Spa}(R_\infty, R_\infty^+)$, where (R_∞, R_∞^+) is the completion of $\varinjlim_j (R_j, R_j^+)$.

Fix a pseudo-uniformizer $\varpi \in R$. We claim that the Čech complex

$$0 \to R^+/\varpi \to R_\infty^+/\varpi \to \cdots$$

is almost exact. The complex

$$0 \to R^+/\varpi \to R_j^+/\varpi \to \cdots$$

is almost exact, because $H^i(X_{\text{ét}}, \mathcal{O}_X^+/\varpi)$ is almost zero for $i > 0$ (and is R^+/ϖ for $i = 0$). Now we can take a direct limit over j. A filtered direct limit of almost exact sequences is almost exact. Thus,

$$0 \to R^+ \to R_\infty^+ \to \cdots$$

is almost exact (all terms are ϖ-torsion free and ϖ-adically complete), which gives the claim about the almost vanishing of the higher cohomology of $\mathcal{O}_X^+$. Now invert ϖ to get that $\mathcal{O}_X$ is a pro-étale sheaf, whose higher cohomology is zero. This implies that $\mathcal{O}_X^+$ is also a sheaf, as $\mathcal{O}_X^+ \subset \mathcal{O}_X$ is the subsheaf given by functions of absolute value at most 1 everywhere.

(2) We can reduce to the case that $X = \mathrm{Spa}(R, R^+)$ and $Y = \mathrm{Spa}(S, S^+)$ are affinoid. Let $\{f_i \colon Y_i \to Y\}$ be a pro-étale cover, on which one has compatible maps $Y_i \to X$. We get a map $R \to H^0(Y_{\text{proét}}, \mathcal{O}_Y) = S$ (as $\mathcal{O}_Y$ is a pro-étale sheaf), and similarly a map $R^+ \to S^+$, as desired. $\square$

8.3   DEFINITION OF DIAMONDS

Recall that our intuitive definition of diamonds involved the tilting functor in case of perfectoid spaces of characteristic 0. For this reason, diamonds will be defined as certain pro-étale sheaves on the category Perf $\subset$ Perfd of perfectoid spaces of characteristic p.

In the following, we will denote the sheaf on Perf represented by a perfectoid space X of characteristic p by X itself.

Definition 8.3.1. A *diamond* is a pro-étale sheaf $\mathcal{D}$ on Perf such that one can write $\mathcal{D} = X/R$ as a quotient of a perfectoid space X of characteristic p by an equivalence relation $R \subset X \times X$ such that R is a perfectoid space with $s, t : R \to X$ pro-étale.

Remark 8.3.2. This definition is similar to the definition of algebraic spaces as quotients of schemes by étale equivalence relations.

Let us analyze this definition. What are the conditions on a pair (R, X) such that X/R defines a diamond with induced equivalence relation $R = X \times_{X/R} X$? First, we need that R injects into $X \times X$.

Definition/Proposition 8.3.3 ([Sch17, Proposition 5.3]). Let $f : Y \to X$ be a map of perfectoid spaces. The following conditions are equivalent.

1. For all perfectoid spaces T, the map $\operatorname{Hom}(T, Y) \to \operatorname{Hom}(T, X)$ is injective.
2. For all perfectoid affinoid fields (K, K^+), the map $Y(K, K^+) \to X(K, K^+)$ is injective.
3. For all algebraically closed affinoid fields (C, C^+), the map $Y(C, C^+) \to X(C, C^+)$ is injective.
4. The map $|Y| \to |X|$ is injective, and for all $y \in Y$ with image $x \in X$, the map $K(y) \to K(x)$ is an isomorphism.
5. The map $|Y| \to |X|$ is injective, and for all perfectoid spaces T, the map

$$\operatorname{Hom}(T, Y) \to \operatorname{Hom}(T, X) \times_{C^0(|T|,|X|)} C^0(|T|, |Y|)$$

is a bijection, where $C^0(-, -)$ denotes the set of continuous maps.

The map f is an *injection* if it satisfies these equivalent conditions.

Now we get the following result.

Proposition 8.3.4 ([Sch17, Proposition 11.3]). Let X be a perfectoid space of characteristic p, and let R be a perfectoid space with two pro-étale maps $s, t : R \to X$ such that the induced map $R \to X \times X$ is an injection making R an equivalence relation on X. Then $\mathcal{D} = X/R$ is a diamond and the natural map $R \to X \times_{\mathcal{D}} X$ is an isomorphism.

The following results will allow us to make sense of absolute products like $\operatorname{Spa} \mathbf{Q}_p \times \operatorname{Spa} \mathbf{Q}_p$ in the world of diamonds, as promised in the introduction.

Lemma 8.3.5. The (absolute) product of two perfectoid spaces of characteristic p is again a perfectoid space.

Remark 8.3.6. Note that Perf lacks a final object, so an absolute product is not a fiber product.

Proof. It suffices to show that the product of two affinoid perfectoid spaces of characteristic p is again an affinoid perfectoid space. Let $X = \operatorname{Spa}(A, A^+)$ and $Y = \operatorname{Spa}(B, B^+)$ be two affinoid perfectoid spaces, with pseudo-uniformizers $\varpi \in A$ and $\varpi' \in B$. For each $m, n \geq 1$, we define a topology on $A \otimes_{\mathbf{F}_p} B$, using the ring of definition $(A^\circ \otimes_{\mathbf{F}_p} B^\circ) \left[\frac{\varpi^m \otimes 1}{1 \otimes \varpi'}, \frac{1 \otimes (\varpi')^n}{\varpi \otimes 1} \right]$, equipped with its $\varpi \otimes 1$-adic topology (this coincides with the $1 \otimes \varpi'$-adic topology). Let $(A \otimes_{\mathbf{F}_p} B)^+$ be the integral closure of $A^+ \otimes_{\mathbf{F}_p} B^+$ in $A \otimes_{\mathbf{F}_p} B$. Then $(A \otimes_{\mathbf{F}_p} B, (A \otimes_{\mathbf{F}_p} B)^+)$ is a Huber pair; we let $(C_{m,n}, C_{m,n}^+)$ be its completion. Then $C_{m,n}$ is a perfect complete Tate ring, so that by Proposition 6.1.6, $C_{m,n}$ is a perfectoid ring, and $\mathrm{Spa}(C_{m,n}, C_{m,n}^+)$ is a perfectoid space.

For $m' \geq m$ and $n' \geq n$ there is a natural open immersion $\mathrm{Spa}(C_{m,n}, C_{m,n}^+) \to \mathrm{Spa}(C_{m',n'}, C_{m',n'}^+)$. We claim that the union of these over all pairs (m, n) represents the product $X \times Y$. Indeed, suppose (R, R^+) is a complete Huber pair, and $f \colon (A, A^+) \to (R, R^+)$ and $g \colon (B, B^+) \to (R, R^+)$ are morphisms. Since the sequences $f(\varpi)^m/g(\varpi')$ and $g(\varpi')^n/f(\varpi)$ both approach 0 in R, there must exist $m, n \geq 1$ such that $f(\varpi)^m/g(\varpi')$ and $g(\varpi')^n/f(\varpi)$ both lie in A°. Then by our construction, the homomorphism $(A \otimes_{\mathbf{F}_p} B, A^+ \otimes_{\mathbf{F}_p} B^+) \to (R, R^+)$ factors uniquely through a morphism of Huber pairs $(C_{m,n}, C_{m,n}^+) \to (R, R^+)$. $\qquad\square$

Proposition 8.3.7. Let $\mathcal{D}$ and $\mathcal{D}'$ be diamonds. Then the product sheaf $\mathcal{D} \times \mathcal{D}'$ is also a diamond.

Proof. Let $\mathcal{D} = X/R$ and $\mathcal{D}' = X'/R'$, where X, X', R, R' are perfectoid spaces of characteristic p, and $R \to X \times X$ and $R' \to X' \times X'$ are injections making R (respectively, R') an equivalence relation on X (respectively, X'). Then $R \times R'$ and $X \times X'$ are perfectoid spaces by Lemma 8.3.5, and $R \times R' \to (X \times X') \times (X \times X')$ is an injection (this is clear from Prop 8.3.3(1)) which induces an equivalence relation. Therefore $\mathcal{D} \times \mathcal{D}' \cong (X \times X')/(R \times R')$ is a diamond, by Proposition 8.3.4. $\qquad\square$

8.4 THE EXAMPLE OF $\mathrm{Spd}\,\mathbf{Q}_p$

Following the discussion at the beginning of this lecture, this is defined as the following sheaf on Perf:

$$\mathrm{Spd}\,\mathbf{Q}_p := \mathrm{Spa}(\mathbf{Q}_p^{\mathrm{cycl}})^\flat/\mathbf{Z}_p^\times = \mathrm{Spa}(\mathbf{F}_p((t^{1/p^\infty})))/\mathbf{Z}_p^\times \,,$$

where $\mathbf{Z}_p^\times$ acts on $\mathbf{F}_p((t^{1/p^\infty}))$ via $\gamma(t) = (1+t)^\gamma - 1$ for all $\gamma \in \mathbf{Z}_p^\times$.

Here, we use the following notation. If X is a perfectoid space and G is a profinite group, then one can define what it means for G to act continuously on X; cf. [Sch18]. Equivalently, this is an action of the pro-étale sheaf of groups $\underline{G}$ on X, where $\underline{G}(T)$ is the set of continuous maps from $|T|$ to G, for any $T \in \mathrm{Perf}$. In particular, given a continuous action of G on X, one can define $R = X \times \underline{G}$ (which agrees with the previous definition of $X \times \underline{S}$ for a profinite set S), which comes with two maps $R \rightrightarrows X$, given by the projection to the first component, and the action map. If the induced map $R \to X \times X$ is an injection, then X/R is a diamond that we also denote by $X/\underline{G}$.

Essentially by unraveling the definitions, we find the following description of $\mathrm{Spd}\,\mathbf{Q}_p$; a full proof will be given in the next lecture.

Proposition 8.4.1. If $X = \mathrm{Spa}(R, R^+)$ is an affinoid perfectoid space of characteristic p, then $(\mathrm{Spd}\,\mathbf{Q}_p)(X)$ is the set of isomorphism classes of data of the following shape:

1. A $\mathbf{Z}_p^\times$-torsor $R \to \tilde{R}$: that is, $\tilde{R} = \left(\varinjlim R_n\right)^\wedge$, where R_n/R is finite étale with Galois group $(\mathbf{Z}/p^n\mathbf{Z})^\times$.
2. A topologically nilpotent element unit $t \in \tilde{R}$ such that for all $\gamma \in \mathbf{Z}_p^\times$, $\gamma(t) = (1+t)^\gamma - 1$.

   Moreover, we will prove the following theorem.

Theorem 8.4.2. The category of perfectoid spaces over $\mathbf{Q}_p$ is equivalent to the category of perfectoid spaces X of characteristic p equipped with a structure morphism $X \to \mathrm{Spd}\,\mathbf{Q}_p$.

Lecture 9

Diamonds II

9.1 COMPLEMENTS ON THE PRO-ÉTALE TOPOLOGY

In the previous lecture on the pro-étale topology, participants raised two issues that we would like to address today.

The first issue concerned descent, or more specifically *pro-étale descent for perfectoid spaces*. In the classical world of schemes, there are various results pertaining to descent along an fpqc covering $X' \to X$. For instance, suppose we are given a morphism of schemes $Y' \to X'$ together with a descent datum over $X' \times_X X'$. We may ask if Y' descends to X; that is, if there exists a morphism $Y \to X$ for which $Y' = Y \times_X X'$. Such a Y is unique if it exists: this is because the presheaf $h_Y = \mathrm{Hom}(Y, -)$ on the fpqc site of X is actually a sheaf. If $Y' \to X'$ is affine, then $Y \to X$ exists and is also affine [Sta, Tag 0244]. A concise way of saying this is that the fibered category

$$X \mapsto \{Y/X \text{ affine}\}$$

on the category of schemes is a stack for the fpqc topology.

Proposition 8.2.8 prompts us to ask whether there are similar descent results for perfectoid spaces, using the pro-étale topology.

Question 9.1.1. Is the fibered category

$$X \mapsto \{\text{morphisms } Y \to X \text{ with } Y \text{ perfectoid}\}$$

on the category of perfectoid spaces a stack for the pro-étale topology? That is, if $X' \to X$ is a pro-étale cover, and we are given a morphism $Y' \to X'$ together with a descent datum over $X' \times_X X'$, etc, then does $Y' \to X'$ descend to $Y \to X$? If not, is this true under stronger hypothesis on the morphisms? What if all spaces are assumed to be affinoid perfectoid? (Such a descent is unique up to unique X-isomorphism if it exists, by Proposition 8.2.8.)

The answer to this type of question is in general very negative.

Example 9.1.2 (There is no good notion of affinoid morphism in rigid geometry). Recall that a morphism of schemes $f \colon Y \to X$ is *affine* if for all open affine subsets $U \subset X$, $f^{-1}(U)$ is affine. It is a basic result that this is equivalent to the condition that there exists an open affine cover $\{U_i\}$ of X such that $f^{-1}(U_i)$ is affine for all i. As a result, the functor $X \mapsto \{Y/X \text{ affine}\}$ is a stack on the category of schemes in the Zariski (and even the fpqc) topology.

One might guess that there is a notion of "affinoid morphism" between adic spaces which works the same way. However, we run into the following counterexample. Let K be a nonarchimedean field, and let $X = \operatorname{Spa} K\langle x, y\rangle$. Let $V \subset X$ be $\{(x,y) | |x| = 1 \text{ or } |y| = 1\}$. This is covered by two affinoids, but certainly is not affinoid itself: $H^1(V, \mathcal{O}_V) = \widehat{\bigoplus}_{m,n>0} K x^{-n} y^{-m}$ (analogous to the situation of the punctured plane in classical geometry).

We claim there is a cover $X = \bigcup_i U_i$ by rational subsets U_i such that $U_i \times_X V \subset U_i$ is rational, thus affinoid. Let $\varpi \in K$ be a pseudo-uniformizer, and take

$$
\begin{aligned}
U_0 &= \{|x|, |y| \leq |\varpi|\} \\
U_1 &= \{|y|, |\varpi| \leq |x|\} \\
U_2 &= \{|x|, |\varpi| \leq |y|\}.
\end{aligned}
$$

Then $X = U_0 \cup U_1 \cup U_2$, and

$$
\begin{aligned}
U_0 \times_X V &= \emptyset \\
U_1 \times_X V &= \{|x| = 1\} \subset U_1 \\
U_2 \times_X V &= \{|y| = 1\} \subset U_2
\end{aligned}
$$

are all rational subsets of U_i. One gets a similar example in the perfectoid setting. Thus the functor $X \mapsto \{Y/X \text{ affinoid perfectoid}\}$ is *not* a stack on the category of affinoid perfectoid spaces for the analytic topology, let alone for a finer topology such as the pro-étale topology.

A positive answer to this is however given by the following result, which involves the notion of (strictly) totally disconnected spaces defined later today.[1]

Theorem 9.1.3 ([Sch17, Propositions 9.3, 9.6, 9.7]). Descent along a pro-étale cover $X' \to X$ of a perfectoid space $f : Y' \to X'$ is effective in the following cases.

1. If X, X' and Y' are affinoid and X is totally disconnected.
2. If f is separated and pro-étale and X is strictly totally disconnected.
3. If f is separated and étale.
4. If f is finite étale.

Moreover, the descended morphism has the same properties.

The other issue was that the property of being a pro-étale morphism is not local for the pro-étale topology on the target. That is, suppose $f: X \to Y$ is a

---

[1] For separatedness, see [Sch17, Definition 5.10].

morphism between perfectoid spaces, and suppose there exists a pro-étale cover $Y' \to Y$, such that the base change $X' = X \times_Y Y' \to Y'$ is pro-étale. We cannot conclude that f is pro-étale; cf. Example 9.1.5 below. How can we characterize such an f? It turns out there is a convenient "punctual" criterion.

Proposition 9.1.4 ([Sch17, Lemma 7.19]). Let $f \colon X \to Y$ be a morphism of affinoid perfectoid spaces. The following are equivalent:

1. There exists $Y' \to Y$ which is affinoid pro-étale surjective, such that the base change $X' = X \times_Y Y' \to Y'$ is affinoid pro-étale.
2. For all geometric points $\operatorname{Spa} C \to Y$ of rank 1, the pullback $X \times_Y \operatorname{Spa} C \to \operatorname{Spa} C$ is affinoid pro-étale; equivalently, $X \times_Y \operatorname{Spa} C = \operatorname{Spa} C \times \underline{S}$ for some profinite set S.

Example 9.1.5 (A non-pro-étale morphism which is locally pro-étale). Assume $p \neq 2$, and let
$$Y = \operatorname{Spa} K\langle T^{1/p^\infty}\rangle \, ,$$
and
$$X = \operatorname{Spa} K\langle T^{1/2p^\infty}\rangle \, .$$

Then $X \to Y$ appears to be ramified at 0, and indeed it is not pro-étale. However, consider the following pro-étale cover of Y: let

$$Y' = \varprojlim Y'_n, \ Y'_n = \{x \in Y \mid |x| \leq |\varpi^n|\} \sqcup \coprod_{i=1}^{n} \left\{x \in Y \mid |\varpi|^i \leq |x| \leq |\varpi|^{i-1}\right\}.$$

Let $X' = X \times_Y Y'$. We claim that the pullback $X' \to Y'$ is affinoid pro-étale.

As a topological space, $\pi_0(Y') = \{1, 1/2, 1/3, \ldots, 0\} \subset \mathbf{R}$. The fiber of Y' over $1/i$ is

$$\left\{x \in Y \mid |\varpi|^i \leq |x| \leq |\varpi|^{i-1}\right\} \, ,$$

and the fiber over 0 is just 0. Let

$$X'_n = \{x \mid |x| \leq |\varpi|^n\} \sqcup \coprod_{i=1}^{n} \left\{|\varpi|^i \leq |x| \leq |\varpi|^{i-1}\right\} \times_Y X$$

so that $X'_n \to Y'_n$ is finite étale. Then $X' \cong \varprojlim(X'_n \times_{Y'_n} Y') \to Y'$ is pro-étale.

The proof of Proposition 9.1.4 in general relies on the notion of strictly totally disconnected perfectoid spaces, where the base space is torn apart even more drastically than in this example.

Definition 9.1.6. A perfectoid space X is totally disconnected (resp. strictly totally disconnected) if it is qcqs and every open (resp. étale) cover splits.

Any totally disconnected perfectoid space is affinoid; cf. [Sch17, Lemma

7.5]. The connected components of a totally disconnected space are of the form $\operatorname{Spa}(K, K^+)$ for a perfectoid affinoid field (K, K^+); for strictly totally disconnected spaces, one has in addition that K is algebraically closed; cf. [Sch17, Lemma 7.3, Proposition 7.16], and this property characterizes these spaces. Moreover, for any affinoid perfectoid space X one can find an affinoid pro-étale map $\widetilde{X} \to X$ such that $\widetilde{X}$ is strictly totally disconnected; cf. [Sch17, Lemma 7.18].

A central technique of [Sch17] is the reduction of many statements to the case of strictly disconnected spaces. We give an example of this type in the appendix to the next lecture.

9.2 QUASI-PRO-ÉTALE MORPHISMS

Proposition 9.1.4 suggests enlarging the class of pro-étale morphisms.

Definition 9.2.1. A morphism $f\colon X \to Y$ of perfectoid spaces is *quasi-pro-étale* if for any strictly totally disconnected perfectoid space Y' with a map $Y' \to Y$, the pullback $X' = X \times_Y Y' \to Y'$ is pro-étale.

In general, $f\colon X \to Y$ is quasi-pro-étale if and only if it is so when restricted to affinoid open subsets, in which case it is equivalent to the condition of Proposition 9.1.4.

Definition 9.2.2. Consider the site Perf of perfectoid spaces of characteristic p with the pro-étale topology. A map $f\colon \mathcal{F} \to \mathcal{G}$ of sheaves on Perf is *quasi-pro-étale* if it is locally separated[2] and for all strictly totally disconnected perfectoid spaces Y with a map $Y \to \mathcal{G}$ (i.e., an element of $\mathcal{G}(Y)$), the pullback $\mathcal{F} \times_{\mathcal{G}} Y$ is representable by a perfectoid space X and $X \to Y$ is pro-étale.

Using this definition, we can give an equivalent characterization of diamonds.

Proposition 9.2.3 ([Sch17, Proposition 11.5]). A pro-étale sheaf Y on Perf is a diamond if and only if there is a surjective quasi-pro-étale map $X \to Y$ from a perfectoid space X.

---

[2]This is per [Sch17, Convention 10.2]; if f is not locally separated, one should modify the definition slightly. Note that if $\mathcal{F}$ is representable by a perfectoid space, then f is automatically locally separated, which will be the most relevant case below.

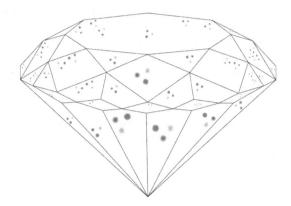

Figure 9.1: We offer here a justification for the terminology "diamond" through an analogy. Suppose $\mathcal{D}$ is a diamond. A geometric point $\operatorname{Spa} C \to \mathcal{D}$ is something like an impurity within a gem which produces a color. This impurity cannot be seen directly, but produces many reflections of this color on the surface of the diamond. Likewise, the geometric point cannot be seen directly, but when we pull it back through a quasi-pro-étale cover $X \to \mathcal{D}$, the result is profinitely many copies of $\operatorname{Spa} C$. Often one can produce multiple such covers $X \to \mathcal{D}$, which result in multiple descriptions of the geometric points of $\mathcal{D}$.

9.3 G-TORSORS

If G is a finite group, we have the notion of G-*torsor* on any topos. This is a map $f \colon \mathcal{F}' \to \mathcal{F}$ with an action $G \times \mathcal{F}' \to \mathcal{F}'$ over $\mathcal{F}$ such that locally on $\mathcal{F}$, one has a G-equivariant isomorphism $\mathcal{F}' \cong \mathcal{F} \times G$.

We also have G-torsors when G is not an abstract group but rather a group object in a topos, for example in the category of pro-étale sheaves on Perf. For any topological space T, we can introduce a sheaf $\underline{T}$ on Perf, by $\underline{T}(X) = C^0(|X|, T)$. As pro-étale covers induce quotient mappings by Proposition 4.3.3, we see that $\underline{T}$ is a pro-étale sheaf. If T is a profinite set, this agrees with the definition of $\underline{T}$ given earlier. If now G is a topological group, then $\underline{G}$ is a sheaf of groups. If $G = \varprojlim_i G_i$ is a profinite group, then in fact $\underline{G} = \varprojlim_i \underline{G_i}$.

Note that $\underline{G}$ is not representable, even if G is finite. The problem is that Perf lacks a final object X (in other words, a base). If it had one, then for finite G the sheaf $\underline{G}$ would be representable by G copies of X. And indeed, $\underline{G}$ becomes representable once we supply the base. If X is a perfectoid space and G is a profinite group, then $X \times \underline{G}$ is representable by a perfectoid space, namely

$$X \times \underline{G} = \varprojlim X \times G/H,$$

where $X \times G/H$ is just a finite disjoint union of copies of X. Finally, note that everything in this paragraph applies when G is a profinite set rather than a group, and that the notation is consistent with the discussion before Remark 8.2.3.

Now a $\underline{G}$-torsor is a morphism $f\colon \mathcal{F}' \to \mathcal{F}$ with an action $\underline{G} \times \mathcal{F}' \to \mathcal{F}'$ such that locally on $\mathcal{F}$ we have a $\underline{G}$-equivariant isomorphism $\mathcal{F}' \cong \mathcal{F} \times \underline{G}$.

Proposition 9.3.1 ([Sch17, Lemma 10.13]). Let $f\colon \mathcal{F}' \to \mathcal{F}$ be a $\underline{G}$-torsor, with G profinite. Then for any affinoid $X = \mathrm{Spa}(B, B^+)$ and any morphism $X \to \mathcal{F}$, the pullback $\mathcal{F}' \times_{\mathcal{F}} X$ is representable by a perfectoid affinoid $X' = \mathrm{Spa}(A, A^+)$. Furthermore, A is the completion of $\varinjlim_H A_H$, where for each open normal subgroup $H \subset G$, A_H/B is a finite étale G/H-torsor in the algebraic sense.

Remark 9.3.2. In fact one can take $A_H = A^H$ to be the ring of elements of A fixed by H. Moreover, [Sch17, Lemma 10.13] is actually a more general result for torsors under *locally* profinite groups G, which is significantly harder.

Proof. (Sketch.) We reduce to the case of G finite this way: If $H \subset G$ is open normal, then $\mathcal{F}'/\underline{H} \to \mathcal{F}$ is a G/H-torsor, and $\mathcal{F}' = \varprojlim_H \mathcal{F}'/\underline{H}$.

The key point now is that the fibered category

$$Y \mapsto \{\text{finite étale } X/Y\}$$

over the category of affinoid perfectoid spaces is a stack for the pro-étale topology, which is part of Theorem 9.1.3. However, it is the easier part of it, so we briefly sketch the argument. If $Y' = \varprojlim Y_i' \to Y$ is an affinoid pro-étale cover, then

$$\{\text{finite étale descent data for } Y' \to Y\}$$
$$= \quad 2\text{-}\varinjlim \{\text{finite étale descent data for } Y_i'/Y\}$$

using Theorem 7.4.8, so we are reduced to showing that étale descent works. For this we can either use a descent to noetherian adic spaces together with a result of Huber, or we can use an argument due to de Jong–van der Put, [dJvdP96], that in general étale descent follows from analytic descent plus classical finite étale descent. □

9.4   THE DIAMOND $\mathrm{Spd}\,\mathbf{Q}_p$

Consider the perfectoid space $\mathrm{Spa}(\mathbf{Q}_p^{\mathrm{cycl}})^\flat$ as a sheaf on Perf. Its (R, R^+)-valued points are the set of continuous homomorphisms $(\mathbf{Q}_p^{\mathrm{cycl}})^\flat \to R$. Since $(\mathbf{Q}_p^{\mathrm{cycl}})^\flat \cong \mathbf{F}_p((t^{1/p^\infty}))$, this is nothing more than the set of topologically nilpotent invertible elements of R.

In the last lecture, we gave the following ad hoc definition of the diamond $\mathrm{Spd}\,\mathbf{Q}_p$.

Definition 9.4.1. $\operatorname{Spd} \mathbf{Q}_p = \operatorname{Spa}(\mathbf{Q}_p^{\mathrm{cycl}})^\flat / \underline{\mathbf{Z}_p^\times}$. That is, $\operatorname{Spd} \mathbf{Q}_p$ is the coequalizer of

$$\underline{\mathbf{Z}_p^\times} \times \operatorname{Spa}(\mathbf{Q}_p^{\mathrm{cycl}})^\flat \rightrightarrows \operatorname{Spa}(\mathbf{Q}_p^{\mathrm{cycl}})^\flat,$$

where one map is the projection and the other is the action.

To see that this is well-behaved, we need to check that

$$\underline{\mathbf{Z}_p^\times} \times \operatorname{Spa}(\mathbf{Q}_p^{\mathrm{cycl}})^\flat \to \operatorname{Spa}(\mathbf{Q}_p^{\mathrm{cycl}})^\flat \times \operatorname{Spa}(\mathbf{Q}_p^{\mathrm{cycl}})^\flat$$

is an injection, which is the content of the next lemma.

Lemma 9.4.2. Let $g\colon \underline{\mathbf{Z}_p^\times} \times \operatorname{Spd} \mathbf{Q}_p^{\mathrm{cycl}} \to \operatorname{Spd} \mathbf{Q}_p^{\mathrm{cycl}} \times \operatorname{Spd} \mathbf{Q}_p^{\mathrm{cycl}}$ be the product of the projection onto the second factor and the group action. Then g is an injection.

Proof. Let (K, K^+) be a perfectoid affinoid field. Then $\mathbf{Z}_p^\times$ acts freely on $\operatorname{Hom}(\mathbf{F}_p((t^{1/p^\infty})), K)$. This implies the result by Proposition 8.3.3. $\qquad\square$

Corollary 9.4.3. The map $\operatorname{Spa} \mathbf{Q}_p^{\mathrm{cycl}} \to \operatorname{Spd} \mathbf{Q}_p$ is a $\underline{\mathbf{Z}_p^\times}$-torsor, and the description of $\operatorname{Spd} \mathbf{Q}_p$ given in Proposition 8.4.1 holds true.

Proof. Indeed, given any map $\operatorname{Spa}(R, R^+) \to \operatorname{Spd} \mathbf{Q}_p$, via pullback we get a $\underline{\mathbf{Z}_p^\times}$-torsor over $\operatorname{Spa}(R, R^+)$, and those are described by Proposition 9.3.1. $\qquad\square$

Theorem 9.4.4. The following categories are equivalent:

- Perfectoid spaces over $\mathbf{Q}_p$.
- Perfectoid spaces X of characteristic p equipped with a "structure morphism" $X \to \operatorname{Spd} \mathbf{Q}_p$.

Before proving Theorem 9.4.4, we observe that both categories are fibered over Perf. For the second category this is obvious. For the first category, the morphism to Perf is $X \mapsto X^\flat$. The existence of pullbacks is Theorem 7.1.4: if $X \to Y$ is a morphism in Perf and $Y^\sharp$ is an untilt of Y, then there exists a unique morphism of perfectoid spaces $X^\sharp \to Y^\sharp$ whose tilt is $X \to Y$.

Our two fibered categories correspond to two presheaves of groupoids on Perf:

- $X \mapsto \operatorname{Untilt}_{\mathbf{Q}_p}(X) = \{(X^\sharp, \iota)\}$, where $X^\sharp$ is a perfectoid space over $\mathbf{Q}_p$ and $\iota\colon X^{\sharp\flat} \cong X$ is an isomorphism.
- $\operatorname{Spd} \mathbf{Q}_p$.

In fact, sections of both presheaves have no nontrivial automorphisms, and so we can think of these as presheaves of sets. (In the second case this is by definition, and in the first case it is again a case of Theorem 7.1.4.)

Exhibiting an isomorphism between the fibered categories is the same as exhibiting an isomorphism between the presheaves. Now $\operatorname{Spd} \mathbf{Q}_p$ is a sheaf for the pro-étale topology on Perf (by definition). We will prove that $\mathrm{Untilt}_{\mathbf{Q}_p}$ is a sheaf as well. In fact, we can prove a bit more. Let Untilt be the presheaf on Perf which assigns to X the set of pairs $(X^\sharp, \iota)$, where $X^\sharp$ is a perfectoid space (of whatever characteristic), and $\iota \colon X^{\sharp\flat} \cong X$ is an isomorphism.

Lemma 9.4.5 ([Sch17, Lemma 15.1 (i)]). Untilt is a sheaf on Perf.

Proof. That is: given a pro-étale cover $\{f_i \colon X_i \to X\}$, we are to show that

$$\mathrm{Untilt}(X) = \mathrm{eq}\left(\prod_i \mathrm{Untilt}(X_i) \rightrightarrows \prod_{i,j} \mathrm{Untilt}(X_i \times_X X_j) \right).$$

As in the proof of Proposition 8.2.8(1), we may assume without loss of generality that $X = \mathrm{Spa}(R, R^+)$ is affinoid, and that the cover consists of a single affinoid $Y = \mathrm{Spa}(S, S^+)$, with $Y \to X$ pro-étale. Then $R \to S$ is an injection and $R^+ = R \cap S^+$ (this follows from the fact that $\mathcal{O}_X$ and $\mathcal{O}_X^+$ are sheaves on $X_{\mathrm{proét}}$.) Let $Z = Y \times_X Y = \mathrm{Spa}(T, T^+)$. Finally, let $Y^\sharp = \mathrm{Spa}(S^\sharp, S^{\sharp+})$ be an untilt of Y. Pullback along the two morphisms $Z \rightrightarrows Y$ gives two untilts $Z_1^\sharp, Z_2^\sharp$ of Z over $Y^\sharp$. We may now translate the sheaf condition as follows: Every descent datum $Z_1^\sharp \cong Z_2^\sharp$ is effective, in the sense that it is induced by a unique untilt $X^\sharp = \mathrm{Spa}(R, R^\sharp)$ of X whose pullback to Y is $Y^\sharp$.

Given such a descent datum, we may identify both untilts of Z with $Z^\sharp = \mathrm{Spa}(T^\sharp, T^{\sharp+})$. We define a pair of topological rings $(R^\sharp, R^{\sharp+})$ by

$$\begin{aligned} R^\sharp &= \mathrm{eq}(S^\sharp \rightrightarrows T^\sharp), \\ R^{\sharp+} &= R^\sharp \cap S^{\sharp+}. \end{aligned}$$

We claim that $(R^\sharp, R^{\sharp+})$ is a perfectoid Huber pair, such that the tilt of $(R^\sharp, R^{\sharp+}) \to (S^\sharp, S^{\sharp+})$ is $(R, R^+) \to (S, S^+)$. It is already clear that $R^{\sharp+} \subset R^\sharp$ is open, integrally closed, and bounded, since the same properties hold for $S^{\sharp+} \subset S^+$.

Choose a pseudouniformizer $\varpi \in R \subset S$, so that $\varpi^\sharp \in S^\sharp$ is a pseudouniformizer for $S^\sharp$. Note that $\varpi^\sharp \in R^\sharp$, and that it is a pseudouniformizer in $R^\sharp$. Since $(\varpi^\sharp)^n S^{\sharp+}$ (for $n = 1, 2, \dots$) constitutes a basis of neighborhoods of 0 in $S^\sharp$, the same is true for the $(\varpi^\sharp)^n R^{\sharp+}$. At this point we can conclude that $(R^\sharp, R^{\sharp+})$ is a uniform Tate Huber pair.

We may choose ϖ in such a way that $(\varpi^\sharp)^p | p$ holds in $S^{\sharp\circ}$. The inclusion $R^{\sharp+} \to S^{\sharp+}$ induces a map

$$\begin{aligned} R^{\sharp+}/(\varpi^\sharp)^p &\to \mathrm{eq}\left(S^{\sharp+}/(\varpi^\sharp)^p \rightrightarrows T^{\sharp+}/(\varpi^\sharp)^p \right) \\ &\cong \mathrm{eq}\left(S^+/\varpi^p \rightrightarrows T^+/\varpi^p \right). \end{aligned}$$

Consider the complex

$$S^+ \xrightarrow{\delta} T^+ \xrightarrow{\delta'} (T')^+ \to \cdots \tag{9.4.1}$$

which computes the Čech cohomology for $\mathcal{O}_X^+$ relative to the cover $Y \to X$. (Thus δ is the difference between the two maps $S^+ \rightrightarrows T^+$, and δ' is the alternating sum of three maps, and so on.) The higher cohomology of this complex is almost zero by Proposition 3.1.8(1). Let $x \in S^+$ be an element whose image in S^+/ϖ^p lies in $\mathrm{eq}(S^+/\varpi^p \rightrightarrows T^+/\varpi^p)$. Then $\delta(x) = \varpi^p y$ for some $y \in T^+$. Since $\delta'(y) = 0$, we have $\varpi y = \delta(x')$ for some $x' \in S^+$. Letting $\tilde{x} = x - \varpi^{p-1}x'$, we see that $\delta(\tilde{x}) = 0$ and therefore $\tilde{x} \in R^+$.

Tracing through this argument shows there is a well-defined map $R^{\sharp+}/(\varpi^{\sharp})^p \to R^+/\varpi^{p-1}$, which induces a map $R^{\sharp+}/\varpi^{\sharp} \to R^+/\varpi$. This map is characterized by the property that

$$
\begin{array}{ccc}
R^{\sharp+}/\varpi^{\sharp} & \longrightarrow & R^+/\varpi \\
\downarrow & & \downarrow \\
S^{\sharp+}/\varpi^{\sharp} & \xrightarrow{\cong} & S^+/\varpi
\end{array}
$$

commutes, where the vertical maps are the inclusions. We claim that $R^{\sharp+}/\varpi^{\sharp} \to R^+/\varpi$ is an isomorphism. Injectivity is obvious from the diagram, and surjectivity follows from observing that if $x \in R^+$, then the class of $x^{\sharp} \in R^{\sharp+}$ modulo $\varpi^{\sharp}$ is a preimage of x mod ϖ.

From the isomorphism $R^{\sharp+}/\varpi^{\sharp} \to R^+/\varpi$ we deduce that $\Phi\colon R^{\sharp+}/(\varpi^{1/p})^{\sharp} \to R^{\sharp+}/\varpi^{\sharp}$ is an isomorphism as well. This in turn implies that $R^{\sharp}$ is perfectoid, and that

$$R^{\sharp\flat+} = \varprojlim_{\Phi} R^{\sharp+}/\varpi^{\sharp} \cong \varprojlim_{\Phi} R^+/\varpi = R^+.$$

Thus $(R^{\sharp}, R^{\sharp+})$ is an untilt of (R, R^+). As for uniqueness: if $(A, A^+) \to (S^{\sharp}, S^{\sharp+})$ is another untilt of $(R, R^+) \to (S, S^+)$, then $(A, A^+) \to (S^{\sharp}, S^{\sharp+})$ must factor through $(R^{\sharp}, R^{\sharp+})$, since this is the equalizer of the two maps into $(T^{\sharp}, T^{\sharp+})$. But then the tilt of $(A, A^+) \to (R^{\sharp}, R^{\sharp+})$ is an isomorphism, so that it must be an isomorphism itself.

$\square$

We can now complete the proof of Theorem 9.4.4. Let $X = \mathrm{Spa}(R, R^+)$ be an affinoid perfectoid space in characteristic p.

If $X^{\sharp} = \mathrm{Spa}(R^{\sharp}, R^{\sharp+})$ is an untilt, let

$$\widetilde{R}^{\sharp} = R^{\sharp} \widehat{\otimes}_{\mathbf{Q}_p} \mathbf{Q}_p^{\mathrm{cycl}},$$

let $\widetilde{R}^{\sharp+}$ be the completion of the integral closure of $R^{\sharp+}$ in $\widetilde{R}^{\sharp}$, and finally let $\widetilde{X}^{\sharp} = \mathrm{Spa}(\widetilde{R}^{\sharp}, \widetilde{R}^{\sharp+})$. Then $\widetilde{X}^{\sharp} \to X^{\sharp}$ is a pro-étale $\underline{\mathbf{Z}_p^{\times}}$-torsor, whose tilt $\widetilde{X} \to X$

is a pro-étale $\mathbf{Z}_p^\times$-torsor equipped with a $\mathbf{Z}_p^\times$-equivariant map $\widetilde{X} \to \mathrm{Spa}(\mathbf{Q}_p^{\mathrm{cycl}})^\flat$. Thus we have produced a morphism $X \to \mathrm{Spd}\,\mathbf{Q}_p$.

In the reverse direction, suppose $\widetilde{X} \to X$ is a pro-étale $\mathbf{Z}_p^\times$-torsor and $\widetilde{X} \to \mathrm{Spa}(\mathbf{Q}_p^{\mathrm{cycl}})^\flat$ a $\mathbf{Z}_p^\times$-equivariant morphism. Then by Theorem 7.1.4 there exists a unique morphism $\widetilde{X}^\sharp \to \mathrm{Spa}\,\mathbf{Q}_p^{\mathrm{cycl}}$, which is also $\mathbf{Z}_p^\times$-equivariant. The equivariance means exactly that $\widetilde{X}^\sharp$ comes equipped with a descent datum along $\widetilde{X} \to X$. Lemma 9.4.5 now produces an untilt $X^\sharp$ of X over $\mathbf{Q}_p$.

We have shown that Untilt is a sheaf. Therefore $\mathrm{Untilt}_{\mathbf{Q}_p}$ is a sheaf as well (since the invertibility of p can be checked locally). Since both $\mathrm{Untilt}_{\mathbf{Q}_p}$ and $\mathrm{Spd}\,\mathbf{Q}_p$ are sheaves on Perf, the above constructions globalize to give an isomorphism between them.

Remark 9.4.6. There is nothing particularly special about $\mathbf{Q}_p^{\mathrm{cycl}}$ here. If $K/\mathbf{Q}_p$ is any perfectoid field which arises as the completion of an algebraic extension, then $\mathrm{Spa}\,K \to \mathrm{Spa}\,\mathbf{Q}_p$ is pro-étale. The proof above shows that untilts of X are in equivalence with pro-étale covers $\widetilde{X} \to X$ together with a morphism $\widetilde{X} \to \mathrm{Spa}\,K^\flat$, which comes equipped with a descent datum relative to $\mathrm{Spa}(K \widehat{\otimes}_{\mathbf{Q}_p} K)^\flat \to \mathrm{Spa}\,K^\flat$.

Lecture 10

Diamonds associated with adic spaces

10.1 THE FUNCTOR $X \mapsto X^\diamond$

Our goal today is to construct a functor

$$\{\text{analytic pre-adic spaces}/\operatorname{Spa}\mathbf{Z}_p\} \quad \to \quad \{\text{diamonds}\}$$
$$X \quad \mapsto \quad X^\diamond$$

forgetting the structure morphism to $\operatorname{Spa}\mathbf{Z}_p$, but retaining topological information.

Definition 10.1.1. Let X be an analytic pre-adic space over $\operatorname{Spa}\mathbf{Z}_p$. Define a presheaf $X^\diamond$ on Perf as follows. For a perfectoid space T in characteristic p, let $X^\diamond(T)$ be the set of isomorphism classes of pairs $(T^\sharp, T^\sharp \to X)$, where $T^\sharp$ is an untilt of T and $T^\sharp \to X$ is a map of pre-adic spaces.
   If $X = \operatorname{Spa}(R, R^+)$, we write $\operatorname{Spd}(R, R^+) = \operatorname{Spa}(R, R^+)^\diamond$.

Remark 10.1.2. Note that if X is perfectoid, then $X^\diamond = X^\flat$ agrees with our prior definition, as perfectoid spaces over X are equivalent to perfectoid spaces over $X^\flat$ by Theorem 7.1.4. Also note that the pairs $(T^\sharp, T^\sharp \to X)$ do not have nontrivial automorphisms, so $X^\diamond$ has some hope of being a sheaf.

The construction above has the following "absolute" analogue.

Definition 10.1.3. Let $\operatorname{Spd}\mathbf{Z}_p = \operatorname{Untilt}$ be the presheaf on Perf which sends S to the set of isomorphism classes of untilts $S^\sharp$ over $\mathbf{Z}_p$ of S.

Remark 10.1.4. An untilt of S over $\mathbf{Z}_p$ is a pair $(S^\sharp, \iota)$, where $S^\sharp$ is a perfectoid space (these are always fibered uniquely over $\operatorname{Spa}\mathbf{Z}_p$) and $\iota\colon S^{\sharp\flat} \to S$ is an isomorphism. These do not have to live in characteristic 0. Indeed, there is always the trivial untilt (S, id_S). This is supposed to correspond to the projection from S onto the closed point of $\operatorname{Spd}\mathbf{Z}_p$.

We warn the reader that $\operatorname{Spd}\mathbf{Z}_p$ is not a diamond. However, it is still a pro-étale sheaf by Lemma 9.4.5. In Lectures 17 and 18, we will see how it fits into the general formalism.

Theorem 10.1.5 ([Sch17, Lemma 15.6]). *The presheaf $X^\diamond$ is a diamond.*

Proof. (Sketch.) First, one checks that $X^\diamond$ is a pro-étale sheaf. As $\operatorname{Spd}\mathbf{Z}_p$ is a pro-étale sheaf as we just mentioned, this reduces to the problem of showing that maps from perfectoid spaces $T^\sharp$ to a fixed pre-adic space X from a pro-étale sheaf; this follows from the argument of Proposition 8.2.8.

Now by Proposition 4.3.1 we may assume $X = \operatorname{Spa}(R, R^+)$ is affinoid, with R a Tate ring. Since $\operatorname{Spa}\mathbf{Z}_p$ is the base, $p \in R$ is topologically nilpotent (but not necessarily a unit).

Lemma 10.1.6 ([Fal02a], [Col02], [Sch17, Lemma 15.3]). *Let R be a Tate ring such that $p \in R$ is topologically nilpotent. Let $\varinjlim R_i$ be a filtered direct limit of algebras R_i finite étale over R, which admits no nonsplit finite étale covers. Endow $\varinjlim R_i$ with the topology making $\varinjlim R_i^\circ$ open and bounded. Let $\widetilde{R}$ be the completion. Then $\widetilde{R}$ is perfectoid.*

Proof. First, find $\varpi \in \widetilde{R}$ a pseudo-uniformizer such that $\varpi^p | p$ in $\widetilde{R}^\circ$. To do this, let $\varpi_0 \in R$ be any pseudo-uniformizer. Let N be large enough so that $\varpi_0 | p^N$. Now look at the equation $x^{p^N} - \varpi_0 x = \varpi_0$. This determines a finite étale $\widetilde{R}$-algebra, and so it admits a solution $x = \varpi_1 \in \widetilde{R}$. Note that $\varpi_1^{p^N} | \varpi_0$ in $\widetilde{R}^\circ$, and ϖ_1 is a unit in $\widetilde{R}$. As $\varpi_1^{p^N} | \varpi_0 | p^N$, we see in particular that $\varpi_1^p | p$, as desired. In fact, taking N larger than necessary, we can even ensure that p/ϖ^p is still topologically nilpotent.

Now we must check that $\Phi\colon \widetilde{R}^\circ/\varpi \to \widetilde{R}^\circ/\varpi$ is surjective. Let $f \in \widetilde{R}^\circ$, and consider the equation $x^p - \varpi^p x - f$. This determines a finite étale $\widetilde{R}$-algebra, and so it admits a solution, giving the desired surjectivity of Frobenius modulo ϖ. □

We can also assume that each R_i is a G_i-torsor over R, compatibly with change in i for an inverse system $\{G_i\}$ of finite groups. Let $G = \varprojlim_i G_i$. The following lemma finishes the proof of the theorem. □

Lemma 10.1.7. $\operatorname{Spd}(R, R^+) = \operatorname{Spd}(\widetilde{R}, \widetilde{R}^+)/\underline{G}$. *Also,* $\operatorname{Spd}(\widetilde{R}, \widetilde{R}^+) \to \operatorname{Spd}(R, R^+)$ *is a $\underline{G}$-torsor. In particular, $\operatorname{Spd}(R, R^+)$ is a diamond.*

Proof. The proof is similar to the case of $\operatorname{Spd}\mathbf{Q}_p$, but we need the fact that for any algebraically closed nonarchimedean field C of characteristic p, the group G acts freely on $\operatorname{Hom}(\widetilde{R}^\flat, C)$, so we verify this. Fix $f\colon \widetilde{R}^\flat \to C$. By the tilting equivalence, this corresponds to a map $f^\sharp\colon \widetilde{R} \to C^\sharp$. More precisely, $\widetilde{R}^\circ = W(\widetilde{R}^{\flat\circ})/I$, where I is G-stable. We get $W(f^\circ)\colon W(\widetilde{R}^{\flat\circ}) \to W(\mathcal{O}_C)$, and then $W(f^\circ) \mod I\colon \widetilde{R}^\circ \to \mathcal{O}_{C^\sharp}$.

Assume there exists $\gamma \in G$ such that

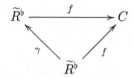

commutes. Apply W and reduce modulo I to obtain

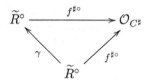

Now invert ϖ. We get that for all i,

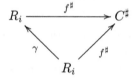

commutes, which shows that $\gamma = 1$. $\qquad\square$

Example 10.1.8. At this point, we can justify the discussion of $\mathrm{Spa}\,\mathbf{Q}_p \times \mathrm{Spa}\,\mathbf{Q}_p$ in the first lecture. Consider the product $\mathrm{Spd}\,\mathbf{Q}_p \times \mathrm{Spd}\,\mathbf{Q}_p$. Let $\mathbf{D}_{\mathbf{Q}_p}$ be the open unit disc, considered as an adic space over $\mathbf{Q}_p$. We can consider $\mathbf{D}_{\mathbf{Q}_p}$ as a subspace of $\mathbf{G}_m$ via $x \mapsto 1 + x$. Let $\widetilde{\mathbf{D}}_{\mathbf{Q}_p} = \varprojlim \mathbf{D}_{\mathbf{Q}_p}$, where the inverse limit is taken with respect to the pth power maps on $\mathbf{G}_m$, and its punctured version $\widetilde{\mathbf{D}}^*_{\mathbf{Q}_p}$.

We claim that there is an isomorphism of diamonds $(\widetilde{\mathbf{D}}^*_{\mathbf{Q}_p})^\diamond / \underline{\mathbf{Z}_p^\times} \cong \mathrm{Spd}\,\mathbf{Q}_p \times \mathrm{Spd}\,\mathbf{Q}_p$ which makes the following diagram commute:

$$
\begin{array}{ccc}
(\widetilde{\mathbf{D}}^*_{\mathbf{Q}_p})^\diamond / \underline{\mathbf{Z}_p^\times} & \longrightarrow & \mathrm{Spd}\,\mathbf{Q}_p \times \mathrm{Spd}\,\mathbf{Q}_p \\
\downarrow & & \downarrow {\scriptstyle \mathrm{pr}_1} \\
\mathrm{Spd}\,\mathbf{Q}_p & \xrightarrow{\ =\ } & \mathrm{Spd}\,\mathbf{Q}_p.
\end{array}
$$

For this it is enough to know that there is an isomorphism

$$
\left(\widetilde{\mathbf{D}}^*_{\mathbf{Q}_p^{\mathrm{cycl}}}\right)^\diamond \cong \mathrm{Spd}\,\mathbf{Q}_p^{\mathrm{cycl}} \times \mathrm{Spd}\,\mathbf{Q}_p^{\mathrm{cycl}}.
$$

which is $\mathbf{Z}_p^\times \times \mathbf{Z}_p^\times$-equivariant. It is not hard to see that $\widetilde{\mathbf{D}}^*_{\mathbf{Q}_p^{\mathrm{cycl}}}$ is a perfectoid space whose tilt is $\widetilde{\mathbf{D}}^*_{(\mathbf{Q}_p^{\mathrm{cycl}})^\flat}$, the corresponding object in characteristic p.

Meanwhile, $\mathrm{Spd}\,\mathbf{Q}_p^{\mathrm{cycl}} \times \mathrm{Spd}\,\mathbf{Q}_p^{\mathrm{cycl}}$ is representable by the perfectoid space $\mathrm{Spa}(\mathbf{Q}_p^{\mathrm{cycl}})^\flat \times \mathrm{Spa}(\mathbf{Q}_p^{\mathrm{cycl}})^\flat$. We have an isomorphism $(\mathbf{Q}_p^{\mathrm{cycl}})^\flat \cong \mathbf{F}_p((t^{1/p^\infty}))$, which when applied to one of the factors in the product gives

$$\mathrm{Spa}(\mathbf{Q}_p^{\mathrm{cycl}})^\flat \times \mathrm{Spa}(\mathbf{Q}_p^{\mathrm{cycl}})^\flat \cong \mathrm{Spa}(\mathbf{Q}_p^{\mathrm{cycl}})^\flat \times \mathrm{Spa}\,\mathbf{F}_p((t^{1/p^\infty})) = \mathbf{D}^*_{(\mathbf{Q}_p^{\mathrm{cycl}})^\flat}$$

as desired. We leave it as an exercise to check that the group actions are compatible.

10.2   EXAMPLE: RIGID SPACES

We want to understand how much information is lost when applying $X \mapsto X^\diamond$. Our intuition is that only topological information is kept. Let us offer some evidence for this intuition. A morphism $f\colon X \to Y$ of adic spaces is a *universal homeomorphism* if all pullbacks of f are homeomorphisms. As in the case of schemes, in characteristic 0 the map f is a universal homeomorphism if and only if it is a homeomorphism and induces isomorphisms on completed residue fields. An example of such an f is the normalization of the cuspidal cubic $y^2 = x^3$ in adic affine space $\mathbf{A}^2_{\mathbf{Q}_p}$ by $t \mapsto (t^2, t^3)$. In keeping with our intuition, universal homeomorphisms induce isomorphisms of diamonds:

Proposition 10.2.1. *Let $f\colon X \to X'$ be a universal homeomorphism of analytic pre-adic spaces over $\mathrm{Spa}\,\mathbf{Q}_p$. Then $f^\diamond\colon X^\diamond \to (X')^\diamond$ is an isomorphism.*

Proof. Let $Y = \mathrm{Spa}(S, S^+)$ be an affinoid perfectoid space, and let $Y \to X'$ be a morphism. We claim there exists a unique factorization

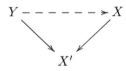

This would imply immediately that $f^\diamond$ is an isomorphism. We have $|f| : |Y| \to |X| = |X'|$. We need also $|f|^* \mathcal{O}_X \to \mathcal{O}_Y$. We have $|f|^* \mathcal{O}_X^+/p^n \to \mathcal{O}_Y^+/p^n$.

Lemma 10.2.2. $\mathcal{O}_{X'}^+/p^n \to \mathcal{O}_X^+/p^n$ *is an isomorphism of sheaves on $|X| = |X'|$.*

Proof. Check on stalks. They are $K(x')^+/p^n \to K(x)^+/p^n$, where x' corresponds to x. But the residue fields are the same, as we have a universal homeomorphism in characteristic 0. $\qquad\qquad\square$

So we have a map $|f|^* \mathcal{O}_X^+/p^n \to \mathcal{O}_Y^+/p^n$, which in the limit induces a map $|f|^* \mathcal{O}_X^+ \to \varprojlim_n \mathcal{O}_Y^+/p^n = \mathcal{O}_Y^+$. The last equality is because Y is perfectoid: $H^i(Y, \mathcal{O}_Y^+/\varpi^n)$ is almost S^+/ϖ^n if $i = 0$ and 0 if $i > 0$. $\qquad\square$

Recall our intuition that $X \mapsto X^\diamond$ keeps only the topological information. Thus a universal homeomorphism, such as the normalization of the cuspidal cubic, gets sent to an isomorphism. At the other extreme, we might ask for a class of X for which $X \mapsto X^\diamond$ is fully faithful. Recall that a ring A is *seminormal* if $t \mapsto (t^2, t^3)$ is a bijection from A onto the set of pairs $(x, y) \in A^2$ with $y^2 = x^3$. A rigid-analytic space X over a nonarchimedean field K is seminormal if locally it is $\mathrm{Spa}(A, A^+)$, where A is seminormal (this definition is well-behaved; see [KL16, (3.7)]). For any rigid space X over a nonarchimedean field K over $\mathbf{Q}_p$, one can find a unique universal homeomorphism $\widetilde{X} \to X$ such that $\widetilde{X}$ is seminormal. Indeed, locally $X = \mathrm{Spa}\, A$, and then $\widetilde{X} = \mathrm{Spa}\, \widetilde{A}$, where $\widetilde{A}$ is the seminormalization of A, which is finite over A (as A is excellent).

Proposition 10.2.3. For any nonarchimedean field K over $\mathbf{Q}_p$, the functor

$$\{\text{seminormal rigid-analytic spaces}/K\} \;\to\; \{\text{diamonds}/\mathrm{Spd}\, K\}$$
$$X \;\mapsto\; X^\diamond$$

is fully faithful.

It is however critical to remember the structure morphism to $\mathrm{Spd}\, K$!

Proof. Let X and Y be rigid spaces over K, and assume that X is seminormal. We claim that
$$\mathrm{Hom}_K(X, Y) \to \mathrm{Hom}_{\mathrm{Spd}\, K}(X^\diamond, Y^\diamond)$$
is bijective. First, we assume that $Y = \mathrm{Spa}\, A$ is affinoid. In that case, the left-hand side is given by the continuous maps from A to $\mathcal{O}_X(X)$, while the right-hand side is similarly given by the continuous maps from A to $\mathcal{O}_{X^\diamond}^\sharp(X^\diamond)$, where $\mathcal{O}_{X^\diamond}^\sharp$ is the pro-étale sheaf that associates to any $T \to X^\diamond$ the sections $\mathcal{O}_{T^\sharp}(T^\sharp)$ of the untilt $T^\sharp$ corresponding to the composite $T \to X^\diamond \to \mathrm{Spd}\, K \to \mathrm{Spd}\, \mathbf{Q}_p$.

Thus, we need to see that the natural map $\mathcal{O}_X(X) \to \mathcal{O}_{X^\diamond}^\sharp(X^\diamond)$ is a homeomorphism, which is [KL16, Theorem 8.2.3]. (In fact, by resolution of singularities and Zariski's main theorem, we can reduce to the case that X is smooth, which follows from [Sch13, Corollary 6.19].)

In general, the underlying topological space of X can be recovered from $X^\diamond$ by the results of the next section, which reduces the general assertion to the affinoid case. $\qquad\square$

10.3 THE UNDERLYING TOPOLOGICAL SPACE OF DIAMONDS

We have seen above that universal homeomorphisms induce isomorphisms on diamonds. However, being a universal homeomorphism is necessary, by Proposition 10.3.7 below.

Definition 10.3.1 ([Sch17, Definition 11.14]). Let $\mathcal{D}$ be a diamond, and choose a presentation $\mathcal{D} = X/R$. The underlying topological space of $\mathcal{D}$ is the quotient $|\mathcal{D}| = |X| / |R|$.

Remark 10.3.2. This is independent of the choice of presentation; cf. [Sch17, Proposition 11.13].

In general, $|\mathcal{D}|$ can be quite pathological. It might not even be T_0, for instance: the quotient of the constant perfectoid space $\mathbf{Z}_p$ over a perfectoid field by the equivalence relation "congruence modulo $\mathbf{Z}$" produces a diamond with underlying topological space $\mathbf{Z}_p/\mathbf{Z}$.

There is a class of $qcqs$ diamonds which avoids such pathologies. To recall the meaning of being qcqs, i.e., quasicompact and quasiseparated, we make a short topos-theoretic digression. Recall from SGA4 the following notions. In any topos, we have a notion of a quasicompact object: this means that any covering family has a finite subcover. An object Z is quasiseparated if for any quasicompact $X, Y \to Z$, the fiber product $X \times_Z Y$ is quasicompact. An object that is both quasicompact and quasiseparated is called qcqs.

If there is a generating family B for the topos (meaning that every object is a colimit of objects in the generating family) consisting of quasicompact objects which is stable under fiber products, then:

1. all objects of B are qcqs,
2. Z is quasicompact if and only if it has a finite cover by objects of B, and
3. Z is quasiseparated if and only if for all $X, Y \in B$ with maps $X, Y \to Z$, the fiber product $X \times_Z Y$ is quasicompact.

In our situation, we consider the topos of sheaves on the pro-étale site of Perf, and we can take B to be the class of all affinoid perfectoid spaces. This is closed under fiber products because maps between analytic adic spaces are adic.

Remark 10.3.3. We warn the reader that B is not stable under direct products. For example if $X = \operatorname{Spa} \mathbf{F}_p((t^{1/p^\infty}))$, then $X \times X = \widetilde{\mathbf{D}}^*_{\mathbf{F}_p((t^{1/p^\infty}))}$ is not quasicompact. This happens in fact always: For any two nonempty perfectoid spaces X and Y, the product $X \times Y$ is not quasicompact.

Proposition 10.3.4. Let $\mathcal{D}$ be a qcqs diamond. Then $|\mathcal{D}|$ is T_0, i.e., for all distinct $x, y \in |\mathcal{D}|$, there is an open subset $U \subset |\mathcal{D}|$ that contains exactly one of x and y.

Proof. (Sketch.) Let $X \to \mathcal{D}$ be a quasi-pro-étale surjection from a strictly totally disconnected space, and let $R = X \times_{\mathcal{D}} X$ which is qcqs pro-étale over X, and so itself a perfectoid space. As $\mathcal{D}$ is qcqs, also R and the map $R \to X$ are qcqs, and so we can apply [Sch17, Lemma 2.7] to $|X| / |R|$. $\qquad\square$

In general, not much more can be said about $|\mathcal{D}|$; however, in Lecture 17, we will introduce a notion of (locally) spatial diamonds that ensures that $|\mathcal{D}|$ is a (locally) spectral space. All diamonds that we will be interested in are locally spatial.

The topological space is closely related with open immersions, as usual.

Definition 10.3.5. A map $\mathcal{G} \to \mathcal{F}$ of pro-étale sheaves on Perf is an *open immersion* if for any map $X \to \mathcal{F}$ from a perfectoid space X, the fiber product $\mathcal{G} \times_{\mathcal{F}} X \to X$ is representable by an open subspace of X. In this case we say that $\mathcal{G}$ is an *open subsheaf* of $\mathcal{F}$.

The following proposition is immediate from the definitions.

Proposition 10.3.6 ([Sch17, Proposition 11.15]). Let $\mathcal{F}$ be a diamond. The category of open subsheaves of $\mathcal{F}$ is equivalent to the category of open immersions into $|\mathcal{F}|$, via $\mathcal{G} \mapsto |\mathcal{G}|$.

In the case of analytic pre-adic spaces, the diamond remembers the underlying topological space.

Proposition 10.3.7 ([Sch17, Lemma 15.6]). Let X be an analytic pre-adic space over $\mathbf{Z}_p$, with associated diamond $X^{\diamond}$. There is a natural homeomorphism $|X^{\diamond}| \cong |X|$.

Proof. (Sketch.) One reduces to the case that $X = \operatorname{Spa}(R, R^+)$ is affinoid, so that $X = \widetilde{X}/\underline{G}$, where $\widetilde{X} = \operatorname{Spa}(\widetilde{R}, \widetilde{R}^+)$ is affinoid perfectoid and a $\underline{G}$-torsor over X. Then $|X| = |\widetilde{X}|/G = |\widetilde{X}^{\flat}|/G = |X^{\diamond}|$, as desired. $\qquad\square$

10.4 THE ÉTALE SITE OF DIAMONDS

Now we want to define an étale site for diamonds, and compare it with the étale site of adic spaces.

Definition 10.4.1 ([Sch17, Definition 10.1 (ii)]). A map $f : \mathcal{G} \to \mathcal{F}$ of pro-étale sheaves on Perf is étale (resp. finite étale) if it is locally separated[1] and for any perfectoid space X with a map $X \to \mathcal{F}$, the pullback $\mathcal{G} \times_{\mathcal{F}} X$ is representable by a perfectoid space Y étale (resp. finite étale) over X.

---

[1] Again, this is per [Sch17, Convention 10.2]; if f is not locally separated, the definition should be slightly changed.

If Y is a diamond, we consider the category $Y_{\text{ét}}$ of diamonds étale over Y (noting that anything étale over Y is automatically itself a diamond—[Sch17, Proposition 11.7]) and turn it into a site by declaring covers to be collections of jointly surjective maps.[2]

For any analytic pre-adic space X, one can define an étale site $X_{\text{ét}}$ by looking at maps that are locally a composite of open embeddings and finite étale maps; cf. [KL15, Definition 8.2.19].

Theorem 10.4.2 ([Sch17, Lemma 15.6])**.** The functor $X \mapsto X^{\diamond}$ from analytic pre-adic spaces over $\mathbf{Z}_p$ to diamonds induces an equivalence of sites $X_{\text{ét}} \cong X_{\text{ét}}^{\diamond}$, restricting to an equivalence $X_{\text{fét}} \cong X_{\text{fét}}^{\diamond}$.

Moreover, for any diamond Y, one can define a quasi-pro-étale site $Y_{\text{qproét}}$ by looking at quasi-pro-étale maps into Y (which again are all diamonds, by [Sch17, Proposition 11.7]). In particular, for any analytic pre-adic space X, we can define $X_{\text{qproét}} = X_{\text{qproét}}^{\diamond}$. This differs from the pro-étale site $X_{\text{proét}}$ defined in [Sch13], because the definition of pro-étale maps used there differs from the ones used here, in that we restricted ourselves to inverse systems of étale maps that are eventually finite étale and surjective, so $X_{\text{proét}}$ is a full subcategory of $X_{\text{qproét}}$, but we also use a stronger notion of covering on $X_{\text{proét}}$. However, one could use $X_{\text{qproét}}$ in place of $X_{\text{proét}}$ in most of [Sch13].[3] In particular $X_{\text{qproét}}$ is a replete topos in the sense of [BS15], and so inverse limits of sheaves behave well. Also, pullback of sheaves along $X_{\text{qproét}} \to X_{\text{ét}}$ is fully faithful:

Proposition 10.4.3 ([Sch17, Proposition 14.8])**.** Let X be an analytic pre-adic space. The pullback functor

$$\nu_X^* : X_{\text{ét}}^{\sim} \to X_{\text{qproét}}^{\sim}$$

from étale sheaves to quasi-pro-étale sheaves on X is fully faithful.

---

[2]In [Sch17, Definition 14.1], this is only considered in case Y is locally spatial, as otherwise it is not clear whether $Y_{\text{ét}}$ contains enough objects.

[3]The exception is around [Sch13, Lemma 8.6], when it becomes necessary to talk about coherent sheaves, which behave better on $X_{\text{proét}}$ than on $X_{\text{qproét}}$, as $X_{\text{proét}}$ has better flatness properties; cf. [Sch13, Lemma 8.7 (ii)].

Appendix to Lecture 10:
Cohomology of local systems

In this appendix, we explain how to use (quasi-)pro-étale descent, and strictly totally disconnected spaces, to prove the following result.[4] Let C be a complete algebraically closed extension of $\mathbf{Q}_p$, and let $f : X \to Y$ be a proper smooth map of rigid-analytic spaces over C, considered as adic spaces.

Theorem 10.5.1. Let $\mathbb{L}$ be an étale $\mathbf{F}_p$-local system on X. Then, for all $i \geq 0$, the higher direct image $R^i f_* \mathbb{L}$ is an étale $\mathbf{F}_p$-local system on Y.

This proves that the extra hypothesis in [Sch13, Theorem 1.10] is always satisfied (as the version with $\mathbf{Z}_p$-coefficients follows formally).

Although this is a very classical statement, its proof will make critical use of (strictly) totally disconnected spaces. Although quite different in the details, the arguments in [Sch17] are often of a similar flavor, and we hope that the reader can get an impression how these wildly ungeometric spaces can be used.

Let us first briefly recall the proof in case $Y = \mathrm{Spa}(C, \mathcal{O}_C)$ is a point; in that case, one has to prove finiteness of $H^i(X, \mathbb{L})$, which is [Sch13, Theorem 1.1]. For this, one first proves that $H^i(X, \mathbb{L} \otimes \mathcal{O}_X^+/p)$ is almost finitely generated. The main idea here is to use the Cartan–Serre technique of choosing two finite affinoid covers $X = \bigcup_{j=1}^n U_j = \bigcup_{j=1}^n V_j$ such that U_j is strictly contained in V_j, and deduces the global finiteness result from a spectral sequence argument and a local finiteness result, which says that the image of

$$H^i(V_j, \mathbb{L} \otimes \mathcal{O}_X^+/p) \to H^i(U_j, \mathbb{L} \otimes \mathcal{O}_X^+/p)$$

is almost finitely generated for all i. Actually, this spectral sequence argument needs $i + 2$ such affinoid covers instead of only 2, when you want to prove finiteness in cohomological degree i; cf. [Sch13, Lemma 5.4]. It will turn out— cf. Lemma 10.5.6 below—that in our variant we will need $\approx 3^i$ affinoid covers! After this, one shows that the map

$$H^i(X, \mathbb{L}) \otimes_{\mathbf{F}_p} \mathcal{O}_C/p \to H^i(X, \mathbb{L} \otimes \mathcal{O}_X^+/p)$$

---

[4]The first author wishes to thank Bhargav Bhatt for some related discussions. The theorem has also been announced by Ofer Gabber, as a consequence of a proof of Poincaré duality.

is an almost isomorphism; together, this formally implies finiteness of $H^i(X, \mathbb{L})$.

In the relative case, we will follow a similar strategy. We already know from [Sch13, Theorem 1.3] that the map

$$R^i f_* \mathbb{L} \otimes \mathcal{O}_Y^+/p \to R^i f_*(\mathbb{L} \otimes \mathcal{O}_X^+/p)$$

of étale sheaves on Y is an almost isomorphism. As we will see in Lemma 10.5.3 below, this reduces us to proving a certain finiteness result for $R^i f_*(\mathbb{L} \otimes \mathcal{O}_X^+/p)$. However, this is still an étale sheaf that does not behave much like a coherent sheaf. However, if $\widetilde{X} = \mathrm{Spa}(R, R^+)$ is a perfectoid space with a pro-étale map $\widetilde{X} \to X$, then the pullback of $Rf_*(\mathbb{L} \otimes \mathcal{O}_X^+/p)$ to the étale site of $\widetilde{X}$ is (almost) given by $M^\bullet \overset{\mathbb{L}}{\otimes}_{R^+/p} \mathcal{O}_{\widetilde{X}}^+/p$ for a complex $M^\bullet$ of R^+/p-modules. However, the tensor product here is derived, so it is hard to isolate individual cohomology groups. Moreover, it is not clear whether R^+/p has any finiteness properties (such as being almost noetherian, or more realistically almost coherent), so it is not clear which finiteness results for the cohomology groups of $M^\bullet$ to hope for.[5]

Surprisingly, by "tearing apart" $\widetilde{X}$ even further and making it (strictly) totally disconnected, these problems disappear. More precisely, in that case the tensor product need not be derived as $\mathcal{O}_{\widetilde{X}}^+/p$ is flat over R^+/p. Moreover, R^+/p is coherent (in particular, almost coherent), and so one can hope that restricting $R^i f_*(\mathbb{L} \otimes \mathcal{O}_X^+/p)$ to the étale site of $\widetilde{X}$ is of the form $M \otimes_{R^+/p} \mathcal{O}_{\widetilde{X}}^+/p$ for some almost finitely presented R^+/p-module M. This will turn out to be the case.

Now we have to explain two parts of the proof:

1. Reduce Theorem 10.5.1 to an almost finite presentation of a different cohomology group, as in Lemma 10.5.3.
2. Prove almost finite generation of that cohomology group, as in Proposition 10.5.10.

By Proposition 10.4.3, it is enough to prove that for any strictly totally disconnected perfectoid space $S = \mathrm{Spa}(R, R^+)$ with a quasi-pro-étale map to Y,[6] with pullback $f_S : X_S \to S$, the higher direct image $R^i f_{S*}\mathbb{L}|_{X_S} = (R^i f_*\mathbb{L})|_S$ is an $\mathbf{F}_p$-local system on S; here this base change equality holds by [Sch17, Corollary 16.10 (ii)]. The general strategy now is to repeat the proof of [Sch13, Theorem 1.1]. By [Sch13, Theorem 1.3] (and [Sch17, Corollary 16.10 (ii)]), we already know that

$$(R^i f_*\mathbb{L})|_S \otimes \mathcal{O}_S^+/p \to R^i f_{S*}(\mathbb{L} \otimes \mathcal{O}_{X_S}^+/p)$$

is an almost isomorphism of étale sheaves on S.

---

[5]One could hope that the complex $M^\bullet$ is "almost perfect," as it will turn out (a posteriori) to be; however, the Cartan–Serre technique for proving finiteness results is fundamentally about cohomology groups, not complexes.

[6]Such S form a basis of the topology for $Y_{\mathrm{qproét}}$; cf. [Sch17, Lemma 7.18].

We need the following ring-theoretic statement.

Proposition 10.5.2. Let $\mathrm{Spa}(R, R^+)$ be a totally disconnected perfectoid space and $\varpi \in R^+$ any pseudouniformizer. Then the ring R^+/ϖ is coherent. In particular, it is almost coherent, i.e., any almost finitely generated submodule of an almost finitely presented module is almost finitely presented.

Here, an R^+/ϖ-module M is almost finitely generated (resp. presented) if for all pseudo-uniformizers $\epsilon \in R^+$, there is a finitely generated (resp. presented) R^+/ϖ-module M_ϵ and maps $M \to M_\epsilon \to M$ whose composites in both directions are both multiplication by ϵ.

Proof. The almost coherence statement follows from coherence by approximating almost finitely generated (resp. presented) modules by finitely generated (resp. presented) modules.

We need to prove that any finitely generated ideal $I \subset R^+/\varpi$ is finitely presented. First, we recall that the basic structure of R^+/ϖ if $\mathrm{Spa}(R, R^+)$ is totally disconnected. Recall that all connected components of $\mathrm{Spa}(R, R^+)$ are of the form $\mathrm{Spa}(K, K^+)$ where K is a perfectoid field and $K^+ \subset K$ is an open and bounded valuation subring. There is a reduction map $\mathrm{Spa}(R, R^+) \to \mathrm{Spec}(R^+/\varpi)$ on topological spaces, which is a homeomorphism in this case (by reduction to the case of $\mathrm{Spa}(K, K^+)$). In particular, any connected component of $\mathrm{Spec}(R^+/\varpi)$ is of the form $\mathrm{Spec}(K^+/\varpi)$ where $K^+ \subset K$ is as above. Let $T = \pi_0 \mathrm{Spa}(R, R^+) = \pi_0 \mathrm{Spec}(R^+/\varpi)$ be the profinite set of connected components; we can and will consider all R^+/ϖ-modules as sheaves on T.

For any $t \in T$, let the connected component of $\mathrm{Spa}(R, R^+)$ be $\mathrm{Spa}(K_t, K_t^+)$; we get an ideal $I_t \subset K_t^+/\varpi$. But K_t^+ is a valuation ring, thus coherent, and then also K_t^+/ϖ is coherent. It follows that I_t is generated by one element $\overline{f_t} \in K_t^+/\varpi$ that (after localization on T) comes from an element $\overline{f} \in R^+/\varpi$ that we lift to $f \in R^+$. If $I_t = 0$, then, as I is finitely generated, I vanishes in a neighborhood. Otherwise, $|f| \geq |\varpi|$ defines an open neighborhood of t, and on this open neighborhood, we have $g = \frac{\varpi}{f} \in R^+$. The map $R^+ \to I$ given by f is surjective at t, and thus in an open neighborhood, to which we pass. Its kernel is then generated by g, and so we see that $I \cong R^+/g$ is finitely presented, as desired. $\square$

This allows us to formulate a module-theoretic criterion.

Lemma 10.5.3. Assume that $H^i(X_S, \mathbb{L} \otimes \mathcal{O}_{X_S}^+/p)$ is an almost finitely presented R^+/p-module. Then $(R^i f_* \mathbb{L})|_S$ is an $\mathbf{F}_p$-local system on S, and therefore $R^i f_* \mathbb{L}$ is an $\mathbf{F}_p$-local system on Y.

Remark 10.5.4. As S is strictly totally disconnected, sheaves on the étale site of S are the same as sheaves on the topological space $|S|$; this will be used below.

Remark 10.5.5. The theory of (almost) finitely presented R^+/p-modules is very different from the theory of finitely presented R-modules. Indeed, the latter become locally free over a Zariski stratification, i.e., a constructible stratification of $\operatorname{Spec} R$, while the former are (almost in a weak sense) locally constant over a constructible stratification of $\operatorname{Spa}(R, R^+)$. These two notions of constructibility are very different: A point of the closed unit disc is Zariski constructible, but is not constructible in the adic sense, as its complement (the punctured unit disc) is not quasicompact! Thus, the almost finite presentation condition in the lemma precludes sections concentrated at one point. This notion of constructibility also plays a critical role in the ℓ-adic cohomology of diamonds for $\ell \neq p$; cf. [Sch17].

Proof. Let $M = H^i(X_S, \mathbb{L} \otimes \mathcal{O}_{X_S}^+/p)$. We start with some general observations on the almost isomorphism

$$(R^i f_* \mathbb{L})|_S \otimes \mathcal{O}_S^+/p \to R^i f_{S*}(\mathbb{L} \otimes \mathcal{O}_{X_S}^+/p) . \tag{10.5.1}$$

Note that

$$R^i f_{S*}(\mathbb{L} \otimes \mathcal{O}_{X_S}^+/p)$$

admits an almost isomorphism from $H^i(X_S, \mathbb{L} \otimes \mathcal{O}_{X_S}^+/p) \otimes_{R^+/p} \mathcal{O}_S^+/p$, by computing the cohomology by an affinoid pro-étale of X_S and applying the base change result [Sch13, Lemma 4.12] on each term, and finally using the fact that $\mathcal{O}_S^+/p$ is flat over R^+/p by [Sch17, Proposition 7.23] to see that the tensor product is underived.

For any $s \in S$ corresponding to a map $\operatorname{Spa}(C(s), C(s)^+) \to S$, the stalk of the left-hand side of Equation (10.5.1) is of the form $H^i(X_{\operatorname{Spa}(C(s), C(s)^+)}, \mathbb{L}) \otimes_{\mathbf{F}_p} C(s)^+/p$, while the stalk of the right-hand side is $M \otimes_{R^+/p} C(s)^+/p$.

By [Sch13, Proposition 2.11], any almost finitely generated $C(s)^+$-module such as $M \otimes_{R^+/p} C(s)^+/p$ has a sequence of elementary divisors. Varying s, one gets a map

$$\gamma_M : S \to \ell_{\geq}^\infty(\mathbb{N})_0$$

where $\ell_{\geq}^\infty(\mathbb{N})_0$ denotes the set of decreasing sequences $(x_0, x_1, \ldots)$ of nonnegative real numbers that converge to 0. The map γ_M is continuous if M is almost finitely presented and the target is equipped with the ℓ^∞-norm; this follows by approximating M by finitely presented modules and using the fact that this approximation has ℓ^∞-bounded effect on the elementary divisors by [Sch13, Proposition 2.11 (i)], and checking continuity in that case.

On the other hand, the elementary divisors of a module of the form $V \otimes_{\mathbf{F}_p} C(s)^+/p$ for an $\mathbf{F}_p$-vector space V are exactly $(1, 1, \ldots, 1, 0, \ldots)$, where 1 occurs $\dim V$ many times. Combining these observations shows that the map sending $s \in S$ to $\dim(R^i f_* \mathbb{L})_s$ is locally constant. After replacing S by an open and closed subset, we may therefore assume that it is constant, equal to n.

Let $s \in S$ be any point, and let $x_1, \ldots, x_n$ be a basis of the stalk $(R^i f_* \mathbb{L})_s$. The sections $x_1, \ldots, x_n$ extend to a small neighborhood, so after shrinking S, we may assume that $x_1, \ldots, x_n$ define global sections. We get maps

$$(\mathcal{O}_S^+/p)^n \to (R^i f_* \mathbb{L})|_S \otimes \mathcal{O}_S^+/p \to R^i f_{S*}(\mathbb{L} \otimes \mathcal{O}_{X_S}^+/p) ,$$

where the second map is an almost isomorphism. Recall that

$$R^i f_{S*}(\mathbb{L} \otimes \mathcal{O}^+_{X_S}/p)$$

admits an almost isomorphism from $M \otimes_{R^+/p} \mathcal{O}^+_S/p$. Thus, the composite map above is almost the same as the map induced from $(R^+/p)^n \to M$ by tensoring with $\mathcal{O}^+_S/p$ over R^+/p. The kernel and cokernel of $(R^+/p)^n \to M$ are again almost finitely presented R^+/p-modules whose elementary divisors vanish at s; on the other hand, the elementary divisors of the kernel and cokernel again only take the values 0 and 1 (by comparing with the kernel and cokernel of the stalks of

$$(\mathcal{O}^+_S/p)^n \to (R^i f_* \mathbb{L})|_S \otimes \mathcal{O}^+_S/p).$$

Thus, these elementary divisors vanish in a neighborhood of s, so after localization we may assume that they are 0. This implies that all localizations of the kernel and cokernel of $(R^+/p)^n \to M$ are almost zero, so that the kernel and cokernel are almost zero. Thus, $(R^+/p)^n \to M$ is an almost isomorphism, and so

$$(\mathcal{O}^+_S/p)^n \to (R^i f_* \mathbb{L})|_S \otimes \mathcal{O}^+_S/p$$

is an almost isomorphism. In particular, it follows that the map $\mathbf{F}^n_p \to (R^i f_* \mathbb{L})|_S$ of sheaves on S is injective, and then bijective by comparing ranks at all points. □

It remains to prove that $H^i(X_S, \mathbb{L} \otimes \mathcal{O}^+_{X_S}/p)$ is an almost finitely presented R^+/p-module. For this, one follows the proof of [Sch13, Lemma 5.8] "replacing almost finitely generated with almost finitely presented everywhere." For the proof, one needs to be able to chase finiteness results through spectral sequences, and so we need a version of [Sch13, Lemma 5.4]. We were (only) able to make this work with exponentially many spectral sequences:

Lemma 10.5.6. Let $E^{p,q}_{*,(i)} \Rightarrow M^{p+q}_{(i)}$, $i = 0, \ldots, (N+1)3^{N+1}$, be upper-right quadrant spectral sequences of R^+/p-modules, together with maps of spectral sequences $E^{p,q}_{*,(i)} \to E^{p,q}_{*,(i+1)}$, $M^{p+q}_{(i)} \to M^{p+q}_{(i+1)}$ for $i = 0, \ldots, (N+1)3^{N+1} - 1$. Assume that for some r, the map $E^{p,q}_{r,(i)} \to E^{p,q}_{r,(i+1)}$ on the rth sheet factors over an almost finitely presented R^+/p-module for all i, p, q. Then the map $M^k_{(0)} \to M^k_{((N+1)3^{N+1})}$ factors over an almost finitely presented R^+/p-module for $k \le N$.

Remark 10.5.7. The statement that a map factors over an almost finitely presented module is slightly weaker than the statement that the image is almost finitely presented; correspondingly, in [Sch18, Lemma 8.8] (where the present lemma was used without proof), the assertion should only be that the map factors over an almost finitely presented module. The rest of [Sch18] remains unaffected; for example, to get the statement of [Sch18, Corollary 8.9], note that if the identity on a module M factors over an almost finitely presented module, then M is almost finitely presented.

Proof. First, we claim that if $E^{p,q}_{r,(i)} \to E^{p,q}_{r,(i+1)}$ factors over an almost finitely presented module $F^{p,q}_{r,(i)}$ for all i, p, q, then the same holds true for the map $E^{p,q}_{r+1,(i)} \to E^{p,q}_{r+1,(i+3)}$. Indeed, first

$$\ker(d : E^{p,q}_{r,(i)} \to E^{p+r,q-r+1}_{r,(i)}) \to \ker(d : E^{p,q}_{r,(i+2)} \to E^{p+r,q-r+1}_{r,(i+2)})$$

factors over

$$G^{p,q}_{r,(i)} = \ker(F^{p,q}_{r,(i)} \to E^{p,q}_{r,(i+1)} \xrightarrow{d} E^{p+r,q-r+1}_{r,(i+1)} \to F^{p+r,q-r+1}_{r,(i+1)}) \,,$$

which is almost finitely presented. Applying a similar argument again to

$$E^{p,q}_{r+1,(i)} = \operatorname{coker}(d : E^{p-r,q+r-1}_{r,(i)} \to \ker(d : E^{p,q}_{r,(i)} \to E^{p+r,q-r+1}_{r,(i)}))$$

gives the desired result.

By induction, we see that

$$E^{p,q}_{r+j,(i)} \to E^{p,q}_{r+j,(i+3^j)}$$

factors over an almost finitely presented module for all i, p, q. In particular, as $E^{p,q}_{N+2,(i)} = E^{p,q}_{\infty,(i)}$ for $p + q \leq N$, it follows that

$$E^{p,q}_{\infty,(i)} \to E^{p,q}_{\infty,(i+3^{N+1})}$$

factors over an almost finitely presented module for all i, p, q with $p + q \leq N$. Using this for $i = i_0 3^{N+1}$, $i_0 = 0, \ldots, N+1$, we get R^+/p-modules

$$M_0 = M^k_{(0)}, M_1 = M^k_{(3^{N+1})}, \ldots, M_{N+1} = M^k_{((N+1)3^{N+1})}$$

that come equipped with increasing filtrations $\operatorname{Fil}^p M_i \subset M_i$, $\operatorname{Fil}^0 M_i = 0$, $\operatorname{Fil}^{N+1} M_i = M_i$, for which $\operatorname{gr}^p M_i \to \operatorname{gr}^p M_{i+1}$ factors over an almost finitely presented module for $i = 0, \ldots, N+1$.

We claim by induction on N that this implies that the map $M_0 \to M_{N+1}$ factors over an almost finitely presented module. If $N = 0$, this is clear. In general, we may apply the inductive hypothesis to $\operatorname{Fil}^N M_1 \to \operatorname{Fil}^N M_{N+1}$; say it factors over an almost finitely presented module Y. On the other hand, $\operatorname{gr}^N M_0 \to \operatorname{gr}^N M_1$ factors over an almost finitely presented module X. Now

look at the diagram

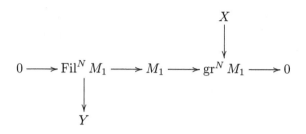

and let Z be obtained from M_1 by taking the pullback along $X \to \mathrm{gr}^N M_1$ and the pushout along $\mathrm{Fil}^N M_1 \to Y$. Then there is an exact sequence $0 \to Y \to Z \to X$, so Z is almost finitely presented, and moreover $M_0 \to M_{N+1}$ factors over Z. $\qquad\square$

The other ingredient one needs is the following.

Lemma 10.5.8. Let

$$V = \mathrm{Spa}(B, B^+) \to \mathbb{B}^n_S = \mathrm{Spa}(R\langle T_1, \ldots, T_n\rangle, R^+\langle T_1, \ldots, T_n\rangle)$$

be a composite of rational embeddings and finite étale maps, and let $U = \mathrm{Spa}(A, A^+) \subset V$ be a rational subset such that U is strictly contained in V relatively to S, i.e. for all affinoid fields (K, K^+) with a map $\mathrm{Spa}(K, \mathcal{O}_K) \to U$ whose projection to S extends to $\mathrm{Spa}(K, K^+)$, there is an extension $\mathrm{Spa}(K, K^+) \to V$. Then the map $B^+/p \to A^+/p$ factors over a finitely presented R^+/p-module.

Proof. By Lemma 10.5.9 below, B^+/p and A^+/p are finitely presented R^+/p-algebras. We claim that in fact the map factors over an R^+/p-algebra that is finitely presented as an R^+/p-module. This can be checked locally on $\mathrm{Spec}(R^+/p)$. Thus, we can assume that $\mathrm{Spec}(R^+/p)$ is connected, in which case $(R, R^+) = (K, K^+)$ is an affinoid field.

In that case K^+ is valuation ring, as is $V = K^+/K^{\circ\circ}$. Let $\overline{A} = A^+ \otimes_{K^+} V$ and $\overline{B} = B^+ \otimes_{K^+} V$. Then $\overline{A}$ and $\overline{B}$ are flat finitely presented V-algebras; let T be the image of $\overline{B}$ in $\overline{A}$. Then $T \subset \overline{A}$ is a flat finitely generated V-algebra, and thus by [RG71, Corollaire 3.4.7], it is finitely presented. We claim that T is in fact finite over V; for this, it is enough to check that it satisfies the valuative criterion of properness, which follows from the assumption, using the fact that all generic points of $\mathrm{Spec}\,T$ lift to $\mathrm{Spec}\,\overline{A}$ (and it is enough to check the valuative criterion for specializations starting at the generic point).

Now we find $f_1, \ldots, f_m \in B^+/p$ such that the ideal generated by $f_1, \ldots, f_m$ in $\overline{B}$ is $\ker(\overline{B} \to \overline{A})$. As the kernel of $A^+/p \to \overline{A}$ is a nilideal, it follows that there is some integer N such that $f_1^N, \ldots, f_m^N$ map to 0 in A^+/p. Therefore, the map $B^+/p \to A^+/p$ factors over the finitely presented R^+/p-algebra $B^+/(p, f_1^N, \ldots, f_m^N)$ which still satisfies the valuative criterion of properness (as its reduced quotient agrees with the one of T), and so is finitely presented as an R^+/p-module. $\square$

Lemma 10.5.9. Let

$$U = \mathrm{Spa}(A, A^+) \to \mathbb{B}_S^n = \mathrm{Spa}(R\langle T_1, \ldots, T_n\rangle, R^+\langle T_1, \ldots, T_n\rangle)$$

be a composite of rational embeddings and finite étale maps. Then A^+/p is a finitely presented R^+/p-algebra.

Proof. This is a consequence of the reduced fiber theorem of Bosch–Lütkebohmert–Raynaud, [BLR95], as follows. We can write $(R, R^+) = \varinjlim_i \widehat{(R_i, R_i^+)}$ as a completed direct limit, where all (R_i, R_i^+) are topologically of finite presentation over $(C, \mathcal{O}_C)$ (so in particular $R_i^+ = R_i^\circ$). Then $S^\diamond = \varprojlim_i S_i^\diamond$ with $S_i = \mathrm{Spa}(R_i, R_i^+)$. By the usual limit arguments—cf. [Sch17, Proposition 11.23]—one can find for large enough i a map $U_i \to \mathbb{B}_{S_i}^n$ that is a composite of rational embeddings and finite étale maps; in particular, U_i is an affinoid rigid space over C. By the finiteness theorem of Grauert and Remmert, we know that $U_i = \mathrm{Spa}(A_i, A_i^+)$ where $A_i^+ = A_i^\circ$ is topologically of finite presentation over R_i^+ (in fact, both are topologically of finite presentation over $\mathcal{O}_C$). By [BLR95, Theorem 2.1], we find that after increasing i (corresponding to a rig-étale cover $S_i' \to S_i$, all of which split over S), the map $\mathrm{Spec}\, A_i^+/p \to \mathrm{Spec}\, R_i^+/p$ has geometrically reduced fibers. In this case, the formation of A_i^+ commutes with base change, i.e., $A_j^+ = A_i^+ \widehat{\otimes}_{R_i^+} R_j^+$ for $j \geq i$ (the p-adic completion of the tensor product). In particular, $A_j^+/p = A_i^+/p \otimes_{R_i^+/p} R_j^+/p$, and by passing to the filtered colimit over j, we see that $A^+/p = A_i^+/p \otimes_{R_i^+/p} R^+/p$. As A_i^+/p is a finitely presented R_i^+/p-algebra, it follows that A^+/p is a finitely presented R^+/p-algebra, as desired. $\square$

Now the proof of [Sch13, Lemma 5.8] goes through (replacing the use of the auxiliary module A at the end of the proof of [Sch13, Lemma 5.6] with the explicit description of $H^i_{\mathrm{cont}}((p^m \mathbb{Z}_p)^n, R^+/p)$ as almost equal to a sum of R_m^+/p^r for varying r (going to 0)), so we deduce the following.[7]

Proposition 10.5.10. The R^+/p-module $H^i(X_S, \mathbb{L} \otimes \mathcal{O}_{X_S}^+/p)$ is almost finitely presented.

By Lemma 10.5.3, this finishes the proof of Theorem 10.5.1.

---

[7] A very similar argument also appears in [Sch18, Section 8].

Lecture 11

Mixed-characteristic shtukas

Today we begin talking about shtukas.

11.1 THE EQUAL CHARACTERISTIC STORY: DRINFELD'S SHTUKAS AND LOCAL SHTUKAS

Let $X/\mathbf{F}_p$ be a smooth projective curve. Let us recall the following definition from Lecture 1.

Definition 11.1.1. Let $S/\mathbf{F}_p$ be a scheme. A *shtuka of rank n over S with legs* $x_1, \ldots, x_m \in X(S)$ is a rank n vector bundle $\mathcal{E}$ over $S \times_{\mathbf{F}_p} X$ together with an isomorphism

$$\varphi_\mathcal{E} \colon (\mathrm{Frob}_S \times 1)^* \mathcal{E}|_{S \times_{\mathbf{F}_p} X \setminus \bigcup_i \Gamma_{x_i}} \cong \mathcal{E}|_{S \times_{\mathbf{F}_p} X \setminus \bigcup_i \Gamma_{x_i}}$$

on $S \times_{\mathbf{F}_p} X \setminus \bigcup_i \Gamma_{x_i}$, where $\Gamma_{x_i} \subset S \times_{\mathbf{F}_p} X$ is the graph of x_i.

In Lecture 1 we mused about the possibility of transporting this definition into mixed characteristic. As we mentioned there, we currently have no hope of replacing X with $\mathrm{Spec}\,\mathbf{Z}$, but rather only $\mathrm{Spec}\,\mathbf{Z}_p$. As a guide to what we should do, let us first discuss the local analogue of Definition 11.1.1. Let $\widehat{X}$ be the formal completion of X at one of its $\mathbf{F}_p$-rational points, so that $\widehat{X} \cong \mathrm{Spf}\,\mathbf{F}_p[\![T]\!]$. We would like to define local shtukas (or $\widehat{X}$-shtukas) over S. Since we expect these to be of an analytic nature, the test object $S/\mathbf{F}_p$ should be an adic formal scheme, or more generally an adic space. Then the legs of such a local shtuka will be elements of $\widehat{X}(S)$, which is to say morphisms $S \to \mathrm{Spf}\,\mathbf{F}_p[\![T]\!]$, where $S/\mathbf{F}_p$ is an adic formal scheme, or more generally an adic space.

Definition 11.1.2. A *local shtuka* of rank n over an adic space $S/\mathrm{Spa}\,\mathbf{F}_p$ with legs $x_1, \ldots, x_m \in \widehat{X}(S)$ is a rank n vector bundle $\mathcal{E}$ over $S \times_{\mathbf{F}_p} \widehat{X}$ together with an isomorphism

$$\varphi_\mathcal{E} \colon (\mathrm{Frob}_S \times 1)^* \mathcal{E}|_{S \times_{\mathbf{F}_p} \widehat{X} \setminus \bigcup_i \Gamma_{x_i}} \cong \mathcal{E}|_{S \times_{\mathbf{F}_p} \widehat{X} \setminus \bigcup_i \Gamma_{x_i}}$$

over $S \times_{\mathbf{F}_p} \widehat{X} \backslash \bigcup_i \Gamma_{x_i}$ that is meromorphic along $\bigcup_i \Gamma_{x_i}$ in the sense of Definition 5.3.5.

Remark 11.1.3. The space S should be sufficiently nice that $S \times_{\mathbf{F}_p} \widehat{X}$ exists and has a good theory of vector bundles. Moreover, each x_i should define a closed Cartier divisor $S \hookrightarrow S \times_{\mathbf{F}_p} \widehat{X}$. This is the case for instance if S is locally of the form $\mathrm{Spa}(R, R^+)$, with R strongly noetherian, or if S is sousperfectoid.

Example 11.1.4. Let $C/\mathbf{F}_p$ be an algebraically closed nonarchimedean field with pseudo-uniformizer ϖ and residue field k, and let $S = \mathrm{Spa}\, C$. Then the product $S \times_{\mathbf{F}_p} \mathrm{Spa}\, \mathbf{F}_p[\![T]\!] = \mathbf{D}_C$ is the rigid open unit disc over C. The legs are given by continuous maps $\mathbf{F}_p[\![T]\!] \to C$, which is to say elements $x_i \in C$ which are topologically nilpotent. We can think of x_i as a C-point of $\mathbf{D}_C$.

Then a shtuka is a rank n vector bundle $\mathcal{E}$ over $\mathbf{D}_C$ together with an isomorphism $\varphi_{\mathcal{E}} \colon \varphi^* \mathcal{E} \to \mathcal{E}$ on $\mathbf{D}_C \backslash \{x_i\}$ that is meromorphic at the x_i. Here, $\varphi \colon \mathbf{D}_C \to \mathbf{D}_C$ sends the parameter T to T, but is the Frobenius on C. Note that $\mathbf{D}_C$ is a classical rigid space over C, but φ is not a morphism of rigid spaces because it is not C-linear. In the case of a single leg at 0, such pairs $(\mathcal{E}, \varphi_{\mathcal{E}})$ are studied in [HP04], where they are called σ-*bundles*.

We could also have taken $S = \mathrm{Spa}\, \mathcal{O}_C$, in which case the product $\mathrm{Spa}\, \mathcal{O}_C \times \widehat{X}$ is $\mathrm{Spa}\, \mathcal{O}_C[\![T]\!]$. This is similar to the space $\mathrm{Spa}\, \mathbf{Z}_p[\![T]\!]$, which we analyzed in §4.2. It contains a unique non-analytic point x_k. Let $\mathcal{Y}$ be the complement of x_k in $\mathrm{Spa}\, \mathcal{O}_C[\![T]\!]$. Once again there is a continuous surjective map $\kappa \colon \mathcal{Y} \to [0, \infty]$, defined by

$$\kappa(x) = \frac{\log |\varpi(\widetilde{x})|}{\log |T(\widetilde{x})|},$$

where $\widetilde{x}$ is the maximal generalization of x. The Frobenius map φ is a new feature of this picture. It satisfies $\kappa \circ \varphi = p\kappa$. See Figure 11.1.

11.2 THE ADIC SPACE "$S \times \mathrm{Spa}\, \mathbf{Z}_P$"

In the mixed characteristic setting, $\widehat{X}$ will be replaced with $\mathrm{Spa}\, \mathbf{Z}_p$. Our test objects S will be drawn from Perf, the category of perfectoid spaces in characteristic p. For an object $S \in \mathrm{Perf}$, a shtuka over S should be a vector bundle over an adic space "$S \times \mathrm{Spa}\, \mathbf{Z}_p$" together with a Frobenius structure. The product is not meant to be taken literally (if so, one would just recover S), but rather it is to be interpreted as a fiber product over a deeper base.

The main idea is that if R is an $\mathbf{F}_p$-algebra, then "$R \otimes \mathbf{Z}_p$" ought to be $W(R)$. As justification for this, note that $W(R)$ is a ring admitting a ring homomorphism $\mathbf{Z}_p \to W(R)$ and also a map $R \to W(R)$ which is not quite a ring homomorphism (it is only multiplicative). Motivated by this, we will define

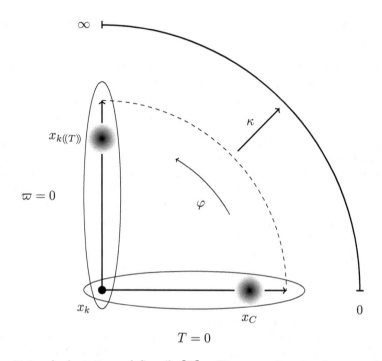

$$T = 0$$

Figure 11.1: A depiction of $\mathrm{Spa}\,\mathcal{O}_C[\![T]\!]$. The two closed subspaces $\mathrm{Spa}\,\mathcal{O}_C$ and $\mathrm{Spa}\,k[\![T]\!]$ appear as the x-axis and y-axis, respectively. Their intersection is the unique non-analytic point x_k of $\mathrm{Spa}\,\mathcal{O}_C[\![T]\!]$. The complement of x_k in $\mathrm{Spa}\,\mathcal{O}_C[\![T]\!]$ is the adic space $\mathcal{Y}$, on which the continuous map $\kappa\colon \mathcal{Y} \to [0,\infty]$ is defined. The automorphism φ of $\mathrm{Spa}\,\mathcal{O}_C[\![T]\!]$ tends to rotate points towards the y-axis (though it fixes both axes).

an analytic adic space[1] $S \,\dot\times\, \mathrm{Spa}\,\mathbf{Z}_p$ and then show that its associated diamond is the appropriate product of sheaves on Perf.

In fact, if $S = \mathrm{Spa}(R, R^+)$, then it is the open subspace $\{\varpi \neq 0\}$ of $\mathrm{Spa}\,R^+$, where ϖ is a pseudouniformizer. As R^+ is perfect, the product $\mathrm{Spa}\,R^+ \,\dot\times\, \mathrm{Spa}\,\mathbf{Z}_p$ should be defined as $\mathrm{Spa}\,W(R^+)$, and then we set

$$S \,\dot\times\, \mathrm{Spa}\,\mathbf{Z}_p := \{[\varpi] \neq 0\} \subset \mathrm{Spa}\,W(R^+)\,.$$

Note that this is independent of the choice of ϖ, as for any other choice ϖ', there is some n such that $\varpi|(\varpi')^n$ and $\varpi'|\varpi^n$.

Proposition 11.2.1. If $S = \mathrm{Spa}(R, R^+)$ is an affinoid perfectoid space of characteristic p, this defines an analytic adic space $S \,\dot\times\, \mathrm{Spa}\,\mathbf{Z}_p$ such that there is a

---

[1]In the lectures, this space was denoted by "$S \times \mathrm{Spa}\,\mathbf{Z}_p$" throughout.

natural isomorphism

$$(S \dot\times \operatorname{Spa} \mathbf{Z}_p)^\diamond = S \times \operatorname{Spd} \mathbf{Z}_p .$$

In particular, there is a natural map of topological spaces

$$\left| S \dot\times \operatorname{Spa} \mathbf{Z}_p \right| = \left| (S \dot\times \operatorname{Spa} \mathbf{Z}_p)^\diamond \right| = \left| S \dot\times \operatorname{Spd} \mathbf{Z}_p \right| \to |S| .$$

For any perfectoid space $S \in \operatorname{Perf}$, one can define an analytic adic space $S \dot\times \operatorname{Spa} \mathbf{Z}_p$ with a natural isomorphism

$$(S \dot\times \operatorname{Spa} \mathbf{Z}_p)^\diamond = S \times \operatorname{Spd} \mathbf{Z}_p$$

which over any open affinoid subspace $U \subset S$ recovers $U \dot\times \operatorname{Spa} \mathbf{Z}_p$ as defined above.

Proof. We will treat the case that $S = \operatorname{Spa}(R, R^+)$ is affinoid; the globalization follows by checking that if $S' \subset S$ is a rational embedding of affinoid perfectoid spaces, then $S' \dot\times \operatorname{Spa} \mathbf{Z}_p \subset S \dot\times \operatorname{Spa} \mathbf{Z}_p$ is a corresponding rational embedding compatible with the displayed identity of diamonds.

It is clear that $S \dot\times \operatorname{Spa} \mathbf{Z}_p$ is analytic as $[\varpi]$ is a topologically nilpotent unit. We claim that it is in fact an adic space. We have a covering of

$$S \dot\times \operatorname{Spa} \mathbf{Z}_p = \operatorname{Spa} W(R^+) \setminus \{[\varpi] = 0\}$$

by affinoid subsets of $\operatorname{Spa} W(R^+)$ of the form $|p| \le \left| [\varpi^{1/p^n}] \right| \ne 0$, for $n = 1, 2, \ldots$.

Let $\operatorname{Spa}(R_n, R_n^+)$ be the rational subset $|p| \le \left| [\varpi^{1/p^n}] \right| \ne 0$. Thus R_n is the ring obtained by $[\varpi]$-adically completing $W(R^+)[p/[\varpi^{1/p^n}]]$ and then inverting $[\varpi]$. One has the following presentation for R_n:

$$R_n = \left\{ \sum_{i \ge 0} [r_i] \left(\frac{p}{[\varpi^{1/p^n}]} \right)^i \;\middle|\; r_i \in R, \; r_i \to 0 \right\} .$$

The ring $R_n \widehat\otimes_{\mathbf{Z}_p} \mathbf{Z}_p[p^{1/p^\infty}]^\wedge$ is a perfectoid ring, and therefore R_n is sousperfectoid: note that $\varpi' = [\varpi^{1/p^{n+1}}]$ serves as a pseudo-uniformizer satisfying the condition of Definition 6.1.1. By Proposition 6.3.4, (R_n, R_n^+) is sheafy.

Now we check the identification of associated diamonds. For an object T of Perf, say $T = \operatorname{Spa}(A, A^+)$, a T-valued point of $(S \dot\times \operatorname{Spa} \mathbf{Z}_p)^\diamond$ is an untilt $T^\sharp$ lying over $S \dot\times \operatorname{Spa} \mathbf{Z}_p$. To give a morphism $T^\sharp \to S \dot\times \operatorname{Spa} \mathbf{Z}_p$ is to give a continuous homomorphism $W(R^+) \to A^{\sharp+}$ such that the image of $[\varpi]$ is invertible in $A^\sharp$. Recall that maps from $W(R^+)$ to $A^{\sharp+}$ are equivalent to maps from R^+ to $(A^{\sharp+})^\flat = A^+$; under this bijection, $[\varpi]$ is invertible in $A^\sharp$ if and only if ϖ is invertible in A. This shows an equivalence between maps from T to $(S \dot\times \operatorname{Spa} \mathbf{Z}_p)^\diamond$ and untilts $T^\sharp$ which come with a map from T to $S = \operatorname{Spa}(R, R^+)$.

In other words,

$$(S \dot{\times} \operatorname{Spa} \mathbf{Z}_p)^\diamond = S \times \operatorname{Spd} \mathbf{Z}_p \, ,$$

as desired.   □

Example 11.2.2. Let $S = \operatorname{Spa} C$, where $C/\mathbf{F}_p$ is an algebraically closed nonar-chimedean field with residue field k. The ring of Witt vectors $W(\mathcal{O}_C)$ is called A_{inf} in Fontaine's theory, [Fon94]. Its spectrum $\operatorname{Spa} W(\mathcal{O}_C)$ is rather like the formal unit disk $\operatorname{Spa} \mathcal{O}_C[\![T]\!]$; cf. Figure 12.1 below.

For any connected open affinoid subspace $U = \operatorname{Spa}(R, R^+) \subset \operatorname{Spa} C \dot{\times} \operatorname{Spa} \mathbf{Z}_p$, the ring R is a principal ideal domain by [Ked16]. Moreover, for any maximal ideal $\mathfrak{m} \subset R$, the quotient $C_\mathfrak{m} = R/\mathfrak{m}$ is an untilt of C, and this induces a bijection between untilts of C and "classical points" of the space $\operatorname{Spa} C \dot{\times} \operatorname{Spa} \mathbf{Z}_p$. We will analyze this relation between untilts and sections of $S \dot{\times} \operatorname{Spa} \mathbf{Z}_p$ in general in the next section.

11.3   SECTIONS OF $(S \dot{\times} \operatorname{Spa} \mathbf{Z}_P)^\diamond \to S$

Even though there is no morphism $S \dot{\times} \operatorname{Spa} \mathbf{Z}_p \to S$, Proposition 11.2.1 shows that there is a morphism $(S \dot{\times} \operatorname{Spa} \mathbf{Z}_p)^\diamond \to S$. The following proposition shows that sections of this morphism behave as expected.

Proposition 11.3.1. Let $S \in \operatorname{Perf}$. The following sets are naturally identified:

1. Sections of $(S \dot{\times} \operatorname{Spa} \mathbf{Z}_p)^\diamond \to S$,
2. Morphisms $S \to \operatorname{Spd} \mathbf{Z}_p$, and
3. Untilts $S^\sharp$ of S.

Moreover, given these data, there is a natural map $S^\sharp \hookrightarrow S \dot{\times} \operatorname{Spa} \mathbf{Z}_p$ of adic spaces over $\mathbf{Z}_p$ that is the inclusion of a closed Cartier divisor.

Thus, the sections make sense as actual subspaces. The statement about Cartier divisors means that $\mathcal{O}_{S \dot{\times} \operatorname{Spa} \mathbf{Z}_p} \to \mathcal{O}_{S^\sharp}$ is surjective and its kernel is locally free of rank 1; cf. Definition 5.3.2 and Definition 5.3.7.

Proof. By Proposition 11.2.1, $(S \dot{\times} \operatorname{Spa} \mathbf{Z}_p)^\diamond = S \times \operatorname{Spd} \mathbf{Z}_p$. Thus a section as in (1) is a section of $S \times \operatorname{Spd} \mathbf{Z}_p \to S$, which is nothing but a morphism $S \to \operatorname{Spd} \mathbf{Z}_p$. Thus (1) and (2) are identified. The sets (2) and (3) are identified by definition of $\operatorname{Spd} \mathbf{Z}_p$.

Given an untilt $S^\sharp = \operatorname{Spa}(R^\sharp, R^{\sharp+})$ as in (3), we have a map $\theta \colon W(R^+) \to R^{\sharp+}$ which sends $[\varpi]$ to a unit in $R^\sharp$. This means that the composite map $S^\sharp \to \operatorname{Spa} R^{\sharp+} \to \operatorname{Spa} W(R^+)$ factors through $S \dot{\times} \operatorname{Spa} \mathbf{Z}_p \subset \operatorname{Spa} W(R^+)$, so we have a map $S^\sharp \to S \dot{\times} \operatorname{Spa} \mathbf{Z}_p$.

It remains to see that this map defines a closed Cartier divisor. As usual let $\xi \in W(R^+)$ be a generator of $\ker \theta$. By Proposition 5.3.8, it suffices to check that for all open affinoid $\mathrm{Spa}(A, A^+) \subset S \dot{\times} \mathrm{Spa}\, \mathbf{Z}_p$, multiplication by ξ defines an injective map $A \to A$ with closed image.

We can define a supremum norm $|\cdot|_\infty : A \to \mathbf{R}_{\geq 0}$ by $|a|_\infty = \sup_x |a(x)|$, where x runs over rank 1 valuations in $\mathrm{Spa}(A, A^+)$, normalized so that $|[\varpi](x)| = 1/p$. Then $|\cdot|_\infty$ induces the topology on A: if $|a|_\infty \leq p^{-N}$, then $\left|a/[\varpi]^N\right|_\infty \leq 1$, so that $a \in [\varpi]^N A^+$.

We claim there exists a constant $c > 0$ such that the inequality

$$|\xi a|_\infty \geq c\, |a|_\infty$$

holds for all $a \in A$. The claim would imply immediately that multiplication by ξ on A is injective. It also implies that $\xi A \subset A$ is closed: if $a_i = \xi b_i$ is a sequence in ξA converging to $a \in A$, then $|b_i - b_j|_\infty \leq c^{-1} |a_i - a_j|_\infty \to 0$, so that the b_i form a Cauchy sequence, from which we conclude that $a \in \xi A$.

To prove the claim, we first make a reduction to the case that $S^\sharp \subset \mathrm{Spa}(A, A^+)$. For this, we consider the intersection of $S^\sharp$ with $\mathrm{Spa}(A, A^+)$, which is a quasicompact open subspace $U^\sharp \subset S^\sharp$ corresponding to a quasicompact open subspace $U \subset S$. The intersection $V = \mathrm{Spa}(A, A^+) \cap U \dot{\times} \mathrm{Spa}\, \mathbf{Z}_p$ contains $U^\sharp$, which is the vanishing locus of ξ. Now V is a constructible subset of $\mathrm{Spa}(A, A^+)$, and so its complement is quasicompact for the constructible topology; it follows that for some n the locus $V' = \{|\xi| \leq |[\varpi]|^n\} \subset \mathrm{Spa}(A, A^+)$ is contained in V. Covering $\mathrm{Spa}(A, A^+)$ by the two rational subsets $\{|\xi| \leq |[\varpi]|^n\}$ and $\{|[\varpi]|^n \leq |\xi|\}$, the desired result is clear on the second, so we can reduce to the first case. In that case, $\mathrm{Spa}(A, A^+) \subset U \dot{\times} \mathrm{Spa}\, \mathbf{Z}_p$ and $U^\sharp \subset \mathrm{Spa}(A, A^+)$, so after replacing U by an affinoid cover, we can indeed assume that $S^\sharp \subset \mathrm{Spa}(A, A^+)$.

By a similar quasicompactness argument, we see that in fact the locus $\{|\xi| \leq |[\varpi]|^n\} \subset S \dot{\times} \mathrm{Spa}\, \mathbf{Z}_p$ is contained in $\mathrm{Spa}(A, A^+)$ for n large enough, and we can reduce to the case $\mathrm{Spa}(A, A^+) = \{|\xi| \leq |[\varpi]|^n\}$ for n large enough (so that this is a quasicompact subspace).

We claim that in this case $|\xi a|_\infty \geq p^{-n}|a|_\infty$ for all $a \in A$. This can be checked fiberwise and on rank 1 points, so we can assume that $S = \mathrm{Spa}(C, \mathcal{O}_C)$ is a geometric point. It is enough to see that the supremum norm on $\{|\xi| \leq |[\varpi]|^n\}$ agrees with the supremum over the set of points that admit specializations to points outside of $\{|\xi| \leq |[\varpi]|^n\}$, i.e. the Shilov boundary. This can be checked on the cover $S \dot{\times} \mathrm{Spa}\, \mathbf{Z}_p[p^{1/p^\infty}]_p^\wedge$ of $S \dot{\times} \mathrm{Spa}\, \mathbf{Z}_p$, and then on its tilt $S \times \mathrm{Spa}\, \mathbf{F}_p[[t^{1/p^\infty}]]$. In this case, we have the perfection of the open unit disc $\mathbb{D}_C$, and $\mathrm{Spa}(A, A^+)$ corresponds to an open affinoid subset of $\mathbb{D}_C$. In this situation, it is well-known that the supremum is achieved on the boundary, giving the result. $\square$

11.4 DEFINITION OF MIXED-CHARACTERISTIC SHTUKAS

By Theorem 5.2.8, one has a good notion of vector bundle for the space $S \dot{\times} \mathrm{Spa}\, \mathbf{Z}_p$.

Definition 11.4.1. Let S be a perfectoid space in characteristic p. Let $x_1, \ldots, x_m \colon S \to \operatorname{Spd} \mathbf{Z}_p$ be a collection of morphisms; for $i = 1, \ldots, m$ let $\Gamma_{x_i} \colon S_i^\sharp \to S \dot{\times} \operatorname{Spa} \mathbf{Z}_p$ be the corresponding closed Cartier divisor. A *(mixed-characteristic) shtuka of rank n over S with legs $x_1, \ldots, x_m$* is a rank n vector bundle $\mathcal{E}$ over $S \dot{\times} \operatorname{Spa} \mathbf{Z}_p$ together with an isomorphism

$$\varphi_\mathcal{E} \colon \operatorname{Frob}_S^*(\mathcal{E})|_{S \dot{\times} \operatorname{Spa} \mathbf{Z}_p \setminus \bigcup_i \Gamma_{x_i}} \to \mathcal{E}|_{S \dot{\times} \operatorname{Spa} \mathbf{Z}_p \setminus \bigcup_i \Gamma_{x_i}}$$

that is meromorphic along $\bigcup_i \Gamma_{x_i}$.

Remark 11.4.2. Here, the final meromorphicity condition is meant with respect to the closed Cartier divisor that is the sum of the m closed Cartier divisors Γ_{x_i}.

In the next lectures, we will analyze the case of shtukas with one leg over $S = \operatorname{Spa} C$, where $C/\mathbf{F}_p$ is an algebraically closed nonarchimedean field. The leg is a map $S \to \operatorname{Spd} \mathbf{Z}_p$. Let us assume that this factors over $\operatorname{Spd} \mathbf{Q}_p$ and thus corresponds to a characteristic 0 untilt $C^\sharp$, an algebraically closed complete nonarchimedean extension of $\mathbf{Q}_p$. In fact, let us change notation, and instead start out with a complete algebraically closed extension $C/\mathbf{Q}_p$, and let $S = \operatorname{Spa} C^\flat$. We have a surjective homomorphism $W(\mathcal{O}_{C^\flat}) \to \mathcal{O}_C$ whose kernel is generated by an element $\xi \in W(\mathcal{O}_{C^\flat})$. Let $\varphi = W(\operatorname{Frob}_{\mathcal{O}_{C^\flat}})$, an automorphism of $W(\mathcal{O}_{C^\flat})$.

It turns out that such shtukas are in correspondence with linear-algebra objects which are essentially shtukas over all of $\operatorname{Spa} W(\mathcal{O}_{C^\flat})$, rather than over just the locus $[\varpi] \neq 0$.

Definition 11.4.3. A *Breuil-Kisin-Fargues module* is a pair (M, φ_M), where M is a finite free $W(\mathcal{O}_{C^\flat})$-module and $\varphi_M \colon (\varphi^* M)[\xi^{-1}] \overset{\sim}{\to} M[\xi^{-1}]$ is an isomorphism.

Remark 11.4.4. Note the analogy to Kisin's work [Kis06], which takes place in the context of a finite totally ramified extension $K/W(k)[1/p]$ (now k is any perfect field of characteristic p). Let ξ generate the kernel of a continuous surjective homomorphism $W(k)[\![u]\!] \to \mathcal{O}_K$. Kisin's φ-modules are pairs (M, φ_M), where M is a finite free $W(k)[\![u]\!]$-module and $\varphi_M \colon (\varphi^* M)[\xi^{-1}] \overset{\sim}{\to} M[\xi^{-1}]$ is an isomorphism. Kisin constructs a fully faithful functor from the category of crystalline representations of $\operatorname{Gal}(\overline{K}/K)$ into the category φ-modules up to isogeny and identifies the essential image.

In fact, much of the theory of mixed-characteristic shtukas is motivated by the structures appearing in (integral) p-adic Hodge theory. In the next lectures, we will concentrate on these relations.

Let (M, φ_M) be a Breuil-Kisin-Fargues module over $W(\mathcal{O}_{C^\flat})$. After passing to $\operatorname{Spa} W(\mathcal{O}_{C^\flat})$ and inverting $[\varpi]$, M gives rise to a vector bundle on $S \dot{\times} \operatorname{Spa} \mathbf{Z}_p \subset W(\mathcal{O}_{C^\flat})$ and therefore a shtuka $\mathcal{E}$. In fact one can go in the other direction:

Theorem 11.4.5 (Fargues). The functor $(M, \varphi_M) \mapsto (\mathcal{E}, \varphi_{\mathcal{E}})$ is an equivalence between the category of Breuil-Kisin-Fargues modules over $W(\mathcal{O}_{C^\flat})$ and the category of shtukas over $\operatorname{Spa} C^\flat$ with one leg at the untilt C of $C^\flat$.

We will prove this theorem in Lectures 12–14. In the course of the proof, we will learn many facts that apply to much more general shtukas.

Remark 11.4.6. In the following lectures, we will make a Frobenius twist, and redefine Breuil-Kisin-Fargues modules by replacing ξ with $\varphi(\xi)$; similarly, we will consider shtukas with a leg at $\varphi^{-1}(x_C)$, where $x_C : \operatorname{Spa} C \hookrightarrow \operatorname{Spa} C^\flat \overset{.}{\times} \operatorname{Spa} \mathbf{Z}_p$ denotes the fixed Cartier divisor. This normalization is necessary to get the comparison with crystalline cohomology in the usual formulation.[2] Moreover, we warn the reader that the definition of Breuil-Kisin-Fargues modules in [BMS18] is slightly different, as there one does not ask that M be finite free (only that M be finitely presented and that $M[\frac{1}{p}]$ be finite free over $A_{\mathrm{inf}}[\frac{1}{p}]$); this is necessary in order to accomodate the possible presence of torsion in the cohomology.

---

[2]Recently, in [BS19], a refinement of crystalline cohomology is constructed that is a Frobenius descent of it; thus, if one works only with prismatic cohomology, the leg will be at x_C and not $\varphi^{-1}(x_C)$.

Lecture 12

Shtukas with one leg

Over the next three lectures, we will analyze shtukas with one leg over a geometric point in detail, and discuss the relation to (integral) p-adic Hodge theory. We will focus on the connection between shtukas with one leg and p-divisible groups, and recover a result of Fargues (Theorem 14.1.1) which states that p-divisible groups are equivalent to one-legged shtukas of a certain kind.

In fact this is a special case of a much more general connection [BMS18] between shtukas with one leg and proper smooth (formal) schemes, which we discuss in the Appendix to Lecture 14.

Throughout, we fix $C/\mathbf{Q}_p$ an algebraically closed nonarchimedean field, with ring of integers $\mathcal{O}_C$ and residue field k.

12.1 p-DIVISIBLE GROUPS OVER $\mathcal{O}_C$

Recall that p-divisible groups over k may be classified by Dieudonné modules. We pose the question here of how to classify p-divisible groups over $\mathcal{O}_C$, and how that classification interacts with the classification over k.

Let G be a p-divisible group over $\mathcal{O}_C$, and let $T_pG = \varprojlim G[p^n](C)$ be its Tate module, a free $\mathbf{Z}_p$-module of finite rank.

Theorem 12.1.1 (The Hodge-Tate exact sequence, [Far]). There is a natural short exact sequence

$$0 \longrightarrow \operatorname{Lie} G \otimes_{\mathcal{O}_C} C(1) \xrightarrow{\alpha_{G^*}^*} T_pG \otimes_{\mathbf{Z}_p} C \xrightarrow{\alpha_G} (\operatorname{Lie} G^*)^* \otimes_{\mathcal{O}_C} C \longrightarrow 0$$

Remark 12.1.2. Here, $C(1)$ denotes a Tate twist of C (which is trivial, as C is algebraically closed). In order to make the presentation more canonical, we will still often write the Tate twist; it becomes important when G is already defined over a subfield of C, so that there are Galois actions around. Tate [Tat67] treated the case where G comes from a p-divisible group over a discrete valuation ring with perfect residue field.

The map α_G is defined as follows. An element of T_pG is really just a morphism of p-divisible groups $f \colon \mathbf{Q}_p/\mathbf{Z}_p \to G$, whose dual is a morphism $f^* \colon G^* \to$ μ_{p^∞}. The derivative of f^* is an $\mathcal{O}_C$-linear map $\operatorname{Lie} f^* \colon \operatorname{Lie} G^* \to \operatorname{Lie} \mu_{p^\infty}$. Note that $\operatorname{Lie} \mu_{p^\infty}$ is canonically free of rank 1, with dual $(\operatorname{Lie} \mu_{p^\infty})^* = \mathcal{O}_C \cdot \frac{dt}{t}$. The dual of $\operatorname{Lie} f^*$ is a map $\mathcal{O}_C \cong (\operatorname{Lie} \mu_{p^\infty})^* \to (\operatorname{Lie} G^*)^*$; let $\alpha_G(f)$ be the image of $1 \in \mathcal{O}_C \cong (\operatorname{Lie} \mu_{p^\infty})^*$.

Remark 12.1.3. It is not at all formal that $\alpha_G \circ \alpha_{G^*}^* = 0$.

Definition 12.1.4. Let $\{(T, W)\}$ be the category of pairs consisting of a finite free $\mathbf{Z}_p$-module T and a C-subvectorspace $W \subset T \otimes_{\mathbf{Z}_p} C$.

Theorem 12.1.1 gives us a functor

$$
\begin{aligned}
\{p\text{-divisible groups}/\mathcal{O}_C\} &\to \{(T, W)\} \\
G &\mapsto (T_p G, \operatorname{Lie} G \otimes_{\mathcal{O}_C} C(1)).
\end{aligned}
$$

Theorem 12.1.5 ([SW13, Theorem B]). This is an equivalence of categories.

Theorem 12.1.5 gives a classification of p-divisible groups over $\mathcal{O}_C$ in terms of linear algebra data, analogous to Riemann's classification of complex abelian varieties.

On the other hand we have the classification of p-divisible groups over k via Dieudonné modules, which we review here. For us, a *Dieudonné module* is a finite free $W(k)$-module M equipped with a φ-linear isomorphism $\varphi_M \colon M[\frac{1}{p}] \cong M[\frac{1}{p}]$. There is a fully faithful embedding $G \mapsto M(G)$ from the category of p-divisible groups over k into the category of Dieudonné modules; the essential image consists of those (M, φ_M) with $M \subset \varphi_M(M) \subset p^{-1}M$.[1] Here $\varphi \colon W(k) \to W(k)$ is induced from the pth power Frobenius map on k.

There is the interesting question of how the classifications over $\mathcal{O}_C$ and k interact. That is, we have a diagram

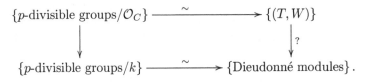

It is not at all clear how to give an explicit description of the arrow labeled "?". If we think of "?" only as a map between sets of isogeny classes of objects, we get the following interpretation. Let (h, d) be a pair of nonnegative integers with $d \leq h$. The set of isomorphism classes of objects (T, W) with $\operatorname{rank} T = h$ and $\dim W = d$ together with a trivialization $T \cong \mathbf{Z}_p^h$ is $\operatorname{Grass}(h, d)(C)$, the set

---

[1]Here, we are using the covariant normalization of Dieudonné theory that sends $\mathbf{Q}_p/\mathbf{Z}_p$ to $(W(k), \varphi)$ and μ_{p^∞} to $(W(k), p^{-1}\varphi)$. This is given as the naive dual of the contravariant normalization (usually, one multiplies φ_M by p so that it preserves M, but this is artificial).

of d-planes in C^h. The set of isogeny classes of Dieudonné modules is identified with the finite set $\mathcal{NP}_{h,d}$ of Newton polygons running between $(0,0)$ and (d,h) whose slopes lie in $[0,1]$. Thus we have a *canonical* $\mathrm{GL}_h(\mathbf{Q}_p)$-equivariant map $\mathrm{Grass}(d,h)(C) \to \mathcal{NP}_{h,d}$. What are its fibers? There is no known explicit answer to this question in general. The preimage of the Newton polygon that is a straight line from $(0,0)$ to (h,d) (corresponding to an isoclinic p-divisible group) is an open subset known as the "admissible locus"; it is contained in the explicit "weakly admissible locus" (cf. [RZ96]), but is in general smaller.

Example 12.1.6. In the cases $d = 0$ and $d = 1$ the map $\mathrm{Grass}(d,h)(C) \to \mathcal{NP}_{h,d}$ can be calculated explicitly. The case $d = 0$ is trivial since both sets are singletons. So consider the case $d = 1$. Let $W \subset C^h$ be a line, so that $W \in \mathbf{P}^{h-1}(C)$. The p-divisible group corresponding to $(\mathbf{Z}_p^h, W)$ is determined by "how rational" W is. To wit, let $H \subset \mathbf{P}^{h-1}(C)$ be the smallest $\mathbf{Q}_p$-linear C-subspace containing W, and let $i = \dim_C H - 1$. Then the p-divisible group associated to $(\mathbf{Z}_p^h, W)$ is isogenous to $(\mathbf{Q}_p/\mathbf{Z}_p)^{\oplus(h-i)} \oplus G_i$, where G_i is a p-divisible formal group of height i and dimension 1. (Since k is algebraically closed, G_i is unique up to isomorphism.) Thus for instance the set of $W \in \mathbf{P}^{h-1}(C)$ which corresponds to a p-divisible group with special fiber G_h is Drinfeld's upper half-space $\Omega^h(C)$, where

$$\Omega^h = \mathbf{P}^{h-1} \setminus \bigcup_{H \subset \mathbf{P}^{h-1}} H$$

with H running over all $\mathbf{Q}_p$-rational hyperplanes.

12.2   SHTUKAS WITH ONE LEG AND P-DIVISIBLE GROUPS: AN OVERVIEW

The functor "?" is hard to describe, but there is a hint for how to proceed: If G is defined over $\mathcal{O}_K$ with $K/\mathbf{Q}_p$ finite, then the relation between the Galois representation $T_p G$ and the Dieudonné module $M(G_k)$ is given by Fontaine's comparison isomorphism:

$$M(G_k)[\tfrac{1}{p}] = (T_p G \otimes_{\mathbf{Z}_p} A_{\mathrm{cris}})^{\mathrm{Gal}(\overline{K}/K)}[\tfrac{1}{p}].$$

Here, A_{cris} is Fontaine's period ring; it is a $W(k)$-algebra which comes equipped with a ϕ-linear endomorphism.

This indicates that we will have to work with period rings. It turns out that the necessary period rings show up naturally in the geometry of $\mathrm{Spa}\, W(\mathcal{O}_{C^\flat})$, and that there is an intimate link to shtukas over $\mathrm{Spa}\, C^\flat$ with one leg at C.

Remark 12.2.1. A notational remark: Fontaine gives the name $\mathcal{R}$ to $\mathcal{O}_{C^\flat}$, and Berger and Colmez call it $\widetilde{\mathbb{E}}^+$, reserving $\widetilde{\mathbb{E}}$ for what we call $C^\flat$. The ring $W(\mathcal{O}_{C^\flat})$ is variously called $W(\mathcal{R})$ and A_{inf}.

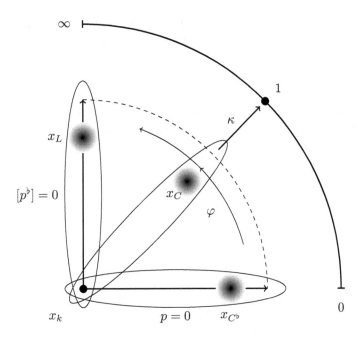

Figure 12.1: A depiction of Spa A_{inf}, where $A_{\mathrm{inf}} = W(\mathcal{O}_{C^\flat})$. The two closed subspaces $p = 0$ and $[p^\flat] = 0$ appear as the x-axis and y-axis, respectively. We have also depicted the closed subspace $p = [p^\flat]$, which cuts out Spa $\mathcal{O}_C$, as a diagonal ellipse. The unique non-analytic point x_k of Spa A_{inf} appears at the origin. Its complement in Spa A_{inf} is the adic space $\mathcal{Y}$, on which the continuous map $\kappa \colon \mathcal{Y} \to [0, \infty]$ is defined. The automorphism φ of Spa A_{inf} rotates points towards the y-axis, as per the equation $\kappa \circ \varphi = p\kappa$.

From now on, we write $A_{\mathrm{inf}} = W(\mathcal{O}_{C^\flat})$. Recall that we have a surjective map $\theta \colon A_{\mathrm{inf}} \to \mathcal{O}_C$. The kernel of θ is generated by a nonzerodivisor $\xi = p - [p^\flat]$, where $p^\flat = (p, p^{1/p}, p^{1/p^2}, \ldots) \in \mathcal{O}_{C^\flat}$. Note that $p^\flat$ is a pseudo-uniformizer of $C^\flat$.

Consider the pre-adic space Spa A_{inf}.[2] We give names to four special points of Spa A_{inf}, labeled by their residue fields:

1. x_k, the unique non-analytic point (recall that k is the residue field of C).
2. $x_{C^\flat}$, which corresponds to $A_{\mathrm{inf}} \to \mathcal{O}_{C^\flat} \to C^\flat$.
3. x_C, which corresponds to $A_{\mathrm{inf}} \to \mathcal{O}_C \to C$ (the first map is θ); moreover, it has φ-translates $\varphi^n(x_C)$ for all $n \in \mathbb{Z}$.
4. x_L, which corresponds to $A_{\mathrm{inf}} \to W(k) \to W(k)[1/p] = L$.

---

[2]It is probably true that Spa A_{inf} is an adic space, but we will not need this.

Let $\mathcal{Y} = \operatorname{Spa} W(\mathcal{O}_{C^\flat}) \setminus \{x_k\}$, an analytic adic space. Then as usual there exists a surjective continuous map $\kappa \colon \mathcal{Y} \to [0, \infty]$, defined by

$$\kappa(x) = \frac{\log |[\varpi](\tilde{x})|}{\log |p(\tilde{x})|},$$

where $\tilde{x}$ is the maximal generalization of x; cf. the discussion in Section 4.2. See Figure 12.1 for a depiction of the various structures associated with $\operatorname{Spa} W(\mathcal{O}_{C^\flat})$. We have:

$$
\begin{aligned}
\kappa(x_{C^\flat}) &= 0, \\
\kappa(x_C) &= 1, \\
\kappa(\varphi^n(x_C)) &= p^n, \\
\kappa(x_L) &= \infty.
\end{aligned}
$$

For an interval $I \subset [0, \infty]$, let $\mathcal{Y}_I$ be the interior of the preimage of $\mathcal{Y}$ under κ. Thus $\mathcal{Y}_{[0,\infty)}$ is the complement in $\mathcal{Y}$ of the point x_L with residue field $L = W(k)[1/p]$. Also note that $\mathcal{Y}_{[0,\infty)} = \operatorname{Spa} C^\flat \dot{\times} \operatorname{Spa} \mathbf{Z}_p$.

The Frobenius automorphism of $\mathcal{O}_{C^\flat}$ induces an automorphism φ of $\operatorname{Spa} A_{\inf}$, which preserves $\mathcal{Y}$ and which satisfies $\kappa \circ \varphi = p\kappa$. Therefore φ sends $\mathcal{Y}_{[a,b]}$ isomorphically to $\mathcal{Y}_{[ap,bp]}$.

We will outline the construction of the functor "?" in three steps:

1. Show that a pair $\{(T, W)\}$ determines a shtuka over $\operatorname{Spa} C^\flat$ with one leg at $\varphi^{-1}(x_C)$,[3] i.e., a vector bundle on $\mathcal{Y}_{[0,\infty)}$ with a certain Frobenius.
2. Show that any shtuka over $\mathcal{Y}_{[0,\infty)}$ extends to all of $\mathcal{Y}$, i.e., it extends over x_L.
3. Show that any shtuka over $\mathcal{Y}$ extends to all of $\operatorname{Spa} A_{\inf}$, i.e. it extends over x_k.

In the end, one has a finite free $A_{\inf}$-module with a certain Frobenius, so we can take the base change along $A_{\inf} \to W(k)$ to get a Dieudonné module. Thus, in total, we get a functor from pairs $\{(T, W)\}$ to Dieudonné modules. The remainder of the lecture concerns Step (1); Step (2) is carried out in Lecture 13, and Step (3) in Lecture 14.

---

[3] It would appear more natural to consider the leg at x_C; however, to get the correct functor (and not a Frobenius twist of it), this normalization is necessary. Again, we mention that if one changes crystalline cohomology into prismatic cohomology as defined in [BS19], this issue disappears.

12.3   SHTUKAS WITH NO LEGS, AND φ-MODULES OVER THE INTEGRAL ROBBA RING

As preparation, we analyze shtukas with no legs. This will be useful also for general shtukas, as we will see that any shtuka whose legs are all in characteristic 0 admits a meromorphic map to a shtuka with no legs.

Definition 12.3.1 (The integral Robba rings). Let $\widetilde{\mathcal{R}}^{\mathrm{int}}$ be the local ring $\mathcal{O}_{\mathcal{Y}, x_{C^\flat}}$. For a rational number $r > 0$, we define $\widetilde{\mathcal{R}}^{\mathrm{int},r}$ to be the ring of global sections of $\mathcal{O}_{\mathcal{Y}_{[0,r]}}$.

Remark 12.3.2.

1. $\widetilde{\mathcal{R}}^{\mathrm{int}}$ is a henselian discrete valuation ring with uniformizer p, residue field $C^\flat$, and completion equal to $W(C^\flat)$.
2. The Frobenius automorphism of A_{inf} induces isomorphisms $\varphi \colon \widetilde{\mathcal{R}}^{\mathrm{int},r} \to \widetilde{\mathcal{R}}^{\mathrm{int},r/p}$ for $r > 0$, and an automorphism of $\widetilde{\mathcal{R}}^{\mathrm{int}}$. In particular, φ^{-1} induces an endomorphism of $\widetilde{\mathcal{R}}^{\mathrm{int},r}$.

Unwinding the definition of κ, we see that $\mathcal{Y}_{[0,r]}$ is the rational subset of $\mathrm{Spa}\, A_{\mathrm{inf}}$ cut out by the conditions $|p(x)|^r \leq |[p^\flat](x)| \neq 0$. Thus

$$\widetilde{\mathcal{R}}^{\mathrm{int},r} = A_{\mathrm{inf}} \langle p/[p^\flat]^{1/r} \rangle [1/[p^\flat]] \,,$$

where $A_{\mathrm{inf}} \langle p/[p^\flat]^{1/r} \rangle$ is the $[p^\flat]$-adic completion of $A_{\mathrm{inf}}[p/[(p^\flat)^{1/r}]]$, and $(p^\flat)^{1/r} \in \mathcal{O}_{C^\flat}$ is any rth root of $p^\flat$. Thus $\widetilde{\mathcal{R}}^{\mathrm{int},r}$ can be described as

$$\widetilde{R}^{\mathrm{int},r} = \left\{ \sum_{n \geq 0} [c_n] p^n \;\middle|\; c_n \in C^\flat,\; c_n (p^\flat)^{n/r} \to 0 \right\}.$$

For $r' < r$, the inclusion of rational subsets $\mathcal{Y}_{[0,r']} \to \mathcal{Y}_{[0,r]}$ allows us to view $\widetilde{R}^{\mathrm{int},r}$ as a subring of $\widetilde{R}^{\mathrm{int},r'}$. Finally,

$$\widetilde{R}^{\mathrm{int}} = \varinjlim \widetilde{R}^{\mathrm{int},r} \text{ as } r \to 0.$$

It will be convenient for us to make the following definition.

Definition 12.3.3. Let R be a ring together with an endomorphism $\varphi \colon R \to R$. A φ-module over R is a finite projective R-module M with an isomorphism $\varphi_M \colon M \otimes_{R,\varphi} R \cong M$. Similarly, if X is an adic space equipped with an endomorphism $\varphi \colon X \to X$, then a φ-module over X is a vector bundle $\mathcal{E}$ on X equipped with an isomorphism $\varphi_{\mathcal{E}} \colon \varphi^* \mathcal{E} \cong \mathcal{E}$.

Note that if φ is an automorphism of R, the map φ_M is equivalent to a φ-semilinear automorphism of M, and we will often make this translation.

The next theorem states that φ-modules over the rings $\widetilde{\mathcal{R}}^{\mathrm{int}}$ and $W(C^\flat)$ are trivial in the sense that one can always find a φ-invariant basis.

Theorem 12.3.4 ([KL15, Theorem 8.5.3]). *The following categories are equivalent:*

1. φ-modules over $\widetilde{\mathcal{R}}^{\mathrm{int}}$;
2. φ-modules over $W(C^\flat)$;
3. finite free $\mathbf{Z}_p$-modules.

The functor from (1) to (2) is base extension, and the functor from (2) to (3) is the operation of taking φ-invariants.

Proof. (Sketch.) The full faithfulness of the functor from (1) to (2) is equivalent to the statement that if M is a φ-module over $\widetilde{\mathcal{R}}^{\mathrm{int}}$, then the map

$$M^{\varphi=1} \to (M \otimes_{\widetilde{\mathcal{R}}^{\mathrm{int}}} W(C^\flat))^{\varphi=1}$$

is an isomorphism. This can be checked by consideration of Newton polygons.

The equivalence between (2) and (3) is a special case of the following fact. Let R be a perfect ring. Then φ-modules over $W(R)$ are equivalent to $\mathbf{Z}_p$-local systems on $\operatorname{Spec} R$. This is sometimes called Artin-Schreier-Witt theory (the case of φ-modules over R and $\mathbf{F}_p$-local systems being due to Artin-Schreier). Given this equivalence, it is clear that the functor from (1) to (2) is essentially surjective, finishing the proof. $\square$

Proposition 12.3.5. *The following categories are equivalent:*

1. Shtukas over $\operatorname{Spa} C^\flat$ with no legs, and
2. φ-modules over $\widetilde{\mathcal{R}}^{\mathrm{int}}$ (in turn equivalent to finite free $\mathbf{Z}_p$-modules by Theorem 12.3.4).

Proof. A shtuka over $\operatorname{Spa} C^\flat$ with no legs is by definition a φ-module on $\mathcal{Y}_{[0,\infty)}$. The localization of $\mathcal{E}$ at $x_{C^\flat}$ is a φ-module over $\mathcal{O}_{\mathcal{Y}, x_{C^\flat}} = \widetilde{\mathcal{R}}^{\mathrm{int}}$.

Going in the other direction, it is convenient to switch to φ^{-1}-modules on both sides. Suppose (M, φ_M) is a φ^{-1}-module over $\widetilde{\mathcal{R}}^{\mathrm{int}}$. Since $\widetilde{\mathcal{R}}^{\mathrm{int}} = \varinjlim \widetilde{\mathcal{R}}^{\mathrm{int},r}$ and φ^{-1} acts compatibly on all rings, the category of φ^{-1}-modules over $\widetilde{\mathcal{R}}^{\mathrm{int}}$ is the colimit of the directed system of categories of φ^{-1}-modules over the $\widetilde{\mathcal{R}}^{\mathrm{int},r}$. This means that we can descend (M, φ_M^{-1}) to a pair $(M_r, \varphi_{M_r}^{-1})$ if r is small enough.

In terms of vector bundles, we have a pair $(\mathcal{E}_r, \varphi_{\mathcal{E}_r}^{-1})$, where $\mathcal{E}_r$ is a vector bundle over $\mathcal{Y}_{[0,r]}$ together with an isomorphism $\varphi_{\mathcal{E}_r}^{-1} \colon (\varphi^{-1})^* \mathcal{E}_r \cong \mathcal{E}_r$. But then $(\varphi^{-1})^* \mathcal{E}_r$ actually defines a φ^{-1}-module over $\mathcal{Y}_{[0,pr]}$, and by continuing, one produces a unique φ^{-1}-module over all of $\mathcal{Y}_{[0,\infty)}$. $\square$

12.4 SHTUKAS WITH ONE LEG, AND B_{dR}-MODULES

The relevance to the study of shtukas with legs comes through the following corollary.

Corollary 12.4.1. Let $(\mathcal{E}, \varphi_{\mathcal{E}})$ be a shtuka over $C^\flat$ with a leg at $\varphi^{-1}(x_C)$. The localization of $\mathcal{E}$ at $x_{C^\flat}$ is a φ-module over $\widetilde{\mathcal{R}}^{\mathrm{int}}$ which by the previous proposition corresponds to a shtuka $(\mathcal{E}', \varphi_{\mathcal{E}'})$ with no legs. There is a unique φ^{-1}-equivariant isomorphism

$$\iota_{\mathcal{E}} : \mathcal{E}|_{\mathcal{Y}_{[0,\infty)}\backslash\bigcup_{n\geq 0}\varphi^n(x_C)} \cong \mathcal{E}'|_{\mathcal{Y}_{[0,\infty)}\backslash\bigcup_{n\geq 0}\varphi^n(x_C)}$$

that induces the given identification at the localization at $x_{C^\flat}$. Moreover, $\iota_{\mathcal{E}}$ is meromorphic along all $\varphi^n(x_C)$.

Remark 12.4.2. There is an obvious version of this corollary for shtukas with any number of legs in characteristic 0, removing all positive φ-translates of all legs. We note that it is not a typo that the isomorphism extends over the leg. Indeed, regarding $\varphi_{\mathcal{E}}$ as a φ-semilinear map, it induces an isomorphism between the localizations of $\mathcal{E}$ at $\varphi^n(x_C)$ and $\varphi^{n-1}(x_C)$, as long as $n \neq 0$. In particular, the localization of $\mathcal{E}$ at $\varphi^{-1}(x_C)$ is identified with all the localizations at $\varphi^{-n}(x_C)$ for $n > 0$.

Proof. We have a φ^{-1}-equivariant isomorphism

$$\mathcal{E}|_{\mathcal{Y}_{[0,r]}} \cong \mathcal{E}'|_{\mathcal{Y}_{[0,r]}}$$

if r is small enough, as we have such an isomorphism at the local ring. On the other hand, if $\mathcal{Y}' = \mathcal{Y}\backslash\bigcup_{n\geq 0}\varphi^n(x_C)$, then there is a natural map $\varphi^{-1} : \mathcal{Y}' \to \mathcal{Y}'$, and isomorphisms $\varphi_{\mathcal{E}}^{-1} : (\varphi^{-1})^*\mathcal{E}|_{\mathcal{Y}'} \cong \mathcal{E}|_{\mathcal{Y}'}$, $\varphi_{\mathcal{E}'}^{-1} : (\varphi^{-1})^*\mathcal{E}'|_{\mathcal{Y}'} \cong \mathcal{E}'|_{\mathcal{Y}'}$. As any quasicompact open subspace of $\mathcal{Y}'$ is carried into $\mathcal{Y}_{[0,r]}$ by φ^{-n} for n large enough, we can extend the isomorphism uniquely, as desired.

Moreover, as $\varphi_{\mathcal{E}}$ is meromorphic, one sees that $\iota_{\mathcal{E}}$ is meromorphic at the $\varphi^n(x_C)$. $\square$

We will now describe $\mathcal{E}$ in terms of $\mathcal{E}'$ and the modification at the points $\varphi^n(x_C)$. Our analysis will involve the completed local ring of $\mathcal{Y}$ at x_C, which is none other than the de Rham period ring B_{dR}^+.

Definition 12.4.3 (The de Rham period ring). Let $B_{dR}^+ = \widehat{\mathcal{O}}_{\mathcal{Y},x_C}$; this agrees with the ξ-adic completion of $W(\mathcal{O}_{C^\flat})[1/p]$. It is a complete discrete valuation ring with residue field C and uniformizer ξ.[4] The map $\theta : A_{\inf} \to \mathcal{O}_C$ extends to a map $B_{dR}^+ \to C$ which we continue to call θ. Let $B_{dR} = B_{dR}^+[\xi^{-1}]$ be the fraction field of B_{dR}^+.

---

[4]To prove these assertions, use that $\operatorname{Spa} C \to \mathcal{Y}$ is a closed Cartier divisor.

We may identify the nth graded piece $\xi^n B_{\mathrm{dR}}^+/\xi^{n+1} B_{\mathrm{dR}}^+$ with the Tate twist $C(n)$.

Remark 12.4.4. The automorphism φ of $\mathcal{Y}$ allows us to identify $\widehat{\mathcal{O}}_{\mathcal{Y},\varphi^n(x_C)}$ with B_{dR}^+ for any $n \in \mathbf{Z}$, and it will be convenient for us to do so. Thus if $\mathcal{E}$ is a vector bundle on an open subset of $\mathcal{Y}$ containing $\varphi^n(x_C)$, its completed stalk $\widehat{\mathcal{E}}_{\varphi^n(x_C)}$ is a B_{dR}^+-module.

Remark 12.4.5. As $\iota_{\mathcal{E}}$ is meromorphic at x_C, we get an isomorphism of completed stalks at x_C after inverting the parameter ξ. That is, we have an isomorphism $\iota_{\mathcal{E},x_C} : \widehat{\mathcal{E}}_{x_C} \otimes_{B_{\mathrm{dR}}^+} B_{\mathrm{dR}} \to \widehat{\mathcal{E}'_{x_C}} \otimes_{B_{\mathrm{dR}}^+} B_{\mathrm{dR}}$. Under the equivalence between shtukas with no legs and finite free $\mathbf{Z}_p$-modules, we have $\mathcal{E}' = T \otimes_{\mathbf{Z}_p} \mathcal{O}_{\mathcal{Y}}$ for some finite free $\mathbf{Z}_p$-module T. Then $\widehat{\mathcal{E}'_{x_C}} \otimes_{B_{\mathrm{dR}}^+} B_{\mathrm{dR}} = T \otimes_{\mathbf{Z}_p} B_{\mathrm{dR}}$. In summary, we get a B_{dR}^+-lattice

$$\Xi = \iota_{\mathcal{E},x_C}(\widehat{\mathcal{E}}_{x_C}) \subset T \otimes_{\mathbf{Z}_p} B_{\mathrm{dR}}$$

encoding the behavior of $\iota_{\mathcal{E}}$ near x_C. This gives a functor from shtukas over $C^\flat$ with one leg at $\varphi^{-1}(x_C)$, towards pairs (T, Ξ) where T is a finite free $\mathbf{Z}_p$-module and $\Xi \subset T \times_{\mathbf{Z}_p} B_{\mathrm{dR}}$ is a B_{dR}^+-lattice.

Proposition 12.4.6. The above functor defines an equivalence between shtukas over $\mathrm{Spa}\, C^\flat$ with one leg at $\varphi^{-1}(x_C)$ and the category of pairs (T, Ξ), where T is a finite free $\mathbf{Z}_p$-module and $\Xi \subset T \otimes_{\mathbf{Z}_p} B_{\mathrm{dR}}$ is a B_{dR}^+-lattice.

Remark 12.4.7. Coming back to p-divisible groups, we note that there is a fully faithful functor $\{(T, W)\} \to \{(T, \Xi)\}$ where Ξ is the unique B_{dR}^+-lattice with

$$T \otimes_{\mathbf{Z}_p} B_{\mathrm{dR}}^+ \subset \Xi \subset \xi^{-1}(T \otimes_{\mathbf{Z}_p} B_{\mathrm{dR}}^+)$$

whose image in

$$\xi^{-1}(T \otimes_{\mathbf{Z}_p} B_{\mathrm{dR}}^+)/(T \otimes_{\mathbf{Z}_p} B_{\mathrm{dR}}^+) \cong T \otimes_{\mathbf{Z}_p} C(-1)$$

is given by $W(-1)$. In particular, given (T, W), the theorem allows us to construct a shtuka over $\mathrm{Spa}\, C^\flat$ with one leg at $\varphi^{-1}(x_C)$.

Proof. This is an easy consequence of Corollary 12.4.1; we describe the inverse functor. Given (T, Ξ), we get the shtuka with no legs $\mathcal{E}' = T \otimes_{\mathbf{Z}_p} \mathcal{O}_{\mathcal{Y}_{[0,\infty)}}$, together with the B_{dR}^+-lattice

$$\Xi \subset T \otimes_{\mathbf{Z}_p} B_{\mathrm{dR}} = \widehat{\mathcal{E}'}_{x_C} \otimes_{B_{\mathrm{dR}}^+} B_{\mathrm{dR}} \ .$$

By the Beauville–Laszlo lemma, Lemma 5.2.9, and as x_C defines a closed Cartier divisor, we can use Ξ to define a vector bundle over $\mathcal{Y}_{[0,\infty)}$ which is isomorphic to $\mathcal{E}'$ away from x_C, and whose completed stalk at x_C is given by Ξ. We can repeat this at $\varphi^n(x_C)$ for all $n \geq 1$, defining a new vector bundle $\mathcal{E}$ on $\mathcal{Y}_{[0,\infty)}$

with a meromorphic isomorphism

$$\iota_{\mathcal{E}} : \mathcal{E}|_{\mathcal{Y}'} \cong \mathcal{E}'|_{\mathcal{Y}'} \,,$$

where $\mathcal{Y}' = \mathcal{Y}_{[0,\infty)} \setminus \bigcup_{n \geq 0} \varphi^n(x_C)$. This meromorphic isomorphism lets us define a meromorphic map $\varphi_{\mathcal{E}} : \varphi^* \mathcal{E}|_{Y'} \to \mathcal{E}|_{\mathcal{Y}'}$ compatibly with $\varphi_{\mathcal{E}'}$. But as we have used the same lattice Ξ at all $\varphi^n(x_C)$, we see that $\varphi_{\mathcal{E}}$ extends over $\varphi^n(x_C)$ for $n > 0$. In other words, $(\mathcal{E}, \varphi_{\mathcal{E}})$ is a shtuka with one leg at $\varphi^{-1}(x_C)$. $\qquad \square$

Lecture 13

Shtukas with one leg II

Today we discuss Step 2 of the plan laid out in Lecture 12. We will show that a shtuka over $\operatorname{Spa} C^\flat$, a priori defined over $\mathcal{Y}_{[0,\infty)} = \operatorname{Spa} A_{\inf} \setminus \{x_k, x_L\}$, actually extends to $\mathcal{Y} = \operatorname{Spa} A_{\inf} \setminus \{x_k\}$. In doing so we will encounter the theory of φ-modules over the Robba ring, due to Kedlaya, [Ked04]. These are in correspondence with vector bundles over the *Fargues-Fontaine curve*, [FF18].

13.1 $\mathcal{Y}$ IS AN ADIC SPACE

As in the previous lecture, let $C/\mathbf{Q}_p$ be an algebraically closed nonarchimedean field, with tilt $C^\flat/\mathbf{F}_p$ and residue field k. Let $A_{\inf} = W(\mathcal{O}_{C^\flat})$, with its $(p, [p^\flat])$-adic topology. We had set $\mathcal{Y} = \mathcal{Y}_{[0,\infty]} = (\operatorname{Spa} A_{\inf}) \setminus \{x_k\}$, an analytic pre-adic space.

Proposition 13.1.1. The space $\mathcal{Y}$ is an adic space.

Proof. From Proposition 11.2.1 applied to $S = \operatorname{Spa} C^\flat$ we know that $S \dot\times \operatorname{Spa} \mathbf{Z}_p = \mathcal{Y}_{[0,\infty)}$ is an adic space, by exhibiting a covering by rational subsets $\operatorname{Spa}(R, R^+)$, where R is sousperfectoid. We apply a similar strategy to the rational subsets $\mathcal{Y}_{[r,\infty]}$ for $r > 0$.

For $r > 0$ rational we have $\mathcal{Y}_{[r,\infty]} = \operatorname{Spa}(R_r, R_r^+)$, where R_r is the ring $W(\mathcal{O}_{C^\flat})\langle (p^\flat)^r/p\rangle[1/p]$. Here $W(\mathcal{O}_{C^\flat})\langle (p^\flat)^r/p\rangle$ is the completion of the ring $W(\mathcal{O}_{C^\flat})[(p^\flat)^r]/p]$ with respect to the $([p^\flat], p)$-adic topology, but since p divides $[(p^\flat)^r]$ in this ring, the topology is just p-adic.

Let $A'_{\inf}$ be the ring $W(\mathcal{O}_{C^\flat})$ equipped with the p-adic topology (rather than the $(p, [p^\flat])$-adic topology, as we have defined $A_{\inf}$). By the observation above, the morphism of adic spaces $\operatorname{Spa} A_{\inf} \to \operatorname{Spa} A'_{\inf}$ induces an isomorphism between $\mathcal{Y}_{[r,\infty]}$ and the rational subset of $\operatorname{Spa} A'_{\inf}$ defined by $\left|[p^\flat](x)\right|^r \le |p(x)| \ne 0$. Therefore to prove the proposition it is enough to show that the rational subset $\{|p(x)| \ne 0\}$ of $\operatorname{Spa} A'_{\inf}$ is an adic space. This rational subset is $\operatorname{Spa}(A'_{\inf}[1/p], A'_{\inf})$.

We claim $A'_{\inf}[1/p]$ is sousperfectoid. Indeed, if $R = A'_{\inf}[1/p]\widehat{\otimes}_{\mathbf{Z}_p}\mathbf{Z}_p[p^{1/p^\infty}]$, then R is a Tate ring with pseudo-uniformizer $p^{1/p}$. Its subring of powerbounded elements is $R^\circ = A'_{\inf} \widehat{\otimes}_{\mathbf{Z}_p} \mathbf{Z}_p[p^{1/p^\infty}]$. Observe that Φ is surjective on $R^\circ/p = \mathcal{O}_{C^\flat} \otimes_{\mathbf{F}_p} \mathbf{F}_p[x^{1/p^\infty}]/x = \mathcal{O}_{C^\flat}[x^{1/p^\infty}]/x$. Thus R is perfectoid. By Proposition 6.3.4, $(A'_{\inf}[1/p], A'_{\inf})$ is sheafy. $\qquad\square$

Remark 13.1.2. The same proof shows that $\operatorname{Spa} W(R)[1/p]$ is an adic space, where R is any (discrete) perfect ring and $W(R)$ has the p-adic topology.

Let us also mention here the following recent result of Kedlaya; the result does not extend to all of $\mathcal{Y}$, as the local ring at x_L is highly nonnoetherian.

Theorem 13.1.3 ([Ked16, Theorem 4.10]). The adic space $\mathcal{Y}_{[0,\infty)}$ is strongly noetherian.

13.2   THE EXTENSION OF SHTUKAS OVER X_L

The main theorem of this lecture concerns the extension of φ-modules from $\mathcal{Y}_{[r,\infty)}$ to $\mathcal{Y}_{[r,\infty]}$, where $0 \le r < \infty$.

Theorem 13.2.1 ([FF18, Théorème 11.1.9, Corollary 11.1.13]). For $0 \le r < \infty$, the restriction functor from φ-modules over $\mathcal{Y}_{[r,\infty]}$ to φ-modules over $\mathcal{Y}_{[r,\infty)}$ is an equivalence.

Remark 13.2.2. In particular, suppose $(\mathcal{E}, \varphi_\mathcal{E})$ is a shtuka over $\operatorname{Spa} C^\flat$ with legs $x_1, \ldots, x_n$. Thus $\mathcal{E}$ is a vector bundle over $\mathcal{Y}_{[0,\infty)}$ and $\varphi_E \colon \varphi^*\mathcal{E} \dashrightarrow \mathcal{E}$ is an isomorphism away from the x_i. Suppose $r > 0$ is greater than $\kappa(x_i)$ for $i = 1, \ldots, n$, so that $\mathcal{E}|_{\mathcal{Y}_{[r,\infty)}}$ is a φ-module over $\mathcal{Y}_{[r,\infty)}$. By the theorem, $\mathcal{E}|_{\mathcal{Y}_{[r,\infty)}}$ extends uniquely to a φ-module over $\mathcal{Y}_{[r,\infty]}$. This can be glued to the given shtuka to obtain a vector bundle $\widehat{\mathcal{E}}$ on $\mathcal{Y}$ together with an isomorphism $\varphi_{\widehat{\mathcal{E}}} \colon \varphi^*\widehat{\mathcal{E}} \to \widehat{\mathcal{E}}$ on $\mathcal{Y} \backslash \bigcup_i \Gamma_{x_i}$.

We only offer some ideas of the proof below.

13.3   FULL FAITHFULNESS

We now sketch a proof that the functor described in Theorem 13.2.1 is fully faithful. This part is more general, and works if C is any perfectoid field (not necessarily algebraically closed).

Suppose I is an interval of the form $[r, \infty)$ or $[r, \infty]$ with $r > 0$.

Lemma 13.3.1. Let $r' > r$, and let $I' = I \cap [r', \infty]$. The restriction functor from φ-modules over $\mathcal{Y}_I$ to φ-modules over $\mathcal{Y}_{I'}$ is an equivalence.

Proof. The inverse functor is given by pullback under φ^{-n} for n large enough.

□

For full faithfulness of the restriction functor in Theorem 13.2.1, it suffices to prove the following proposition about global sections, by applying it to an appropriate internal Hom.

Proposition 13.3.2 ([FF18, Proposition 4.1.3, Théorème 11.1.12]). Let $r > 0$, and let $\mathcal{E}$ be a φ-module over $\mathcal{Y}_{[r,\infty]}$. Restriction induces an isomorphism

$$H^0(\mathcal{Y}_{[r,\infty]}, \mathcal{E})^{\varphi=1} \xrightarrow{\sim} H^0(\mathcal{Y}_{[r,\infty)}, \mathcal{E})^{\varphi=1}.$$

Proof. By [FF18, Théorème 11.1.12], there is a φ-equivariant isomorphism $\mathcal{E} \cong M \otimes_L \mathcal{O}_{\mathcal{Y}_{[r,\infty]}}$ for some φ-module M over $L = W(k)[\frac{1}{p}]$; here, we fixed a section $k \to \mathcal{O}_{C^\flat}$ of the projection.[1] By the Dieudonné–Manin classification (cf. Remark 13.4.2 below), one can assume that M is a simple isocrystal of slope $\lambda = \frac{d}{h}$, $h \geq 1$. In that case, we have to prove that

$$H^0(Y_{[r,\infty]}, \mathcal{O}_{Y_{[r,\infty]}})^{\varphi^h = p^d} = H^0(Y_{[r,\infty)}, \mathcal{O}_{Y_{[r,\infty)}})^{\varphi^h = p^d} .$$

Applying Lemma 13.3.1, this is equivalent to

$$H^0(Y_{(0,\infty]}, \mathcal{O}_{Y_{(0,\infty]}})^{\varphi^h = p^d} = H^0(Y_{(0,\infty)}, \mathcal{O}_{Y_{(0,\infty)}})^{\varphi^h = p^d} .$$

Following [FF18], we put $B^+ = H^0(Y_{(0,\infty]}, \mathcal{O}_{Y_{(0,\infty]}})$ and $B = H^0(Y_{(0,\infty)}, \mathcal{O}_{Y_{(0,\infty)}})$, so we have to prove that $(B^+)^{\varphi^h = p^d} = B^{\varphi^h = p^d}$. This is exactly [FF18, Propositions 4.1.1, 4.1.2, 4.1.3]. Let us sketch the argument. Take any $f \in B^{\varphi^h = p^d}$; we have to show that it extends over x_L. Consider the *Newton polygon* of f: if

$$f = \sum_{i \in \mathbf{Z}} [a_i] p^i, \ a_i \in C^\flat,$$

let $\mathrm{Newt}(f)$ be the convex hull of the polygon in $\mathbf{R}^2$ joining the points $(\mathrm{val}(a_i), i)$ for $i \in \mathbf{Z}$. Here val is a valuation on $C^\flat$, written additively. Then $\mathrm{Newt}(f)$ is independent of the expression of f as a series (which may not be unique). We have that f extends to $\mathcal{Y}_{[r,\infty]}$ if and only if $\mathrm{Newt}(f)$ lies on the right of the y-axis; cf. [FF18, Proposition 1.10.7]. Now

$$\mathrm{Newt}(\varphi^h(f)) = \mathrm{Newt}(p^d f) = d + \mathrm{Newt}(f).$$

But $\mathrm{Newt}(\varphi^h(f))$ is $\mathrm{Newt}(f)$ but scaled by p^h in the val-axis. If $\mathrm{Newt}(f)$ goes to the left of the y-axis, then $\mathrm{Newt}(\varphi^h(f))$ would go further to the left, a contradiction. □

---

[1]One could also carry out the following arguments without the choice of such a trivialization, but it simplifies the argument.

13.4   ESSENTIAL SURJECTIVITY

For essential surjectivity in Theorem 13.2.1, the strategy is to classify all φ-modules over $\mathcal{Y}_{[r,\infty)}$ and show by inspection that each one extends over $\mathcal{Y}_{[r,\infty]}$.

Recall that $L = W(k)[1/p]$. If we choose an embedding $k \hookrightarrow \mathcal{O}_{C^\flat}$ which reduces to the identity modulo the maximal ideal of $\mathcal{O}_{C^\flat}$, we obtain an embedding $L \hookrightarrow A_{\inf}[1/p]$.

Theorem 13.4.1 ([Ked04]). Let $(\mathcal{E}, \varphi_{\mathcal{E}})$ be a φ-module over $\mathcal{Y}_{[r,\infty)}$. Then there exists a φ-module (M, φ_M) over L such that $(\mathcal{E}, \varphi_E) \cong (M, \varphi_M) \otimes_L \mathcal{O}_{\mathcal{Y}_{[r,\infty)}}$. (Here we have fixed an embedding $L \to A_{\inf}[1/p]$.)

Remark 13.4.2.

1. As a result, the φ-module $\widehat{\mathcal{E}} = M \otimes_L \mathcal{O}_{\mathcal{Y}_{[r,\infty]}}$ extends $\mathcal{E}$ to $\mathcal{Y}_{[r,\infty]}$.
2. φ-modules over L are by definition the same as isocrystals over k . The category of isocrystals over k admits a *Dieudonné-Manin classification*: it is semisimple, with simple objects M_λ classified by $\lambda \in \mathbf{Q}$. For a rational number $\lambda = d/h$ written in lowest terms with $h > 0$, the rank of M_λ is h, and φ_{M_λ} can be expressed in matrix form as

$$\varphi_{M_\lambda} = \begin{pmatrix} 0 & 1 & & & \\ & 0 & 1 & & \\ & & \ddots & \ddots & \\ & & & 0 & 1 \\ p^d & & & & 0 \end{pmatrix}.$$

   In fact, the formulation of Kedlaya's theorem is slightly different; let us explain the translation, which is also explained in detail in [FF18, Section 11.2], especially [FF18, Proposition 11.2.20, Corollaire 11.2.22]. The usual "Frobenius pullback" trick shows that φ-modules over $\mathcal{Y}_{[r,\infty)}$ for $r > 0$ can be extended arbitrarily close to 0 in the sense that the restriction map

$$\{\varphi\text{-modules over } \mathcal{Y}_{(0,\infty)}\} \xrightarrow{\sim} \{\varphi\text{-modules over } \mathcal{Y}_{[r,\infty)}\}$$

is an equivalence; cf. Lemma 13.3.1.

Definition 13.4.3 (The extended Robba rings, [KL15, Defn. 4.2.2]). Let $\widetilde{\mathcal{R}}^r = H^0(\mathcal{Y}_{(0,r]}, \mathcal{O}_\mathcal{Y})$, and let $\widetilde{\mathcal{R}} = \varinjlim \widetilde{\mathcal{R}}^r$.

Thus $\widetilde{\mathcal{R}}$ is the ring of functions defined on some punctured disc of small (and unspecified) radius around $x_{C^\flat}$. Note that φ induces an automorphism of $\widetilde{\mathcal{R}}$ (but not of any $\widetilde{\mathcal{R}}^r$). A similar Frobenius pullback trick shows that the category of φ-modules over $\mathcal{Y}_{(0,\infty)}$ is equivalent to the category of φ-modules over $\widetilde{\mathcal{R}}$. Kedlaya's theorem is usually stated in terms of φ-modules over $\widetilde{\mathcal{R}}$.

13.5   THE FARGUES-FONTAINE CURVE

Another perspective on these objects is offered by the Fargues–Fontaine curve. As φ acts properly discontinuously on $\mathcal{Y}_{(0,\infty)}$ (as follows from $\kappa \circ \varphi = p\kappa$), it makes sense to form the quotient.

Definition 13.5.1. The *adic Fargues-Fontaine curve* is the quotient $\mathcal{X}_{\mathrm{FF}} = \mathcal{Y}_{(0,\infty)}/\varphi^{\mathbf{Z}}$.

Now φ-modules over $\mathcal{Y}_{(0,\infty)}$ are visibly the same as vector bundles over $\mathcal{X}_{FF}$, so we see that vector bundles on $\mathcal{X}_{\mathrm{FF}}$ are also equivalent to the categories considered in the previous section.

Fargues–Fontaine also defined a scheme version of their curve. The curve $\mathcal{X}_{FF}$ comes equipped with a natural line bundle $\mathcal{O}(1)$, corresponding to the φ-module on $\mathcal{Y}_{(0,\infty)}$ whose underlying line bundle is trivial and for which $\varphi_{\mathcal{O}(1)}$ is $p^{-1}\varphi$. Let $\mathcal{O}(n) = \mathcal{O}(1)^{\otimes n}$, and let

$$P = \bigoplus_{n\geq 0} H^0(\mathcal{X}_{FF}, \mathcal{O}(n)),$$

a graded ring. The nth graded piece is $H^0(\mathcal{Y}_{(0,\infty)}, \mathcal{O}_{\mathcal{Y}})^{\varphi=p^n} = B^{\varphi=p^n}$, where $B = H^0(\mathcal{Y}_{(0,\infty)}, \mathcal{O}_{\mathcal{Y}})$. Note that by Proposition 13.2.1 this is the same as $(B^+)^{\varphi=p^n}$, where $B^+ = H^0(\mathcal{Y}_{(0,\infty]}, \mathcal{O}_{\mathcal{Y}})$.

These spaces can be reformulated in terms of the crystalline period rings of Fontaine. Let A_{crys} be the p-adic completion of the divided power envelope of the surjection $A_{\inf} \to \mathcal{O}_C$. By a general fact about divided power envelopes for principal ideals in flat $\mathbf{Z}$-algebras, this divided power envelope is the same as the $A_{\inf}$-subalgebra of $A_{\inf}[1/p]$ generated by $\xi^n/n!$ for $n \geq 1$. Let $B^+_{\mathrm{crys}} = A_{\mathrm{crys}}[1/p]$. An element of B^+_{crys} may be written

$$\sum_{n\geq 0} a_n \frac{\xi^n}{n!},\ a_n \in W(\mathcal{O}_{C^\flat})[1/p],\ a_n \to 0\ p\text{-adically}.$$

Using the estimate $n/(p-1)$ for the p-adic valuation of $n!$, one can show that such series converge in $H^0(\mathcal{Y}_{[r,\infty]}, \mathcal{O}_{\mathcal{Y}})$ for $r = \frac{1}{p-1}$. Thus we have an embedding $B^+_{\mathrm{crys}} \subset H^0(\mathcal{Y}_{[r,\infty]}, \mathcal{O}_{\mathcal{Y}})$; it can be shown that

$$(B^+_{\mathrm{crys}})^{\varphi=p^n} = (B^+)^{\varphi=p^n} = B^{\varphi=p^n}\ ,$$

cf. [FF18, Corollaire 11.1.14].

The full crystalline period ring B_{crys} is defined by inverting the element $t \in B^+_{\mathrm{crys}}$, where $t = \log[\varepsilon]$, $\varepsilon = (1, \zeta_p, \zeta_{p^2}, \dots) \in \mathcal{O}_{C^\flat}$. We have $\varphi(t) = \log[\varepsilon^p] = \log[\varepsilon]^p = pt$, so that t is a section of $\mathcal{O}(1)$. Also, $\theta(t) = 0$, so that t has a zero at $x_C \in \mathcal{X}_{\mathrm{FF}}$; in fact this is the only zero.

Definition 13.5.2 (The schematic Fargues-Fontaine curve). Let $X_{\mathrm{FF}} = \operatorname{Proj} P$.

The map $\theta \colon B \to C$ determines a distinguished point $x_C \in X_{\mathrm{FF}}$ with residue field C.

Theorem 13.5.3 ([FF18]).

1. X_{FF} is a regular noetherian scheme of Krull dimension 1 which is locally the spectrum of a principal ideal domain.
2. In fact, $X_{\mathrm{FF}} \setminus \{x_C\}$ is an affine scheme $\operatorname{Spec} B_e$, where $B_e = B_{\mathrm{crys}}^{\varphi=1}$ is a principal ideal domain.
3. We have $\widehat{\mathcal{O}}_{X_{\mathrm{FF}},x_C} = B_{\mathrm{dR}}^+$. Thus vector bundles over X_{FF} correspond to B-pairs (M_e, M_{dR}^+) in the sense of Berger, [Ber08], consisting of finite projective modules over B_e and B_{dR}^+ respectively, with an isomorphism over B_{dR}.
4. The set $|X_{\mathrm{FF}}|$ of closed points of X_{FF} is identified with the set of characteristic 0 untilts of $C^\flat$ modulo Frobenius. This identification sends $x \in |X_{\mathrm{FF}}|$ to its residue field. In particular, residue fields of X_{FF} at closed points are algebraically closed nonarchimedean fields C' such that $(C')^\flat \cong C^\flat$.

Moreover, [FF18] shows that there is a Dieudonné-Manin classification for vector bundles over X_{FF}, just as in Kedlaya's theory.

Theorem 13.5.4 ([FF18]). There is a faithful and essentially surjective functor from isocrystals over k to vector bundles over X_{FF}, which sends M to the vector bundle associated to the graded P-module $\bigoplus_{n \geq 0} (B_{\mathrm{crys}}^+ \otimes M)^{\varphi = p^n}$. This functor induces a bijection on the level of isomorphism classes.

For $\lambda \in \mathbf{Q}$, we write $\mathcal{O}(\lambda)$ for the image of the simple isocrystal $M_{-\lambda}$ under the functor described in Theorem 13.5.4. If $\lambda = d/h$ is in lowest terms with $h > 0$, then $\mathcal{O}(\lambda)$ has rank h and degree d. If $\lambda = n \in \mathbf{Z}$, this definition is consistent with how we have previously defined the line bundle $\mathcal{O}(n)$. In general, if $\mathcal{E}$ is a vector bundle corresponding to $\bigoplus_i M_{\lambda_i}$ under Theorem 13.5.4, the rational numbers λ_i are called the *slopes* of $\mathcal{E}$. A vector bundle with only one slope λ is called *semistable of slope 0*.

Corollary 13.5.5. The category of vector bundles on X_{FF} that are semistable of slope 0 is equivalent to the category of finite-dimensional $\mathbf{Q}_p$-vector spaces.

Proof. The functors are given by $\mathcal{E} \mapsto H^0(X_{\mathrm{FF}}, \mathcal{E})$ and $V \mapsto V \otimes_{\mathbf{Q}_p} \mathcal{O}_{X_{\mathrm{FF}}}$, respectively. To see that they are inverse, one only needs to see that all vector bundles on X_{FF} that are semistable of slope 0 are trivial, which follows from the classification of vector bundles, and that $H^0(X_{\mathrm{FF}}, \mathcal{O}_{X_{\mathrm{FF}}}) = \mathbf{Q}_p$. $\square$

The global sections of $\mathcal{O}(\lambda)$ are

$$H^0(X_{\mathrm{FF}}, \mathcal{O}(\lambda)) = \begin{cases} \text{big}, & \lambda > 0, \\ \mathbf{Q}_p, & \lambda = 0, \\ 0, & \lambda < 0. \end{cases}$$

In the "big" case, the space of global sections is a "finite-dimensional Banach Space" (with a capital S) in the sense of Colmez[2] [Col02], which is a mixture of finite-dimensional $\mathbf{Q}_p$-vector spaces and finite-dimensional C-vector spaces; cf. Lecture 15. For example if $\lambda = 1$, we have an exact sequence

$$0 \to \mathbf{Q}_p t \to H^0(X_{\mathrm{FF}}, \mathcal{O}(1)) \to C \to 0.$$

Furthermore, $H^1(X_{\mathrm{FF}}, \mathcal{O}(\lambda)) = 0$ if $\lambda \geq 0$, and $H^1(X_{\mathrm{FF}}, \mathcal{O}(-1))$ is isomorphic to the quotient $C/\mathbf{Q}_p$.

Theorem 13.5.6 ([KL15, Theorem 8.7.7], "GAGA for the curve"). Vector bundles over X_{FF} and vector bundles over $\mathcal{X}_{\mathrm{FF}}$ are equivalent.

There is a map of locally ringed spaces $\mathcal{X}_{\mathrm{FF}} \to X_{\mathrm{FF}}$, so one really does have a functor from vector bundles over X_{FF} to vector bundles over $\mathcal{X}_{\mathrm{FF}}$. The following theorem will be critical to the proof of the local Drinfeld lemma.

Theorem 13.5.7 ([FF18, Théorème 8.6.1]). $X_{\mathrm{FF}, \overline{\mathbf{Q}}_p}$ is simply connected; i.e. any finite étale cover is split. Equivalently, the categories of finite étale covers of X_{FF} and of $\mathbf{Q}_p$ are equivalent. Moreover, the categories of finite étale covers of X_{FF} and $\mathcal{X}_{\mathrm{FF}}$ are equivalent.

Proof. The following argument due to Fargues-Fontaine also gives a proof that $\mathbf{P}^1$ is simply connected over an algebraically closed field, which avoids using the Riemann-Hurwitz formula.

We need to show that $A \mapsto \mathcal{O}_{X_{\mathrm{FF}}} \otimes_{\mathbf{Q}_p} A$ is an equivalence from finite étale $\mathbf{Q}_p$-algebras to finite étale $\mathcal{O}_{X_{\mathrm{FF}}}$-algebras. Suppose $\mathcal{E}$ is a finite étale $\mathcal{O}_{X_{\mathrm{FF}}}$-algebra. By Theorem 13.5.4, the underlying vector bundle of $\mathcal{E}$ is isomorphic to $\bigoplus_i \mathcal{O}(\lambda_i)$ for some $\lambda_1, \ldots, \lambda_s \in \mathbf{Q}$. The étaleness provides a perfect trace pairing on $\mathcal{E}$, hence a self-duality of the underlying vector bundle, which implies that $\sum_i \lambda_i = 0$. Let $\lambda = \max \lambda_i$, so that $\lambda \geq 0$.

Assume $\lambda > 0$. The multiplication map $\mathcal{E} \otimes \mathcal{E} \to \mathcal{E}$ restricts to a map $\mathcal{O}(\lambda) \otimes \mathcal{O}(\lambda) \to \mathcal{E}$, which gives a global section of $\mathcal{E} \otimes \mathcal{O}(-\lambda)^{\otimes 2}$. But the latter has negative slopes, implying that $H^0(X_{\mathrm{FF}}, \mathcal{E} \otimes \mathcal{O}(-2\lambda)) = 0$. It follows that $f^2 = 0$ for every $f \in H^0(X_{\mathrm{FF}}, \mathcal{O}(\lambda)) \subset H^0(X_{\mathrm{FF}}, \mathcal{E})$. But since $\mathcal{E}$ is étale over $\mathcal{O}_{X_{\mathrm{FF}}}$, its ring of global sections is reduced, so in fact $H^0(X_{\mathrm{FF}}, \mathcal{O}(\lambda)) = 0$ and $\lambda < 0$, a contradiction. Therefore $\lambda = 0$, and so $\lambda_i = 0$ for all i, meaning that $\mathcal{E}$ is trivial. But the category of trivial vector bundles on X_{FF} is equivalent to the category of finite dimensional vector spaces, given by $\mathcal{E} \mapsto H^0(X_{\mathrm{FF}}, \mathcal{E})$. Thus $H^0(X_{\mathrm{FF}}, \mathcal{E})$ is a finite étale $\mathbf{Q}_p$-algebra, which gives us the functor in the other direction.

The equivalence of finite étale covers of X_{FF} and $\mathcal{X}_{\mathrm{FF}}$ follows from Theorem 13.5.6 by encoding both as sheaves of finite étale algebras (and interpreting finite étale as locally free with perfect trace pairing). □

---

[2]These are now known as Banach-Colmez spaces.

Lecture 14

Shtukas with one leg III

We continue in the usual setup: Let $C/\mathbf{Q}_p$ be an algebraically closed nonarchimedean field, $C^\flat$ its tilt, k its residue field, $L = W(k)[1/p]$, and $A_{\inf} = W(\mathcal{O}_{C^\flat})$. There is the map $\theta: A_{\inf} \to \mathcal{O}_C$ whose kernel is generated by $\xi = p - [p^\flat]$. We have the associated adic space $\operatorname{Spa} A_{\inf}$ containing its analytic locus $\mathcal{Y} = \operatorname{Spa} A_{\inf} \backslash \{x_k\}$. Moreover, we have the complete discrete valuation ring B_{dR}^+ and its fraction field B_{dR}, where B_{dR}^+ is the ξ-adic completion of $W(\mathcal{O}_{C^\flat})[1/p]$. Finally, we have the Fargues-Fontaine curve X_{FF} defined in Definition 13.5.2, with a distinguished point ∞ corresponding to the untilt C of $C^\flat$.

14.1 FARGUES' THEOREM

Our goal today is to complete the proof of the following theorem of Fargues.

Theorem 14.1.1 (Fargues). The following categories are equivalent:

1. Shtukas over $\operatorname{Spa} C^\flat$ with one leg at $\varphi^{-1}(x_C)$, i.e., vector bundles $\mathcal{E}$ on $\mathcal{Y}_{[0,\infty)}$ together with an isomorphism $\varphi_{\mathcal{E}} : (\varphi^* \mathcal{E})|_{\mathcal{Y}_{[0,\infty)} \backslash \varphi^{-1}(x_C)} \cong \mathcal{E}|_{\mathcal{Y}_{[0,\infty)} \backslash \varphi^{-1}(x_C)}$ that is meromorphic at $\varphi^{-1}(x_C)$.
2. Pairs (T, Ξ), where T is a finite free $\mathbf{Z}_p$-module, and $\Xi \subset T \otimes_{\mathbf{Z}_p} B_{\mathrm{dR}}$ is a B_{dR}^+-lattice.
3. Quadruples $(\mathcal{F}, \mathcal{F}', \beta, T)$, where $\mathcal{F}$ and $\mathcal{F}'$ are vector bundles on the Fargues-Fontaine curve X_{FF} such that $\mathcal{F}$ is trivial, $\beta: \mathcal{F}|_{X_{\mathrm{FF}} \backslash \{\infty\}} \xrightarrow{\sim} \mathcal{F}'|_{X_{\mathrm{FF}} \backslash \{\infty\}}$ is an isomorphism, and $T \subset H^0(X_{\mathrm{FF}}, \mathcal{F})$ is a $\mathbf{Z}_p$-lattice.
4. Vector bundles $\widetilde{\mathcal{E}}$ on $\mathcal{Y}$ together with an isomorphism $\varphi_{\widetilde{\mathcal{E}}} : (\varphi^* \widetilde{\mathcal{E}})|_{\mathcal{Y} \backslash \varphi^{-1}(x_C)} \cong \widetilde{\mathcal{E}}|_{\mathcal{Y} \backslash \varphi^{-1}(x_C)}$.
5. Breuil-Kisin-Fargues modules over $A_{\inf}$, i.e., finite free $A_{\inf}$-modules M together with an isomorphism $\varphi_M : (\varphi^* M)[\frac{1}{\varphi(\xi)}] \cong M[\frac{1}{\varphi(\xi)}]$.

Moreover, restricting to the full subcategories given (in case (2)) by the condition

$$T \otimes_{\mathbf{Z}_p} B_{\mathrm{dR}}^+ \subset \Xi \subset \xi^{-1}(T \otimes_{\mathbf{Z}_p} B_{\mathrm{dR}}^+),$$

the corresponding full subcategories are equivalent to the category of p-divisible groups over $\mathcal{O}_C$.

Proof. The equivalence between (1) and (2) is Theorem 12.4.6. Let us explain the equivalence between (2) and (3). By Corollary 13.5.5, the datum of a trivial vector bundle $\mathcal{F}$ is equivalent to the datum of a finite-dimensional $\mathbf{Q}_p$-vector space; together with T, this data is equivalent to a finite free $\mathbf{Z}_p$-module T. Now by the Beauville–Laszlo lemma, Lemma 5.2.9, the datum of $\mathcal{F}'$ and β is equivalent to the datum of a B_{dR}^+-lattice in

$$\widehat{\mathcal{F}}_\infty \otimes_{B_{\mathrm{dR}}^+} B_{\mathrm{dR}} = T \otimes_{\mathbf{Z}_p} B_{\mathrm{dR}} .$$

This shows that (2) and (3) are equivalent.

Let us also briefly discuss the relation between (1) and (3). Suppose $(\mathcal{E}, \varphi_{\mathcal{E}})$ is a shtuka over $\mathrm{Spa}\, C^\flat$ with one leg at $\varphi^{-1}(x_C)$. This means that $\mathcal{E}$ is a vector bundle on $\mathcal{Y}_{[0,\infty)}$ and $\varphi_{\mathcal{E}} \colon \varphi^* \mathcal{E} \to \mathcal{E}$ is an isomorphism away from $\varphi^{-1}(x_C)$. The vector bundles $\mathcal{F}$ and $\mathcal{F}'$ on $\mathcal{X}_{\mathrm{FF}} = \mathcal{Y}_{(0,\infty)}/\varphi^{\mathbf{Z}}$ come from descending $(\mathcal{E}, \varphi_{\mathcal{E}})$ "on either side" of x_C, respectively, as we now explain. Namely, if $r < 1$ resp. $r > 1$, then $\mathcal{E}|_{\mathcal{Y}_{(0,r]}}$ resp. $\mathcal{E}_{\mathcal{Y}_{[r,\infty)}}$ is a vector bundle on which φ^{-1} resp. φ becomes an isomorphism. As $\mathcal{Y}_{(0,r]} \to \mathcal{X}_{\mathrm{FF}}$ resp. $Y_{[r,\infty)} \to \mathcal{X}_{\mathrm{FF}}$ is surjective, we can thus descend it to a vector bundle $\mathcal{F}$ resp. $\mathcal{F}'$ on $\mathcal{X}_{\mathrm{FF}}$, which by Theorem 13.5.6 is equivalent to a vector bundle on X_{FF}. As $\mathcal{E}$ extends to $\mathcal{Y}_{[0,r]}$ without a leg in characteristic p, Proposition 12.3.4 implies that $\mathcal{F}$ is trivial, and comes with a distinguished $\mathbf{Z}_p$-lattice. Moreover, that $\mathcal{F}$ and $\mathcal{F}'$ both come from $\mathcal{E}$ is encoded in the modification β.

In the last lecture, we proved the equivalence between (1) and (4). By Theorem 14.2.1 below, categories (4) and (5) are equivalent. The final sentence follows from description (2) and Theorem 12.1.5. $\qquad\square$

14.2   EXTENDING VECTOR BUNDLES OVER THE CLOSED POINT OF $\mathrm{Spec}\, A_{\mathrm{inf}}$

We have seen that to complete the proof of Fargues' theorem, it remains to prove the following result, where $\mathcal{Y} = \mathrm{Spa}\, A_{\mathrm{inf}} \setminus \{x_k\}$.

Theorem 14.2.1 (Kedlaya, [Ked19b, Theorem 3.6]). There is an equivalence of categories between:

1. Finite free A_{inf}-modules, and
2. Vector bundles on $\mathcal{Y}$.

One should think of this as being an analogue of a classical result: If $(R, \mathfrak{m})$ is a 2-dimensional regular local ring, then finite free R-modules are equivalent to vector bundles on $(\mathrm{Spec}\, R) \setminus \{\mathfrak{m}\}$. In fact the proof we give for Theorem 14.2.1 works in that setup as well.

Question 14.2.2. One can ask whether Theorem 14.2.1 extends to torsors under groups other than GL_n. If $G/\mathbf{Q}_p$ is a connected reductive group with parahoric model $\mathcal{G}/\mathbf{Z}_p$, is it true that $\mathcal{G}$-torsors over A_{inf} are the same as $\mathcal{G}$-torsors on $\operatorname{Spec} A_{\mathrm{inf}} \backslash \{\mathfrak{m}\}$, and on $\mathcal{Y}$? The analogue for 2-dimensional regular local rings is treated in [CTS79, Theorem 6.13] if $\mathcal{G}$ is reductive, and in some cases when $\mathcal{G}$ is parahoric in [KP18, Section 1.4].

First we prove the algebraic version of Theorem 14.2.1.

Lemma 14.2.3. Let $\mathfrak{m} \in \operatorname{Spec} A_{\mathrm{inf}}$ be the unique closed point, and let $Y = \operatorname{Spec}(A_{\mathrm{inf}}) \backslash \{\mathfrak{m}\}$. Then $\mathcal{E} \mapsto \mathcal{E}|_Y$ is an equivalence between vector bundles on $\operatorname{Spec} A_{\mathrm{inf}}$ (that is, finite free A_{inf}-modules) and vector bundles on Y.

Proof. Let $R = A_{\mathrm{inf}}$, and let

$$
\begin{aligned}
R_1 &= R[1/p] \\
R_2 &= R[1/[p^\flat]] \\
R_{12} &= R[1/p[p^\flat]].
\end{aligned}
$$

Then Y is covered by $\operatorname{Spec} R_1$ and $\operatorname{Spec} R_2$, with overlap $\operatorname{Spec} R_{12}$. Thus the category of vector bundles on Y is equivalent to the category of triples (M_1, M_2, h), where M_i is a finite projective R_i-module for $i = 1, 2$, and $h \colon M_1 \otimes_{R_1} R_{12} \overset{\sim}{\to} M_2 \otimes_{R_2} R_{12}$ is an isomorphism of R_{12}-modules. We wish to show that the obvious functor $M \mapsto (M \otimes_R R_1, M \otimes_R R_2, h_M)$ from finite free R-modules to such triples is an equivalence.

For full faithfulness, suppose we are given finite free R-modules M and M' and a morphism of triples $(M \otimes_R R_1, M \otimes_R R_2, h_M) \to (M' \otimes_R R_1, M' \otimes_R R_2, h_{M'})$. The matrix coefficients of such a morphism lie in $R_1 \cap R_2 = R$, and thus the morphism extends uniquely to a morphism $M \to M'$.

For essential surjectivity, suppose we are given a triple (M_1, M_2, h). Using h we may identify both $M_1 \otimes_{R_1} R_{12}$ and $M_2 \otimes_{R_2} R_{12}$ with a common R_{12}-module M_{12}. Consider the map $M_1 \oplus M_2 \to M_{12}$ defined by $(x, y) \mapsto x - y$. Let M be the kernel, an R-module. For $i = 1, 2$, the projection map $\operatorname{pr}_i \colon M_1 \oplus M_2 \to M_i$ induces a map $\operatorname{pr}_i \colon M \otimes_R R_i \to M_i$, which is an isomorphism. Indeed, in geometric terms, let $j \colon \operatorname{Spec} R \backslash \{\mathfrak{m}\} \to \operatorname{Spec} R$ be the inclusion; then j_* preserves quasicoherent sheaves and $j^* j_* = \mathrm{id}$.

We now must show that M is a finite free R-module, given that its localizations to R_1 and R_2 are finite projective. First we present some generalities concerning finite projective modules over Tate rings. (Even though R is not Tate, its localizations R_1 and R_2 are both Tate, with pseudo-uniformizers p and $[p^\flat]$, respectively.)

Let A be a Tate ring, let $f \in A$ be a topologically nilpotent unit, and let $A_0 \subset A$ be a ring of definition containing f. Then A_0 has the f-adic topology and $A = A_0[f^{-1}]$ (Proposition 2.2.6(2)). If M is a finite projective A-module, it comes with a canonical topology. This may be defined by writing M as a direct summand of A^n and giving M the induced subspace (or equivalently, quotient space) topology. We gather a few facts:

1. If A is complete, then so is M.
2. An A_0-submodule $N \subset M$ is open if and only if $N[f^{-1}] = M$.
3. An A_0-submodule $N \subset M$ is bounded if and only if it is contained in a finitely generated A_0-submodule.
4. If A_0 is f-adically separated and complete, then an open and bounded A_0-submodule $N \subset M$ is also f-adically separated and complete.
5. Let A' be a Tate ring containing A as a topological subring. If $X \subset M \otimes_A A'$ is a bounded subset, then $X \cap M \subset M$ is also bounded.

(For the last point: suppose that f is a pseudo-uniformizer of A, and thus also of A'. Let $A_0' \subset A'$ be a ring of definition containing f; then $A_0 := A \cap A_0'$ is a ring of definition for A. Use a presentation of M as a direct summand of a free module to reduce to the case that M is free, and then to the case that $M = A$. By Proposition 2.2.6(3), boundedness of $X \subset A'$ means that $X \subset f^{-n} A_0'$ for some n, and therefore $X \cap A \subset f^{-n} A_0$ is bounded.)

Thus all open and bounded submodules of M differ by bounded f-torsion, in the sense that if $M_0, M_0' \subset M$ are open and bounded, then there exists n such that $f^n M_0 \subset M_0' \subset f^{-n} M_0$.

Now we return to the situation of the lemma. Endow R_1 with the p-adic topology making R a ring of definition. Then R_1 is Tate and $p \in R_1$ is a topologically nilpotent unit. We claim that $M \subset M_1$ is an open and bounded R_1-submodule. Since $M \otimes_R R_1 = M_1$, point (2) above (applied to $A_0 = R$ and $A = R_1$) shows that M is open. For boundedness, endow R_{12} with the p-adic topology making R_2 a ring of definition. Then $R_1 \subset R_{12}$ is a topological subring. Since $M_2 \subset M_{12} = M_2 \otimes_{R_2} R_{12}$ is bounded, we can apply (5) to conclude that $M = M_2 \cap M_1 \subset M_1$ is bounded as well.

Thus $M \subset M_1$ is open and bounded. Since R_1 is p-adically separated and complete, point (4) shows that M is p-adically complete. It is also p-torsion free, since M_1 is. It follows that in order to prove that M is finite free, it is enough to prove that M/p is finite free over $R/p = \mathcal{O}_{C^\flat}$.

We claim that the inclusion $M \hookrightarrow M_2$ induces an injection $M/p \hookrightarrow M_2/p = M_2 \otimes_{R_2} C^\flat$. Assume $m \in M$ maps to $0 \in M_2/p$. Write $m = pm_2$, with $m_2 \in M_2$. Then $m' := (m/p, m_2) \in \ker(M_1 \oplus M_2 \to M_{12}) = M$, so that $m = pm'$, giving the claim.

Thus M/p is an $\mathcal{O}_{C^\flat}$-submodule of a $C^\flat$-vector space of finite dimension d, and we want to show that it is actually free of rank d over $\mathcal{O}_{C^\flat}$. Note that if K is a discretely valued nonarchimedean field, then any open and bounded $\mathcal{O}_K$-submodule of $K^{\oplus d}$ is necessarily finite free of rank d. However, the same statement is false when K is not discretely valued: the maximal ideal $\mathfrak{m}_K$ of $\mathcal{O}_K$ is open and bounded in K, but it is not even finitely generated.

Lemma 14.2.4. Let V be a valuation ring with fraction field K and residue field k and let $\Lambda \subset K^d$ be any V-submodule. Then

$$\dim_k(\Lambda \otimes_V k) \leq d,$$

with equality if and only if $\Lambda \cong V^d$.

Proof. Let $x_1, \ldots, x_e \in \Lambda$ be elements whose reductions $\bar{x}_1, \ldots \bar{x}_e \in \Lambda \otimes_V k$ are linearly independent. We claim that $x_1, \ldots, x_e$ are linearly independent in K. Indeed, if $\sum a_i x_i = 0$ with $a_i \in K$, then by clearing common denominators, we may assume that all $a_i \in V$ and one of them is invertible. But then $\sum \bar{a}_i \bar{x}_i = 0$ in $\Lambda \otimes_V k$ where not all $\bar{a}_i$ are nonzero, which is a contradiction.

In particular, we have $e \leq d$, and $\dim_k(\Lambda \otimes_V k) \leq d$. If one has $e = d$, then the elements $x_1, \ldots, x_e$ induce an injective map $V^d \to \Lambda$. If $x \in \Lambda$ is any other element, there are unique $a_i \in K$ such that $x = \sum a_i x_i$. If one of the a_i is not in V, then by multiplying by a common denominator, we see that $\sum a'_i x_i = a' x$ where all $a'_i \in V$, one of them invertible, and a' in the maximal ideal of V. But then $\sum \bar{a}'_i \bar{x}_i = 0$, which is a contradiction. Thus, $\Lambda \cong V^d$, as desired. $\square$

Our goal was to show that M/p is a finite free $\mathcal{O}_{C^\flat}$-module. By Lemma 14.2.4 it suffices to show that $\dim_k(M/p \otimes_{\mathcal{O}_{C^\flat}} k) = \dim_k(M \otimes_R k)$ is at least d. Let T be the image of $M \otimes_R W(k)$ in $M_1 \otimes_{R_1} W(k)[1/p] \cong L^{\oplus d}$. Since M is open and bounded in M_1, T is open and bounded in $L^{\oplus d}$, so $T \cong W(k)^{\oplus d}$, which implies that $M \otimes_R k$ surjects onto $T \otimes_R k \cong k^{\oplus d}$, and we conclude. $\square$

Remark 14.2.5. In the proof, we did not use that $\mathcal{O}_{C^\flat}$ is a valuation ring of rank 1. In fact, the same proof shows the following slightly more general result, also contained in [Ked19b].

Proposition 14.2.6. Let K be a perfectoid field of characteristic p with an open and bounded valuation subring $K^+ \subset K$, and let $\varpi \in K$ be a pseudouniformizer. Then the category of vector bundles on $\operatorname{Spec} W(K^+) \backslash \{p = [\varpi] = 0\}$ is equivalent to the category of finite free $W(K^+)$-modules.

14.3   PROOF OF THEOREM 14.2.1

We can now give the proof of Theorem 14.2.1, which is the statement that finite free A_{inf}-modules are in equivalence with vector bundles on $\mathcal{Y}$.

For full faithfulness: We have $\mathcal{Y} = \mathcal{Y}_{[0,1]} \cup \mathcal{Y}_{[1,\infty]}$, with intersection $\mathcal{Y}_{\{1\}}$. Let $\operatorname{Spa} S_1 = \mathcal{Y}_{[0,1]}$ and $\operatorname{Spa} S_2 = \mathcal{Y}_{[1,\infty]}$, with $Y_{\{1\}} = \operatorname{Spa} S_{12}$. Then S_1, S_2, and S_{12} are complete Tate rings:

1. The ring $S_1 = A_{\mathrm{inf}}\langle [p^\flat]/p \rangle[1/p]$ has ring of definition $A_{\mathrm{inf}}\langle [p^\flat]/p \rangle$ and pseudo-uniformizer p,
2. The ring $S_2 = A_{\mathrm{inf}}\langle p/[p^\flat] \rangle[1/[p^\flat]]$ has ring of definition $A_{\mathrm{inf}}\langle p/[p^\flat] \rangle$ and pseudo-uniformizer $[p^\flat]$,

3. Finally, the ring $S_{12} = A_{\mathrm{inf}}\langle p/[p^\flat], [p^\flat]/p\rangle[1/p[p^\flat]]$ has ring of definition $A_{\mathrm{inf}}\langle p/[p^\flat], [p^\flat]/p\rangle$; both p and $[p^\flat]$ are pseudo-uniformizers.

The ring S_{12} contains S_1 and S_2 as topological subrings. The intersection $S_1 \cap S_2$ inside S_{12} is A_{inf}: this is [KL15, Lemma 5.2.11(c)]. This gives full faithfulness.

We turn to essential surjectivity. A vector bundle $\mathcal{E}$ on $\mathcal{Y}$ is the same, by Theorem 5.2.8, as finite projective S_i-modules $\mathcal{E}_i$, for $i = 1, 2$, which glue over S_{12}. We want to produce a vector bundle on the scheme Y. For this, consider again $R_1 = A_{\mathrm{inf}}[1/p]$, endowed with the p-adic topology on A_{inf}. Then $\mathrm{Spa}\,R_1$ is covered by open subsets $\{|[p^\flat]| \leq |p| \neq 0\} = \mathrm{Spa}\,S_1$ and $\{|p| \leq |[p^\flat]| \neq 0\} =: \mathrm{Spa}\,S_2'$, where

$$S_2' = A_{\mathrm{inf}}[p/[p^\flat]]_p^\wedge[1/p].$$

Lemma 14.3.1. The natural map gives an isomorphism

$$A_{\mathrm{inf}}[p/[p^\flat]]_p^\wedge \cong A_{\mathrm{inf}}[p/[p^\flat]]_{(p,[p^\flat])}^\wedge.$$

Proof. It is enough to observe that $A_{\mathrm{inf}}[p/[p^\flat]]_p^\wedge$ is already $[p^\flat]$-adically complete. To check this, we calculate that for all $n \geq 1$, $A_{\mathrm{inf}}[p/[p^\flat]]/p^n$ is already $[p^\flat]$-adically complete as follows. In the $n = 1$ case we have

$$
\begin{aligned}
A_{\mathrm{inf}}[p/[p^\flat]]/p &= (A_{\mathrm{inf}}/p)[T]/(p - [p^\flat]T) \\
&= \mathcal{O}_{C^\flat}[T]/p^\flat T = \mathcal{O}_{C^\flat} \oplus \bigoplus_{i \geq 1}(\mathcal{O}_{C^\flat}/p^\flat)T^i,
\end{aligned}
$$

which is indeed $[p^\flat]$-adically complete. $\square$

After inverting p in Lemma 14.3.1 we find an isomorphism $S_2' \cong S_2[1/p]$ (which is not a homeomorphism). Under this isomorphism, $\mathcal{E}_2[1/p]$ can be considered as a vector bundle on S_2'. Then $\mathcal{E}_1$, $\mathcal{E}_2[1/p]$ and $\mathcal{E}_{12}$ define a gluing datum for a vector bundle on $\mathrm{Spa}\,R_1$. Since R_1 is sheafy (as in the proof of Proposition 13.1.1), they glue to give a finite projective R_1-module M_1.

Now apply the Beauville–Laszlo lemma, Lemma 5.2.9, to the scheme $\mathrm{Spec}\,A_{\mathrm{inf}} \setminus \{x_k\}$ (or rather an open affine subset containing $x_{C^\flat}$, e.g., $\mathrm{Spec}\,A_{\mathrm{inf}}[1/[p^\flat]]$) and the element $p \in A_{\mathrm{inf}}$. We find that M_1 and the $W(C^\flat)$-lattice

$$\mathcal{E}_2 \otimes_{S_2} W(C^\flat) \subset \mathcal{E}_2 \otimes_{S_2} W(C^\flat)[\tfrac{1}{p}] \cong M_1 \otimes_{R_1} W(C^\flat)[\tfrac{1}{p}]$$

determine a vector bundle $\mathcal{E}'$ on $\mathrm{Spec}\,A_{\mathrm{inf}} \setminus \{x_k\}$. By Lemma 14.2.3, we see that there is a finite free A_{inf}-module M such that $M[\tfrac{1}{p}] = M_1$ and such that $M \otimes_{A_{\mathrm{inf}}} W(C^\flat) = \mathcal{E}_2 \otimes_{S_2} W(C^\flat)$, compatibly with the identifications over $W(C^\flat)[\tfrac{1}{p}]$. As $M[\tfrac{1}{p}] = M_1$, we find in particular that $M \otimes_{A_{\mathrm{inf}}} S_1 = \mathcal{E}_1$. On the other hand, applying Lemma 5.2.9 to S_2 and the element $p \in S_2$, we see that $M \otimes_{A_{\mathrm{inf}}} S_2 = M_2$ as this is true after inverting p and after p-adic completion, in a compatible way. Moreover, these two identifications match after base extension to S_{12}, as desired. This finishes the proof of Theorem 14.2.1, and thus of Theorem 14.1.1.

14.4   DESCRIPTION OF THE FUNCTOR "?"

This concludes the program outlined in §12.2. Let us now explain the applications of this formalism to p-divisible groups. The following is a restatement of part of Theorem 14.1.1 in the case of p-divisible groups.

Theorem 14.4.1. The category of p-divisible groups G over $\mathcal{O}_C$ is equivalent to the category of finite free A_{inf}-modules M together with a φ-linear isomorphism $\varphi_M : M[\frac{1}{\xi}] \cong M[\frac{1}{\varphi(\xi)}]$ such that

$$M \subset \varphi_M(M) \subset \tfrac{1}{\varphi(\xi)} M .$$

For applications of this theorem, one needs to understand how this relates to the crystalline classification of the special fiber of G. There is in fact a crystalline classification of $G_{\mathcal{O}_C/p}$. Recall that A_{crys} is the p-adic completion of $A_{\mathrm{inf}}[\frac{\xi^n}{n!}]$, which is also the universal p-adically complete PD thickening of $\mathcal{O}_C/p$, and $B_{\mathrm{crys}}^+ = A_{\mathrm{crys}}[\frac{1}{p}]$.

Theorem 14.4.2 ([SW13, Theorem A]). The category of p-divisible groups over $\mathcal{O}_C/p$ embeds fully faithfully into the category of finite free A_{crys}-modules M_{crys} together with an isomorphism

$$\varphi_{\mathrm{crys}} : M_{\mathrm{crys}} \otimes_{A_{\mathrm{crys}},\varphi} B_{\mathrm{crys}}^+ \cong M_{\mathrm{crys}}[\tfrac{1}{p}] .$$

The compatibility between these different classifications is given by the following theorem that is proved in the appendix to this lecture as an application of the results in integral p-adic Hodge theory in [BMS18].

Theorem 14.4.3. Let G be a p-divisible group over $\mathcal{O}_C$ with associated Breuil-Kisin-Fargues module (M, φ_M). Then the crystalline Dieudonné module of $G_{\mathcal{O}_C/p}$ is given by $M \otimes_{A_{\mathrm{inf}}} A_{\mathrm{crys}}$ with its induced Frobenius $\varphi_M \otimes \varphi$.

In particular, by passing further from $\mathcal{O}_C/p$ to k, we get the following corollary, answering the question in Lecture 12.

Corollary 14.4.4. Under the equivalence between p-divisible groups G over $\mathcal{O}_C$ and Breuil-Kisin-Fargues modules (M, φ_M) and the equivalence between p-divisible groups over k and Dieudonné modules, the functor $G \mapsto G_k$ of passage to the special fiber corresponds to the functor $(M, \varphi_M) \mapsto (M \otimes_{A_{\mathrm{inf}}} W(k), \varphi_M \otimes \varphi)$.

We note that there are similar results for Breuil-Kisin modules for p-divisible groups over $\mathcal{O}_K$, where K is a finite extension of $W(k)[\frac{1}{p}]$. In that setting, the case $p \neq 2$ was handled by Kisin, [Kis06], and the case $p = 2$ was completed by T. Liu, [Liu13].

Appendix to Lecture 14:
Integral p-adic Hodge theory

The goal of this appendix is to establish Proposition 14.4.3. We will not give a streamlined argument; instead, we want to explain how the various different (equivalent) linear-algebraic categories of Theorem 14.1.1 arise from the cohomology of proper smooth rigid-analytic spaces or formal schemes, or p-divisible groups.

We will consider the following three kinds of geometric objects.

- (Rigid) A proper smooth rigid-analytic variety X over C.
- (Formal) A proper smooth formal scheme $\mathfrak{X}$ over $\mathcal{O}_C$.
- (p-Div) A p-divisible group G over $\mathcal{O}_C$.

Note that the generic fiber of $\mathfrak{X}$ is an object X of the first kind. Moreover, if $\mathfrak{X}$ is a formal abelian variety (i.e., it is an abelian variety modulo any power of p), then $G = \mathfrak{X}[p^\infty]$ is a p-divisible group, and up to direct factors, every p-divisible group arises in this way by Proposition 14.9.4 below.

At least when the objects are algebraic and defined over a finite extension of $W(k)[\frac{1}{p}]$, classical constructions recalled below naturally associate the following data to these objects:

- (Rigid) For any $i = 0, \ldots, 2 \dim X$, an object "$H^i(X)$" of category (2) of Theorem 14.1.1.
- (Formal) For any $i = 0, \ldots, 2 \dim \mathfrak{X}$, an object "$H^i(\mathfrak{X})$" of categories (2) and (3) of Theorem 14.1.1, and (slightly indirectly) an object of category (4).
- (p-Div) An object of the categories (2), (3), and (4) of Theorem 14.1.1.

We note that while it is easy to establish the equivalence of categories (1), (2), and (3) in Theorem 14.1.1 (and their equivalence holds over general base spaces, not just algebraically closed nonarchimedean fields), we regard it as an open problem to give a direct cohomological construction of objects of the categories (1) and (3) in the rigid case. We do *not* expect that one can naturally associate an object of categories (4) and (5) to a rigid space; more precisely, such a construction should not work in families, as the extension steps in the proof of Theorem 14.1.1 relied critically on the assumption that the base is an algebraically closed nonarchimedean field.

In the formal case, [BMS18] gives a direct construction of a Breuil-Kisin-

Fargues module, and thus of objects of all other categories (as the functors from (5) to the other categories are given by naive specializations). In the case of p-divisible groups, a similarly direct construction is still missing, although one can bootstrap from the case of formal schemes by using abelian varieties with a given p-divisible group, as we will do below.

14.6 COHOMOLOGY OF RIGID-ANALYTIC SPACES

Let X/C be a proper smooth rigid-analytic variety. Here we will show that one can associate pairs (T, Ξ) to the p-adic cohomology of X.

For any $i = 0, \ldots, 2 \dim X$, we have the étale cohomology group

$$H^i(X_{\text{ét}}, \mathbf{Z}_p) = \varprojlim H^i(X_{\text{ét}}, \mathbf{Z}/p^n\mathbf{Z})$$

(which could also be directly defined as $H^i(X_{\text{proét}}, \underline{\mathbf{Z}_p})$). By [Sch13, Theorem 1.1], it is a finitely generated $\mathbf{Z}_p$-module. As we want to work with finite free $\mathbf{Z}_p$-modules, we take the quotient by its torsion subgroup to get $T = H^i(X_{\text{ét}}, \mathbf{Z}_p)/(\text{torsion})$.

To get a pair (T, Ξ), it remains to construct a natural B_{dR}^+-lattice in $T \otimes_{\mathbf{Z}_p} B_{\text{dR}}$.

Theorem 14.6.1 ([BMS18, Theorem 13.1, Theorem 13.8]). There is a natural B_{dR}^+-lattice $\Xi \subset H^i_{\text{ét}}(X, \mathbf{Z}_p) \otimes_{\mathbf{Z}_p} B_{\text{dR}}$. If $X = Y_C$ for a smooth proper rigid-analytic variety Y over a discretely valued field $K \subset C$ with perfect residue field, then $\Xi = H^i_{\text{dR}}(Y/K) \otimes_K B_{\text{dR}}^+$, embedded into $H^i_{\text{ét}}(X, \mathbf{Z}_p) \otimes_{\mathbf{Z}_p} B_{\text{dR}}$ via the comparison isomorphism. In general, $\Xi \otimes_{B_{\text{dR}}^+} C$ is canonically isomorphic to $H^i_{\text{dR}}(X/C)$.

The construction of Ξ is a version of crystalline cohomology relative to the pro-thickening $B_{\text{dR}}^+ \to C$.

14.7 COHOMOLOGY OF FORMAL SCHEMES

Now let $\mathfrak{X}/\mathcal{O}_C$ be a proper smooth formal scheme. In addition to the cohomology groups associated with its generic fiber, we also have its special fiber $\mathfrak{X}_k/k$, which has crystalline cohomology groups

$$H^i(\mathfrak{X}_k/W(k))$$

that are finitely generated $W(k)$-modules equipped with a Frobenius endomorphism that becomes an isomorphism after inverting p.

In fact, one has a finer invariant given as the crystalline cohomology of the base change $\mathfrak{X}_{\mathcal{O}_C/p}$ of $\mathfrak{X}$ to $\mathcal{O}_C/p$. Namely, as recalled above, $\mathcal{O}_C/p$ has a universal p-adically complete divided power thickening $A_{\mathrm{crys}} \to \mathcal{O}_C/p$. Then the

$$H^i(\mathfrak{X}_{\mathcal{O}_C/p}/A_{\mathrm{crys}})$$

are A_{crys}-modules equipped with a Frobenius endomorphism. In fact, $H^i(\mathfrak{X}_{\mathcal{O}_C/p}/A_{\mathrm{crys}})[\frac{1}{p}]$ is a φ-module over $B_{\mathrm{crys}}^+ = A_{\mathrm{crys}}[\frac{1}{p}]$. By [FF18, Corollaire 11.1.14], this is equivalent to a φ-module over B^+, or a φ-module over B, or a vector bundle $\mathcal{F}'$ on the Fargues-Fontaine curve. To compare this with the cohomologies of the generic fiber amounts to a comparison between étale and crystalline cohomology. This is given by the crystalline comparison isomorphism. If $\mathfrak{X}$ is a scheme that is defined over a finite extension of $W(k)$, this result is due to Tsuji, [Tsu99], with other proofs given by Faltings, Niziol, and others.

Theorem 14.7.1 ([BMS18, Theorem 14.5 (i)]). There is a natural φ-equivariant isomorphism

$$H^i(\mathfrak{X}_{\mathcal{O}_C/p}/A_{\mathrm{crys}}) \otimes B_{\mathrm{crys}} \cong H^i(X_{\text{ét}}, \mathbf{Z}_p) \otimes_{\mathbf{Z}_p} B_{\mathrm{crys}} .$$

In other words, the φ-modules over B coming from $H^i(\mathfrak{X}_{\mathcal{O}_C/p}/A_{\mathrm{crys}})$ and $H^i(X_{\text{ét}}, \mathbf{Z}_p)$ respectively become isomorphic after inverting t; geometrically, this means that we have two vector bundles $\mathcal{F}'$ and $\mathcal{F}$ on the Fargues-Fontaine curve that are isomorphic away from ∞. Together with the lattice $T = H^i(X_{\text{ét}}, \mathbf{Z}_p)/$ (torsion), this defines an object of category (3).

Proposition 14.7.2 ([BMS18, Theorem 14.5 (i)]). Under the equivalence of Theorem 14.1.1, this object of category (3) corresponds to the object of category (2) constructed from the generic fiber X. Equivalently, the B_{dR}^+-lattice

$$H^i(\mathfrak{X}_{\mathcal{O}_C/p}/A_{\mathrm{crys}}) \otimes_{A_{\mathrm{crys}}} B_{\mathrm{dR}}^+ \subset H^i(\mathfrak{X}_{\mathcal{O}_C/p}/A_{\mathrm{crys}}) \otimes B_{\mathrm{dR}} \cong H^i(X_{\text{ét}}, \mathbf{Z}_p) \otimes_{\mathbf{Z}_p} B_{\mathrm{dR}}$$

agrees with the B_{dR}^+-lattice $\Xi \subset H^i(X_{\text{ét}}, \mathbf{Z}_p) \otimes_{\mathbf{Z}_p} B_{\mathrm{dR}}$ constructed in Theorem 14.6.1.

In the discretely valued case, the proposition is a basic compatibility between the crystalline and de Rham comparison isomorphisms.

Finally, we claim that we can actually produce an object of category (4). Namely, we have an object of categories (2) and (3), which by an easy argument was equivalent to an object of category (1), i.e., a shtuka over $\operatorname{Spa} C^\flat$ with one leg at $\varphi^{-1}(x_C)$. To extend this to a vector bundle on $\mathcal{Y}$ with a similar Frobenius, we had to extend a φ-module over $\mathcal{Y}_{[r,\infty)}$ to $\mathcal{Y}_{[r,\infty]}$ as in Theorem 13.2.1. This was achieved abstractly only through the difficult classification result for φ-modules over $\mathcal{Y}_{[r,\infty)}$, Theorem 13.4.1. However, here we already have a φ-module over B_{crys}^+, which by base extension gives a φ-module over $\mathcal{Y}_{[r,\infty]}$ for any large enough r.

More precisely, we get the following result, using for simplicity the associated Breuil-Kisin-Fargues module in its formulation.

Corollary 14.7.3. Let $\mathfrak{X}/\mathcal{O}_C$ be a proper smooth formal scheme, and fix some integer i. Consider the Breuil-Kisin-Fargues module (M_i, φ_{M_i}) associated with the pair (T_i, Ξ_i), where $T_i = H^i(X_{\text{ét}}, \mathbf{Z}_p)/(\text{torsion})$ and $\Xi_i \subset T_i \otimes_{\mathbf{Z}_p} B_{\text{dR}}$ is the natural B_{dR}^+-lattice of Theorem 14.6.1. Then there is a natural φ-equivariant isomorphism

$$H^i(\mathfrak{X}_{\mathcal{O}_C/p}/A_{\text{crys}})[\tfrac{1}{p}] \cong M_i \otimes_{A_{\text{inf}}} B_{\text{crys}}^+ \ .$$

Proof. We need to compare two φ-modules over B_{crys}^+. But by [FF18, Corollaire 11.1.14], those are equivalent to vector bundles on the Fargues-Fontaine curve, and we have a comparison of those by the previous proposition. $\qquad\square$

Note that both sides of the isomorphism in the previous proposition have natural integral structures, and one can wonder whether they match. This happens, at least under torsion-freeness hypothesis, by the main results of [BMS18], which will be discussed below; cf. Corollary 14.9.2.

14.8   *p*-DIVISIBLE GROUPS

Let G be a p-divisible group over $\mathcal{O}_C$. We can associate with it the finite free $\mathbf{Z}_p$-module $T = T_p G = \varprojlim G[p^n](\mathcal{O}_C)$, where the transition maps are multiplication by p. Moreover, the Hodge-Tate filtration, Theorem 12.1.1, defines a subspace $\text{Lie}\, G \otimes_{\mathcal{O}_C} C \subset T \otimes_{\mathbf{Z}_p} C(-1)$, which in turn can be used to define the B_{dR}^+-lattice Ξ such that

$$T \otimes_{\mathbf{Z}_p} B_{\text{dR}}^+ \subseteq \Xi \subset \xi^{-1}(T \otimes_{\mathbf{Z}_p} B_{\text{dR}}^+) \ ,$$

and whose image in $\xi^{-1}(T \otimes_{\mathbf{Z}_p} B_{\text{dR}}^+)/(T \otimes_{\mathbf{Z}_p} B_{\text{dR}}^+) = T \otimes_{\mathbf{Z}_p} C(-1)$ is given by $\text{Lie}\, G \otimes_{\mathcal{O}_C} C$.

On the other hand, Dieudonné theory applied to $G_{\mathcal{O}_C/p}$ defines a finite free A_{crys}-module M_{crys} equipped with a Frobenius φ that becomes an isomorphism after inverting p. Again, we have a comparison isomorphism.

Proposition 14.8.1 ([Fal99]). There is a natural φ-equivariant isomorphism

$$M_{\text{crys}} \otimes_{A_{\text{crys}}} B_{\text{crys}} \cong T \otimes_{\mathbf{Z}_p} B_{\text{crys}} \ .$$

The induced B_{dR}^+-lattice

$$M_{\text{crys}} \otimes_{A_{\text{crys}}} B_{\text{dR}}^+ \subset M_{\text{crys}} \otimes_{A_{\text{crys}}} B_{\text{dR}} \cong T \otimes_{\mathbf{Z}_p} B_{\text{dR}}$$

is given by Ξ.

Proof. The proof is simple, so let us recall it. We will in fact construct a φ-

equivariant map

$$T \otimes_{\mathbf{Z}_p} A_{\mathrm{crys}} \to M_{\mathrm{crys}} \;;$$

to check that this induces the desired isomorphism, one can apply this map also to the dual p-divisible group (and dualize) to obtain an inverse up to multiplication by t. To construct such a map, it suffices to construct a map $T \to M_{\mathrm{crys}}^{\varphi=1}$. Now given any element of T, we have equivalently a map $f : \mathbf{Q}_p/\mathbf{Z}_p \to G$. We can evaluate this map on Dieudonné modules to get a φ-equivariant map $A_{\mathrm{crys}} \to M_{\mathrm{crys}}$, which when evaluated on 1 gives the desired φ-invariant element of M_{crys}.

The comparison with Ξ follows directly from the explicit definitions. $\square$

As for formal schemes, we get the following corollary.

Corollary 14.8.2. Let G be a p-divisible group over $\mathcal{O}_C$, and let (M, φ_M) be the Breuil-Kisin-Fargues module associated under Theorem 14.1.1 to the pair $(T = T_pG, \Xi)$. Then there is a natural φ-equivariant isomorphism

$$M_{\mathrm{crys}}[\tfrac{1}{p}] \cong M \otimes_{A_{\mathrm{inf}}} B_{\mathrm{crys}}^{+} \;.$$

To finish the proof of Theorem 14.4.3, it is thus sufficient to prove the following result.

Proposition 14.8.3. The isomorphism $M_{\mathrm{crys}}[\tfrac{1}{p}] \cong M \otimes_{A_{\mathrm{inf}}} B_{\mathrm{crys}}^{+}$ induces an isomorphism of subobjects $M_{\mathrm{crys}} \cong M \otimes_{A_{\mathrm{inf}}} A_{\mathrm{crys}}$.

This proposition will be proved in the next subsection, as a consequence of a similar result for formal schemes in [BMS18].

14.9   THE RESULTS OF [BMS18]

The following theorem sums up the main theorem of [BMS18]. Here $\mu = [\epsilon] - 1 \in A_{\mathrm{inf}}$, where $\epsilon \in \mathcal{O}_{C^\flat}$ is given by $(1, \zeta_p, \zeta_{p^2}, \ldots)$ for a compatible system of primitive p-power roots of unity ζ_{p^r}.

Theorem 14.9.1 ([BMS18, Theorem 1.8, Theorem 14.5 (iii)]). Let $\mathfrak{X}$ be a proper smooth formal scheme over $\mathcal{O}_C$ with generic fiber X. Then there is a perfect complex of A_{inf}-modules $R\Gamma_{A_{\mathrm{inf}}}(\mathfrak{X})$ with a φ-semilinear action, with the following comparisons.

1. A φ-equivariant isomorphism

$$R\Gamma_{A_{\mathrm{inf}}}(\mathfrak{X}) \overset{\mathbb{L}}{\otimes}_{A_{\mathrm{inf}}} A_{\mathrm{crys}} \cong R\Gamma_{\mathrm{crys}}(\mathfrak{X}_{\mathcal{O}_C/p}/A_{\mathrm{crys}}) \;.$$

2. A φ-equivariant isomorphism

$$R\Gamma_{A_{\mathrm{inf}}}(\mathfrak{X}) \otimes_{A_{\mathrm{inf}}} A_{\mathrm{inf}}[1/\mu] \cong R\Gamma(X_{\text{ét}}, \mathbf{Z}_p) \otimes_{\mathbf{Z}_p} A_{\mathrm{inf}}[1/\mu] \;.$$

After base extension to $A_{\mathrm{crys}}[1/\mu] = B_{\mathrm{crys}}$, the resulting φ-equivariant isomorphism

$$R\Gamma_{\mathrm{crys}}(\mathfrak{X}_{\mathcal{O}_C/p}/A_{\mathrm{crys}}) \otimes B_{\mathrm{crys}} \cong R\Gamma(X_{\text{ét}}, \mathbf{Z}_p) \otimes_{\mathbf{Z}_p} B_{\mathrm{crys}}$$

agrees with the isomorphism from Theorem 14.7.1.

Moreover, all cohomology groups $H^i_{A_{\mathrm{inf}}}(\mathfrak{X})$ are finitely presented A_{inf}-modules that are free over $A_{\mathrm{inf}}[\frac{1}{p}]$ after inverting p, and if $H^i_{\mathrm{crys}}(\mathfrak{X}_k/W(k))$ is p-torsion free for some i, then $H^i_{A_{\mathrm{inf}}}(\mathfrak{X})$ is free over A_{inf}, and consequently (by (ii)), $H^i(X_{\text{ét}}, \mathbf{Z}_p)$ is p-torsion free.

The proof proceeds by giving a direct, but rather elaborate, cohomological construction of $R\Gamma_{A_{\mathrm{inf}}}(\mathfrak{X})$. As a corollary, we see that integral structures match in Corollary 14.7.3.

Corollary 14.9.2. Assume that $H^j_{\mathrm{crys}}(\mathfrak{X}_k/W(k))$ is p-torsion free for $j = i, i + 1, i + 2$. Then $H^i_{\mathrm{crys}}(\mathfrak{X}_{\mathcal{O}_C/p}/A_{\mathrm{crys}})$ is p-torsion free, and one has an equality

$$H^i_{\mathrm{crys}}(\mathfrak{X}_{\mathcal{O}_C/p}/A_{\mathrm{crys}}) = M_i \otimes_{A_{\mathrm{inf}}} A_{\mathrm{crys}}$$

as subobjects of

$$H^i_{\mathrm{crys}}(\mathfrak{X}_{\mathcal{O}_C/p}/A_{\mathrm{crys}})[\tfrac{1}{p}] \cong M_i \otimes_{A_{\mathrm{inf}}} B^+_{\mathrm{crys}} \;,$$

where M_i is the Breuil-Kisin-Fargues module associated with the ith cohomology of X as in Corollary 14.7.3.

Proof. By [BMS18, Lemma 4.9 (iii)], all $H^*_{A_{\mathrm{inf}}}(\mathfrak{X})$ have Tor-dimension ≤ 2 over A_{inf}; moreover, the assumption on torsion freeness in crystalline cohomology implies that $H^j_{A_{\mathrm{inf}}}(\mathfrak{X})$ is finite free over A_{inf} for $j = i, i+1, i+2$. Thus, the Tor spectral sequence for the φ-equivariant quasi-isomorphism

$$R\Gamma_{A_{\mathrm{inf}}}(\mathfrak{X}) \overset{\mathbb{L}}{\otimes}_{A_{\mathrm{inf}}} A_{\mathrm{crys}} \cong R\Gamma_{\mathrm{crys}}(\mathfrak{X}_{\mathcal{O}_C/p}/A_{\mathrm{crys}})$$

degenerates in degree i to a φ-equivariant isomorphism

$$H^i_{A_{\mathrm{inf}}}(\mathfrak{X}) \otimes_{A_{\mathrm{inf}}} A_{\mathrm{crys}} \cong H^i_{\mathrm{crys}}(\mathfrak{X}_{\mathcal{O}_C/p}/A_{\mathrm{crys}}) \;.$$

This implies that the right-hand side is p-torsionfree (in fact, finite free over A_{crys}). Moreover, this isomorphism is compatible with the given one after inverting p, which gives the desired result, noting that necessarily $H^i_{A_{\mathrm{inf}}}(\mathfrak{X})$ is the Breuil-Kisin-Fargues module M_i. $\qquad\square$

We need to know that in case $G = \mathfrak{A}[p^\infty]$ for a (formal) abelian variety $\mathfrak{A}$

over $\mathcal{O}_C$, the comparison isomorphism from Proposition 14.8.1 agrees with the comparison isomorphism from Theorem 14.7.1.

Proposition 14.9.3. Let $\mathfrak{A}$ be a formal abelian variety over $\mathcal{O}_C$ with generic fiber A, and with p-divisible group $G = \mathfrak{A}[p^\infty]$. Under the natural duality between $H^1(A_{\text{ét}}, \mathbf{Z}_p)$ and $T_p G$, and between $H^1_{\text{crys}}(\mathfrak{A}_{\mathcal{O}_C/p}/A_{\text{crys}})$ and $M_{\text{crys}}(G_{\mathcal{O}_C/p})$, the comparison isomorphisms of Theorem 14.7.1 and Proposition 14.8.1 correspond.

Proof. Fix an element of $T_p G$, given by a morphism $f : \mathbf{Q}_p/\mathbf{Z}_p \to G$, or equivalently $f : \mathbf{Q}_p/\mathbf{Z}_p \to \mathfrak{A}$; we can assume that it is not divisible by p, i.e. injective. This induces a map $H^1(A_{\text{ét}}, \mathbf{Z}_p) \to \mathbf{Z}_p$ under the duality between $H^1(A_{\text{ét}}, \mathbf{Z}_p)$ and $T_p A = T_p G$. The map in Theorem 14.7.1 comes as the composite of the identifications

$$H^i_{\text{crys}}(\mathfrak{A}_{\mathcal{O}_C/p}/A_{\text{crys}}) \cong H^i_{A_{\text{inf}}}(\mathfrak{A}) \otimes_{A_{\text{inf}}} A_{\text{crys}}$$

and

$$H^i_{A_{\text{inf}}}(\mathfrak{A}) \otimes_{A_{\text{inf}}} A_{\text{inf}}[\tfrac{1}{\mu}] \cong H^i(A_{\text{ét}}, \mathbf{Z}_p) \otimes_{\mathbf{Z}_p} A_{\text{inf}}[\tfrac{1}{\mu}] \ .$$

From f, we will construct a map $H^1_{A_{\text{inf}}}(\mathfrak{A}) \to A_{\text{inf}}$ that is compatible with the maps $H^1(A_{\text{ét}}, \mathbf{Z}_p) \to \mathbf{Z}_p$ and $H^1_{\text{crys}}(\mathfrak{A}_{\mathcal{O}_C/p}/A_{\text{crys}}) \to A_{\text{crys}}$ constructed from f via the displayed identifications.

It is enough to do this modulo p^n for all n. Look at the map $f_n = f|_{\frac{1}{p^n}\mathbf{Z}_p/\mathbf{Z}_p}$: $\frac{1}{p^n}\mathbf{Z}_p/\mathbf{Z}_p \hookrightarrow \mathfrak{A}$; let $\mathfrak{A} \to \mathfrak{A}_n$ be the corresponding quotient. Then $\mathfrak{A}$ is a $\mathbf{Z}/p^n\mathbf{Z}$-torsor over $\mathfrak{A}_n$, which induces a natural map

$$H^1_{A_{\text{inf}}}(\mathfrak{A}) \to H^2(\mathbf{Z}/p^n\mathbf{Z}, H^0_{A_{\text{inf}}}(\mathfrak{A})) = A_{\text{inf}}/p^n$$

as part of the corresponding Hochschild-Serre spectral sequence, and similarly for all other cohomology theories. Using the functoriality of the comparison isomorphisms in [BMS18], this implies the result, as in the étale specialization, that this differential in the Hochschild-Serre spectral sequence recovers the map on étale cohomology detailed above, and on the crystalline specialization it recovers the desired map on crystalline cohomology; both of these assertions are purely internal to étale resp. crystalline cohomology. $\square$

Now Proposition 14.8.3 follows from Corollary 14.9.2, Proposition 14.9.3 and the next proposition that shows that any p-divisible group comes from an abelian variety, up to direct factors.

Proposition 14.9.4. Let G be a p-divisible group over $\mathcal{O}_C$, with Serre dual G^*. Then there is a formal abelian variety $\mathfrak{A}$ over $\mathcal{O}_C$ such that there is an isomorphism of p-divisible groups

$$\mathfrak{A}[p^\infty] \cong G \times G^* \ .$$

Keeping track of polarizations, one could also arrange $\mathfrak{A}$ to be algebraic.

Proof. As the Newton polygon of $G_k \times G_k^*$ is symmetric, we can find an abelian variety A_k over k with $A_k[p^\infty] \cong G_k \times G_k^*$, by [Oor01]. By [SW13, Theorem 5.1.4 (i)], we can, up to changing A_k by a quasi-isogeny, lift to an abelian variety $A_{\mathcal{O}_C/p}$ such that $A_{\mathcal{O}_C/p}[p^\infty] \cong G_{\mathcal{O}_C/p} \times G_{\mathcal{O}_C/p}^*$. Now by Serre-Tate deformation theory, deforming $A_{\mathcal{O}_C/p}$ to characteristic 0 is equivalent to deforming its p-divisible group, so we find a formal abelian variety $\mathfrak{A}$ over $\mathcal{O}_C$ with $\mathfrak{A}[p^\infty] \cong G \times G^*$. $\qquad\square$

Lecture 15

Examples of diamonds

In this lecture, we discuss some interesting examples of diamonds. So far, the only example we have encountered (other than $X^\diamond$ for X an analytic adic space) is the self-product of copies of $\operatorname{Spd} \mathbf{Q}_p$. Let us first study this self-product a little more.

15.1 THE SELF-PRODUCT $\operatorname{Spd} \mathbf{Q}_p \times \operatorname{Spd} \mathbf{Q}_p$

We already encountered $\operatorname{Spd} \mathbf{Q}_p \times \operatorname{Spd} \mathbf{Q}_p$ in previous lectures. It is useful to keep in mind that a diamond $\mathcal{D}$ can have multiple "incarnations," by which we mean that there are multiple presentations of $\mathcal{D}$ as $X^\diamond/\underline{G}$, where X is an analytic adic space over $\operatorname{Spa} \mathbf{Z}_p$, and G is a profinite group. In the case of $\operatorname{Spd} \mathbf{Q}_p \times \operatorname{Spd} \mathbf{Q}_p$, there are (at least) the following two incarnations:

1. $X = \widetilde{\mathbf{D}}^*_{\mathbf{Q}_p}$, $G = \mathbf{Z}_p^\times$.
2. $X = \operatorname{Spa} A_{\inf} \setminus \{p[p^\flat] = 0\}$ with $A_{\inf} = W(\mathcal{O}_{\mathbf{C}_p^\flat})$, and $G = G_{\mathbf{Q}_p}$.

The first incarnation was discussed in Example 10.1.8. (In fact, since the roles of the two factors $\operatorname{Spd} \mathbf{Q}_p$ can be switched, this should already count as two incarnations.) Recall that $\widetilde{\mathbf{D}}^*_{\mathbf{Q}_p} = \varprojlim \mathbf{D}^*_{\mathbf{Q}_p}$ (with transition map $T \mapsto (1+T)^p - 1$). This has an action of $\mathbf{Z}_p^\times$ via $T \mapsto (1+T)^a - 1$. Then we have an isomorphism of diamonds

$$(\widetilde{\mathbf{D}}^*_{\mathbf{Q}_p})^\diamond / \underline{\mathbf{Z}_p^\times} \cong \operatorname{Spd} \mathbf{Q}_p \times \operatorname{Spd} \mathbf{Q}_p \ .$$

The Frobenius on the second factor of $\operatorname{Spd} \mathbf{Q}_p \times \operatorname{Spd} \mathbf{Q}_p$ corresponds to the automorphism $T \mapsto (1+T)^p - 1$; in fact, we have an action of $\mathbf{Q}_p^\times$ on $\widetilde{\mathbf{D}}^*_{\mathbf{Q}_p}$ where p acts as $T \mapsto (1+T)^p - 1$. The case $n = 2$ of Drinfeld's Lemma for diamonds (Theorem 16.3.1) which will be proved in the next lecture says that

Theorem 15.1.1.

$$\pi_1((\widetilde{\mathbf{D}}^*_{\mathbf{Q}_p})^\diamond / \underline{\mathbf{Q}_p^\times}) \cong G_{\mathbf{Q}_p} \times G_{\mathbf{Q}_p} \ .$$

Moreover, for any complete and algebraically closed field $C/\mathbf{Q}_p$, we have

$$\pi_1((\widetilde{\mathbf{D}}_C^*)^\diamond/\mathbf{Q}_p^\times) \cong G_{\mathbf{Q}_p}.$$

Thus, rather surprisingly, $G_{\mathbf{Q}_p}$ can be realized as a *geometric fundamental group*; cf. [Wei17]. Let us sketch the proof, which proceeds by switching to a different incarnation of $\mathrm{Spd}\,\mathbf{Q}_p \times \mathrm{Spd}\,\mathbf{Q}_p$. Recall from Proposition 11.2.1 that $\mathrm{Spa}\,\mathbf{C}_p^\flat \dot\times \mathrm{Spa}\,\mathbf{Z}_p$ is the analytic adic space $\mathcal{Y}_{[0,\infty)} = \mathrm{Spa}\,A_{\inf}\setminus\{[p^\flat]\} = 0$ with $A_{\inf} = W(\mathcal{O}_{\mathbf{C}_p^\flat})$, and that its associated diamond is $\mathrm{Spa}\,\mathbf{C}_p^\flat \times \mathrm{Spd}\,\mathbf{Z}_p$. After inverting p, we find that the diamond associated to $\mathcal{Y}_{(0,\infty)} = \mathrm{Spa}\,A_{\inf}\setminus\{p[p^\flat] = 0\}$ is $\mathrm{Spa}\,\mathbf{C}_p^\flat \times \mathrm{Spd}\,\mathbf{Q}_p$. It follows from this that

$$\mathcal{Y}_{(0,\infty)}^\diamond / G_{\mathbf{Q}_p} = \mathrm{Spd}\,\mathbf{Q}_p \times \mathrm{Spd}\,\mathbf{Q}_p \ .$$

Recall that the Fargues-Fontaine curve was defined as $\mathcal{X}_{\mathrm{FF}} = \mathcal{Y}_{(0,\infty)}/\varphi^{\mathbf{Z}}$. By Theorem 13.5.7, we know that $\pi_1(\mathcal{X}_{\mathrm{FF}}) = G_{\mathbf{Q}_p}$. On the other hand,

$$\pi_1(\mathcal{X}_{\mathrm{FF}}) = \pi_1(\mathcal{X}_{\mathrm{FF}}^\diamond) = \pi_1((\mathrm{Spd}\,\mathbf{C}_p \times \mathrm{Spd}\,\mathbf{Q}_p)/\mathrm{Frob}_{\mathrm{Spd}\,\mathbf{Q}_p}^{\mathbf{Z}}) = \pi_1((\widetilde{\mathbf{D}}_{\mathbf{C}_p}^*)^\diamond/\mathbf{Q}_p^\times) \ ,$$

and so Theorem 15.1.1 follows.

By Theorem 15.1.1, to each finite extension $F/\mathbf{Q}_p$ of degree n there must correspond a connected finite étale n-fold cover of $\widetilde{\mathbf{D}}_C^*/\mathbf{Q}_p^\times$; it is natural to ask what this cover is. The following discussion appears in [Wei17, Section 4.5]. Let $\varpi \in \mathcal{O}_F$ be a uniformizer, and let $\mathrm{LT}/\mathcal{O}_F$ be a *Lubin-Tate formal $\mathcal{O}_F$-module law*: this is a formal scheme isomorphic to $\mathrm{Spf}\,\mathcal{O}_F[\![T]\!]$ equipped with an $\mathcal{O}_F$-module structure, with the property that multiplication by ϖ sends T to a power series congruent to T^q modulo ϖ (here $q = \#\mathcal{O}_F/\varpi$). Then we can form the geometric generic fiber LT_C: this is an $\mathcal{O}_F$-module object in the category of adic spaces, whose underlying adic space is once again the open unit disc $\mathbf{D}_C$.

Now consider the inverse limit $\varprojlim_\varpi \mathrm{LT}_C^\diamond$. It is given by $\widetilde{\mathrm{LT}}_C^\diamond$, where $\widetilde{\mathrm{LT}}_C$ is an F-vector space object in the category of perfectoid spaces over C, whose underlying adic space is the perfectoid open unit disc $\widetilde{\mathbf{D}}_C$.

One can define a *norm map* $N_{F/\mathbf{Q}_p}\colon \mathrm{LT}_C \to \mathbf{D}_C$, which is a morphism of pointed adic spaces. Its construction goes as follows. Let $\breve{F}$ be the completion of the maximal unramified extension of F; in fact the norm map is defined over $\mathcal{O}_{\breve{F}}$. We have that $\mathrm{LT}_{\mathcal{O}_{\breve{F}}}$ is a p-divisible group of height n and dimension 1. By [Hed14], the nth exterior power of $\mathrm{LT}_{\mathcal{O}_{\breve{F}}}$ exists as a p-divisible group of dimension 1 and height 1, and so is isomorphic to $\widehat{\mathbf{G}}_{m,\mathcal{O}_{\breve{F}}}$, the formal multiplicative group. Thus we have an alternating map $\lambda\colon \mathrm{LT}_{\mathcal{O}_{\breve{F}}}^n \to \widehat{\mathbf{G}}_{m,\mathcal{O}_{\breve{F}}}$. Let $\alpha_1,\dots,\alpha_n$ be a basis for $\mathcal{O}_F/\mathbf{Z}_p$, and let $N(x) = \lambda(\alpha_1 x,\dots,\alpha_n x)$, so that N is a morphism $\mathrm{LT}_{\mathcal{O}_{\breve{F}}} \to \widehat{\mathbf{G}}_{m,\mathcal{O}_{\breve{F}}}$. This induces the desired norm map $N_{F/\mathbf{Q}_p}\colon \mathrm{LT}_C \to \mathbf{D}_C$ after passing to geometric generic fibers.

By construction we have $N_{F/\mathbf{Q}_p}(\alpha x) = N_{F/\mathbf{Q}_p}(\alpha)N_{F/\mathbf{Q}_p}(x)$ for all $\alpha \in \mathcal{O}_F$. Therefore $N_{F/\mathbf{Q}_p}$ can be used to define a norm map $\widetilde{\mathrm{LT}}_C \to \widetilde{\mathbf{D}}_C$, which is equivariant for the norm map $F \to \mathbf{Q}_p$; then

$$\widetilde{\mathrm{LT}}_C^* / \underline{F^\times} \to \widetilde{\mathbf{D}}_C^* / \underline{\mathbf{Q}_p^\times}$$

is the desired finite étale cover.

15.2 BANACH-COLMEZ SPACES

Another important class of diamonds, which in fact were one of the primary motivations for their definition, is the category of Banach-Colmez spaces, [Col02]. Recently, le Bras has reworked their theory in terms of perfectoid spaces, [LB18]. In particular, he shows that the following definition recovers Colmez's original definition.

Definition 15.2.1. Fix an algebraically closed nonarchimedean field $C/\mathbf{Q}_p$. The category of Banach-Colmez spaces over C is the thick abelian subcategory of the category of pro-étale sheaves of $\underline{\mathbf{Q}_p}$-modules on Perfd_C generated by $\underline{\mathbf{Q}_p}$ and $\mathbf{G}_{a,C}^\diamond$.

Remark 15.2.2. This is similar to a category considered by Milne in characteristic p, [Mil76]. There he considers the category of sheaves of abelian groups on the category of perfect schemes generated by $\mathbf{F}_p$ and $\mathbf{G}_a$. This notably contains truncated Witt vectors $W_r(\mathbf{G}_a)$, so there are nonsplit extensions of $\mathbf{G}_a$ by $\mathbf{G}_a$. Also, one has Artin-Schreier extensions in this category.

Classically, there are no nontrivial extensions of $\mathbf{G}_a$ by $\mathbf{G}_a$ in the category of rigid spaces of characteristic 0. However, we will see that such extensions exist in the category of Banach-Colmez spaces.

In the following, we will identify $\mathrm{Perfd}_C \cong \mathrm{Perf}_{C^\flat}$. In particular, we will say that a pro-étale sheaf on Perfd_C is a diamond if the corresponding pro-étale sheaf on $\mathrm{Perf}_{C^\flat}$ is.

Let us construct several examples.

First, in the spirit of the previous section, consider the open unit disc $\mathbf{D}_C$ as a subgroup of $\mathbf{G}_m$ via $T \mapsto 1 + T$. The logarithm defines an exact sequence

$$0 \to \mu_{p^\infty} \to \mathbf{D}_C \to \mathbf{G}_{a,C} \to 0$$

of étale sheaves on the category of adic spaces over C. After passing to the inverse limit over multiplication by p, we get an exact sequence

$$0 \to \underline{\mathbf{Q}_p} \to \widetilde{\mathbf{D}}_C \to \mathbf{G}_{a,C} \to 0$$

of pro-étale sheaves on Perfd_C. Therefore, $\widetilde{\mathbf{D}}_C^{\diamond}$ is a Banach-Colmez space.

More generally, assume that G is any p-divisible group over $\mathcal{O}_C$, and consider G as a formal scheme. Then its generic fiber G_C is a disjoint union of open unit balls; in fact, if G is connected, it is simply a unit ball of dimension $\dim G$. We have an exact sequence

$$0 \to G_C[p^{\infty}] \to G_C \to \mathrm{Lie}\, G \otimes_{\mathcal{O}_C} \mathbf{G}_{a,C} \to 0$$

of étale sheaves on adic spaces over C. This induces an exact sequence

$$0 \to \underline{V_p G} \to \widetilde{G}_C \to \mathrm{Lie}\, G \otimes_{\mathcal{O}_C} \mathbf{G}_{a,C} \to 0$$

of pro-étale sheaves on Perfd_C, where

$$\widetilde{G}_C = \varprojlim_p G_C^{\diamond}$$

is the universal cover of G_C in the sense of [SW13].[1] Thus, the universal cover $\widetilde{G}_C$ of a p-divisible group G is a Banach-Colmez space.

In [SW13], the pro-étale sheaves $\widetilde{G}_C$ have been described in terms of Fontaine's period rings. Let $\mathrm{Spa}(R, R^+)$ be an affinoid perfectoid space over C; we wish to understand the (R, R^+)-valued points of $\widetilde{G}_C$. These are given by

$$\varprojlim_p \varprojlim_n G(R^+/p^n)\,,$$

where the outer inverse limit is over multiplication by p. Exchanging the two limits, this is the same as

$$\varprojlim_n \varprojlim_p G(R^+/p^n)\,.$$

If we set $\widetilde{G}(S) = \varprojlim_p G(S)$ for any p-torsion R-algebra S, this defines a formal scheme over $\mathrm{Spf}\,\mathcal{O}_C$ whose generic fiber agrees with $\widetilde{G}_C$, and the points are given by

$$\varprojlim_n \widetilde{G}(R^+/p^n)\,.$$

But now [SW13, Proposition 3.1.3 (ii)] says that the maps $\widetilde{G}(R^+/p^n) \to \widetilde{G}(R^+/p)$ are isomorphisms. Thus, the answer is simply

$$\widetilde{G}(R^+/p) = \mathrm{Hom}_{R^+/p}(\mathbf{Q}_p/\mathbf{Z}_p, G)[\tfrac{1}{p}]\,.$$

Now we use the following result about Dieudonné theory over semiperfect rings

---

[1]This is somewhat of a misnomer, as usually $\widetilde{G}_C$ will have nontrivial π_1. The name is meant to suggest that it is a p-adic analogue of the universal cover of a torus by an affine space.

like R^+/p.

Theorem 15.2.3 ([SW13, Theorem A]). *Let (R, R^+) be a perfectoid Tate-Huber pair, and fix a pseudouniformizer $\varpi \in R^+$ dividing p. Let $A_{\mathrm{crys}}(R^+/\varpi)$ be the p-adically completed divided power envelope of $A_{\inf}(R^+) \to R^+/\varpi$. The category of p-divisible groups G over R^+/ϖ embeds fully faithfully into the category of finite projective $A_{\mathrm{crys}}(R^+/\varpi)$-modules M equipped with a φ-linear map $\varphi_M : M \to M[\frac{1}{p}]$, via sending G to its crystal evaluated on the PD thickening $A_{\mathrm{crys}}(R^+/\varpi) \to R^+/\varpi$.*

Thus, we finally find that if M is the Dieudonné module of G over $A_{\mathrm{crys}}(\mathcal{O}_C/p)$, then

$$\widetilde{G}_C(R, R^+) \cong \widetilde{G}(R^+/p) = (M \otimes_{A_{\mathrm{crys}}(\mathcal{O}_C/p)} B^+_{\mathrm{crys}}(R^+/p))^{\varphi=1} .$$

Here $B^+_{\mathrm{crys}}(R^+/p) = A_{\mathrm{crys}}(R^+/p)[\frac{1}{p}]$.

Example 15.2.4. In the case $G = \mu_{p^\infty}$, we have $\widetilde{G}_C \cong \widetilde{\mathbf{D}}_C$, and Theorem 15.2.3 gives

$$\widetilde{\mathbf{D}}_C(R, R^+) \cong B^+_{\mathrm{crys}}(R^+/p)^{\varphi=p} .$$

We can spell out this isomorphism explicitly. The tilting equivalence gives $\widetilde{\mathbf{D}}_C(R, R^+) = \widetilde{\mathbf{D}}_{C^\flat}(R^\flat, R^{\flat+})$. Since $R^\flat$ is perfect, the latter group is simply $\mathbf{D}_{C^\flat}(R^\flat, R^{\flat+}) = 1 + R^{\circ\circ}$, where $R^{\circ\circ}$ is the set of topological nilpotent elements in R. The above isomorphism sends an element $x \in 1 + R^{\circ\circ}$ to $\log[x]$.

These examples of Banach-Colmez spaces are represented by perfectoid spaces. This is not true in general, and further examples can be constructed by considering more general φ-modules over B^+_{crys}.

For this, let us discuss a few facts about the category of Dieudonné modules over $B^+_{\mathrm{crys}} = B^+_{\mathrm{crys}}(\mathcal{O}_C/p)$, i.e. the category of finite projective B^+_{crys}-modules M equipped with an isomorphism

$$\varphi_M : \varphi^* M \cong M .$$

Any such (M, φ_M) defines a vector bundle on the Fargues-Fontaine curve, as in Lecture 13.

Theorem 15.2.5 ([FF18]). *Via this functor, the category of Dieudonné modules (M, φ_M) over B^+_{crys} is equivalent to the category of vector bundles $\mathcal{E}$ on the Fargues-Fontaine curve $\mathcal{X}_{\mathrm{FF}}$. Under this equivalence, for any affinoid perfectoid space $S = \mathrm{Spa}(R, R^+)$ over C, one has*

$$(M \otimes_{B^+_{\mathrm{crys}}} B^+_{\mathrm{crys}}(R^+/p))^{\varphi=1} = H^0(\mathcal{X}_{\mathrm{FF},S}, \mathcal{E}|_{\mathcal{X}_{\mathrm{FF},S}}) .$$

Here, we use the relative Fargues-Fontaine curve $\mathcal{X}_{\mathrm{FF},S} = \mathcal{X}_{\mathrm{FF},S^\flat}$, so let us quickly define this. Let S be a perfectoid space of characteristic p. We have the adic space

$$\mathcal{Y}_{(0,\infty)}(S) = S \dot{\times} \operatorname{Spa} \mathbf{Q}_p = \{p \neq 0\} \subset S \dot{\times} \operatorname{Spa} \mathbf{Z}_p$$

on which there is a Frobenius automorphism φ coming from the Frobenius on S. This acts freely and totally discontinuously. Indeed, this can be checked if $S = \operatorname{Spa}(R, R^+)$ is affinoid, and then a choice of pseudouniformizer $\varpi \in R$ defines a continuous map

$$\kappa : \mathcal{Y}_{(0,\infty)}(S) \to (0,\infty)$$

given by

$$\kappa(x) = \frac{\log |[\varpi](\tilde{x})|}{\log |p(\tilde{x})|} ,$$

where $\tilde{x}$ is the maximal generalization of x. Then $\kappa \circ \varphi = p\kappa$, implying the desired properties of the φ-action.

Definition 15.2.6. The relative Fargues-Fontaine curve is the quotient

$$\mathcal{X}_{\mathrm{FF},S} = \mathcal{Y}_{(0,\infty)}(S)/\varphi^{\mathbf{Z}} .$$

This defines a sousperfectoid adic space over $\mathbf{Q}_p$. For any perfectoid field $K/\mathbf{Q}_p$, the base change $\mathcal{X}_{\mathrm{FF},S} \times_{\operatorname{Spa} \mathbf{Q}_p} \operatorname{Spa} K$ is a perfectoid space. Moreover, there is a simple formula for the diamond.

Proposition 15.2.7. There is a natural isomorphism

$$\mathcal{X}^{\diamond}_{\mathrm{FF},S} \cong S/\varphi^{\mathbf{Z}} \times \operatorname{Spd} \mathbf{Q}_p .$$

Proof. This follows immediately from Proposition 11.2.1.   □

Coming back to the discussion of Banach-Colmez spaces, we can change perspective even further, and fix a vector bundle $\mathcal{E}$ on $\mathcal{X}_{\mathrm{FF}}$. Even more generally, we allow coherent sheaves on $\mathcal{X}_{\mathrm{FF}}$. Any such $\mathcal{E}$ decomposes as a direct sum $\mathcal{E}_{\geq 0} \oplus \mathcal{E}_{<0}$, where the torsion-free quotient of $\mathcal{E}_{\geq 0}$ has only nonnegative Harder-Narasimhan slopes, while $\mathcal{E}_{<0}$ is a vector bundle whose Harder-Narasimhan polygon has only negative slopes. As

$$H^0(\mathcal{X}_{\mathrm{FF},S}, \mathcal{E}_{<0}|_{\mathcal{X}_{\mathrm{FF},S}}) = 0$$

for all $S \in \operatorname{Perfd}_C$, we restrict attention to the nonnegative part.

Definition 15.2.8. Let $\mathcal{E}$ be a coherent sheaf on $\mathcal{X}_{\mathrm{FF}}$ whose torsion-free quotient has only nonnegative slopes. The Banach-Colmez space $\mathcal{BC}(\mathcal{E})$ associated with $\mathcal{E}$ is the pro-étale sheaf

$$\mathcal{BC}(\mathcal{E}) : S \in \operatorname{Perfd}_C \mapsto H^0(\mathcal{X}_{\mathrm{FF},S}, \mathcal{E}|_{\mathcal{X}_{\mathrm{FF},S}}) .$$

We will prove that $\mathcal{BC}(\mathcal{E})$ is indeed a Banach-Colmez space by going through the classification.

Example 15.2.9. 1. The case $\mathcal{E} = \mathcal{O}_{\mathcal{X}_{\mathrm{FF}}}$. Then $\mathcal{BC}(\mathcal{E}) = \mathbf{Q}_p$.

2. The case $\mathcal{E} = i_{\infty*}C$, where $i_\infty : \mathrm{Spa}\, C \to \mathcal{X}_{\mathrm{FF}}$. Then $\overline{\mathcal{BC}(\mathcal{E})} = \mathbf{G}_{a,C}$.

3. The case $\mathcal{E} = \mathcal{O}_{\mathcal{X}_{\mathrm{FF}}}(\lambda)$, with $0 < \lambda = \frac{s}{r} \leq 1$. In that case $\mathcal{BC}(\mathcal{E})$ is representable by the universal cover of the simple p-divisible formal group G_λ of slope λ, by the above discussion. In particular, $\mathcal{BC}(\mathcal{E})$ is representable by the perfection of an open unit ball of dimension given by the denominator of λ. Also, there is a short exact sequence

$$0 \to \mathcal{O}_{\mathcal{X}_{\mathrm{FF}}}^s \to \mathcal{O}_{\mathcal{X}_{\mathrm{FF}}}(\lambda) \to i_{\infty*}C^r \to 0$$

of coherent sheaves on $\mathcal{X}_{\mathrm{FF}}$, which on global sections induces the sequence

$$0 \to \underline{\mathbf{Q}_p}^s \to \mathcal{BC}(\mathcal{O}_{\mathcal{X}_{\mathrm{FF}}}(\lambda)) \to \mathbf{G}_{a,C}^r \to 0 \,,$$

which agrees with the logarithm sequence for $\widetilde{G}_{\lambda,C}$.

4. The case $\mathcal{E} = \mathcal{O}_{\mathcal{X}_{\mathrm{FF}}}(\lambda)$, with $\lambda = \frac{s}{r} > 1$. In that case, one has a short exact sequence

$$0 \to \mathcal{O}_{\mathcal{X}_{\mathrm{FF}}}(\lambda - 1) \to \mathcal{O}_{\mathcal{X}_{\mathrm{FF}}}(\lambda) \to i_{\infty*}C^r \to 0 \,,$$

which induces on global sections a short exact sequence

$$0 \to \mathcal{BC}(\mathcal{O}_{\mathcal{X}_{\mathrm{FF}}}(\lambda - 1)) \to \mathcal{BC}(\mathcal{O}_{\mathcal{X}_{\mathrm{FF}}}(\lambda)) \to \mathbf{G}_{a,C}^r \to 0 \,.$$

In particular, $\mathcal{BC}(\mathcal{O}_{\mathcal{X}_{\mathrm{FF}}}(\lambda))$ is a Banach-Colmez space. One could think of $\mathcal{BC}(\mathcal{O}(\lambda))$ as the "universal cover of a formal group of slope λ". As the sequence is pro-étale locally split, one deduces that it is a diamond.

5. The case of the torsion sheaf $\mathcal{E} = B_{\mathrm{dR}}^+/\xi^n$ supported at ∞. There is a short exact sequence

$$0 \to \mathcal{O}_{\mathcal{X}_{\mathrm{FF}}} \to \mathcal{O}_{\mathcal{X}_{\mathrm{FF}}}(n) \to \mathcal{E} \to 0 \,,$$

which induces a short exact sequence

$$0 \to \mathbf{Q}_p \to \mathcal{BC}(\mathcal{O}_{\mathcal{X}_{\mathrm{FF}}}(n)) \to \mathcal{BC}(\mathcal{E}) \to 0 \,,$$

showing that $\mathcal{BC}(\mathcal{E})$ is also a Banach-Colmez space. Alternatively, one can write $\mathcal{E}$ as a successive extension of $i_{\infty*}C$, writing $\mathcal{BC}(\mathcal{E})$ as a successive extension of $\mathbf{G}_{a,C}$. In particular, if $n = 2$, one gets a nonsplit short exact sequence

$$0 \to \mathbf{G}_{a,C} \to \mathcal{BC}(B_{\mathrm{dR}}^+/\xi^2) \to \mathbf{G}_{a,C} \to 0 \,.$$

As the sequence is pro-étale locally split, one sees again that $\mathcal{BC}(B_{\mathrm{dR}}^+/\xi^2)$ is a diamond. Although this shows that $\mathcal{BC}(B_{\mathrm{dR}}^+/\xi^2)$ is built from rigid spaces, it is not itself a rigid space!

6. The previous case, for a different untilt of $C^\flat$, which works similarly.

Let us analyze the case of $\mathcal{E} = B_{\mathrm{dR}}^+/\xi^n$ further. For this, we recall the construction of the de Rham period ring. Let (R, R^+) be a perfectoid Tate-Huber pair. We have the surjective homomorphism $\theta \colon W(R^{\flat+}) \to R^+$, whose kernel is generated by a non-zero-divisor ξ. Let $\varpi^\flat \in R^\flat$ be a pseudouniformizer such that $\varpi = (\varpi^\flat)^\sharp$ satisfies $\varpi^p | p$.

We get a surjection $\theta \colon W(R^{\flat+})[[\varpi^\flat]^{-1}] \to R = R^+[\varpi^{-1}]$. Let $B_{\mathrm{dR}}^+(R)$ be the ξ-adic completion of $W(R^{\flat+})[[\varpi^\flat]^{-1}]$. This comes with a canonical filtration $\mathrm{Fil}^i B_{\mathrm{dR}}^+ = \xi^i B_{\mathrm{dR}}^+$, whose associated gradeds are $\mathrm{gr}^i B_{\mathrm{dR}}^+ \cong \xi^i R$.

Philosophically, we think of $\mathrm{Spf}\, B_{\mathrm{dR}}^+(R)$ as the completion of "$\mathrm{Spec}\,\mathbf{Z} \times \mathrm{Spec}\, R$" along the graph of $\mathrm{Spec}\, R \to \mathrm{Spec}\,\mathbf{Z}$. The construction of $B_{\mathrm{dR}}^+(R)$ encompasses the following two important cases:

Example 15.2.10. 1. Let $R = C/\mathbf{Q}_p$ be algebraically closed and complete. Then $B_{\mathrm{dR}}^+(C)$ is the usual de Rham period ring of Fontaine.
2. If R has characteristic p, then we can take $\xi = p$, and $B_{\mathrm{dR}}^+(R) = W(R)$.

We see that the pro-étale sheaf $(R, R^+) \mapsto B_{\mathrm{dR}}^+(R)/\xi^n$ defines the Banach-Colmez space $\mathcal{BC}(B_{\mathrm{dR}}^+/\xi^n)$. The rings $B_{\mathrm{dR}}^+(R)$ will play an important role in the discussion of affine Grassmannians. In particular, the local structure of the B_{dR}^+-affine Grassmannian will be closely related to the Banach-Colmez spaces $\mathcal{BC}(B_{\mathrm{dR}}^+/\mathrm{Fil}^n)$.

Le Bras has in fact obtained a description of the category of Banach-Colmez spaces in terms of coherent sheaves on the Fargues-Fontaine curve. To state the result, we need to consider also vector bundles $\mathcal{E}$ all of whose Harder-Narasimhan slopes are negative. In this case, the global sections of $\mathcal{E}$ vanish, but they have an interesting H^1.

Definition 15.2.11. Let $\mathcal{E}$ be a vector bundle on $\mathcal{X}_{\mathrm{FF}}$ all of whose Harder-Narasimhan slopes are negative. The Banach-Colmez space $\mathcal{BC}(\mathcal{E}[1])$ associated with $\mathcal{E}$ is the pro-étale sheaf

$$\mathcal{BC}(\mathcal{E}[1]) \colon S \in \mathrm{Perfd}_C \mapsto H^1(\mathcal{X}_{\mathrm{FF},S}, \mathcal{E}|_{\mathcal{X}_{\mathrm{FF},S}}) \,.$$

One needs to verify that this is indeed a pro-étale sheaf, which we omit here. Again, we check that this is indeed a Banach-Colmez space by going into the classification. We can reduce to the case $\mathcal{E} = \mathcal{O}_{\mathcal{X}_{\mathrm{FF}}}(\lambda)$, with $\lambda = \frac{s}{r} < 0$. In that case, one has a short exact sequence

$$0 \to \mathcal{O}_{\mathcal{X}_{\mathrm{FF}}}(\lambda) \to \mathcal{O}_{\mathcal{X}_{\mathrm{FF}}}(\lambda + 1) \to i_{\infty*}C^r \to 0$$

of coherent sheaves on $\mathcal{X}_{\mathrm{FF}}$. If $\lambda \geq -1$, this induces a short exact sequence

$$0 \to \mathcal{BC}(\mathcal{O}_{\mathcal{X}_{\mathrm{FF}}}(\lambda + 1)) \to \mathbf{G}_{a,C}^r \to \mathcal{BC}(\mathcal{O}_{\mathcal{X}_{\mathrm{FF}}}(\lambda)[1]) \to 0$$

on cohomology, showing again that $\mathcal{BC}(\mathcal{O}_{\mathcal{X}_{\mathrm{FF}}}(\lambda)[1])$ is a Banach-Colmez space.

Note that if $\lambda = -1$, this shows that

$$\mathcal{BC}(\mathcal{O}_{\mathcal{X}_{\mathrm{FF}}}(-1)) = \mathbf{G}_{a,C}/\underline{\mathbf{Q}_p} \ ,$$

which is a rather interesting space. If $\lambda < -1$, one gets a short exact sequence

$$0 \to \mathbf{G}_{a,C}^r \to \mathcal{BC}(\mathcal{O}_{\mathcal{X}_{\mathrm{FF}}}(\lambda)[1]) \to \mathcal{BC}(\mathcal{O}_{\mathcal{X}_{\mathrm{FF}}}(\lambda+1)[1]) \to 0 \ ,$$

which once again shows that $\mathcal{BC}(\mathcal{O}_{\mathcal{X}_{\mathrm{FF}}}(\lambda)[1])$ is a Banach-Colmez space.

Finally, we can state the main theorem of Le Bras which shows that the examples above exhaust all Banach-Colmez spaces.

Theorem 15.2.12 ([LB18]). *All Banach-Colmez spaces are diamonds. The category of Banach-Colmez spaces is equivalent to the full subcategory of the derived category of coherent sheaves on $\mathcal{X}_{\mathrm{FF}}$ of objects of the form $\mathcal{E}_{\geq 0} \oplus \mathcal{E}_{<0}[1]$, where the torsion-free quotient of $\mathcal{E}_{\geq 0}$ has only nonnegative Harder-Narasimhan slopes, and $\mathcal{E}_{<0}$ is a vector bundle all of whose Harder-Narasimhan slopes are negative. In particular, the category of Banach-Colmez space depends on C only through its tilt $C^\flat$.*

Lecture 16

Drinfeld's lemma for diamonds

In this lecture, we will prove a local analogue of Drinfeld's lemma, thereby giving a first nontrivial argument involving diamonds. This lecture is entirely about fundamental groups. For ease of notation we will omit mention of base points.

16.1 THE FAILURE OF $\pi_1(X \times Y) = \pi_1(X) \times \pi_1(Y)$

It is a basic fact that for connected topological spaces X and Y, the natural map $\pi_1(X \times Y) \to \pi_1(X) \times \pi_1(Y)$ is an isomorphism; let us call this the *Künneth formula for* π_1. Is the same result true if instead X and Y are varieties over a field k, and π_1 is interpreted as the étale fundamental group? The answer is no in general. For instance, suppose k is a non-separably closed field, and $X = Y = \operatorname{Spec} k$. Then $X \times_k Y = \operatorname{Spec} k$ once again, and the diagonal map $\operatorname{Gal}(\overline{k}/k) \to \operatorname{Gal}(\overline{k}/k) \times \operatorname{Gal}(\overline{k}/k)$ is not an isomorphism. So certainly we want to assume that k is separably closed; as perfection does not change the étale site, we can then even assume that k is algebraically closed.

If k is an algebraically closed field of characteristic 0, then the answer becomes yes. If $k = \mathbb{C}$, this can be proved by relating the étale π_1 to the usual π_1; in general, one can prove it by appeal to the Lefschetz principle: X can be descended to a finitely generated field, which can be embedded into $\mathbb{C}$. In fact, one also shows that if k'/k is an extension of algebraically closed fields of characteristic 0, and X is a variety over k, then the natural map $\pi_1(X_{k'}) \to \pi_1(X)$ is an isomorphism. That is, such varieties satisfy *permanence of π_1 under (algebraically closed) base field extension*.

So let us assume that k is an algebraically closed field of characteristic p. Both the Künneth theorem and the permanence of π_1 under base extension hold for *proper* varieties over k. But both properties can fail for non-proper varieties.

Example 16.1.1. Keep the assumption that k is an algebraically closed field of characteristic p. Let $X = \operatorname{Spec} R$ be an affine k-scheme. Then $\operatorname{Hom}(\pi_1(X), \mathbf{F}_p) = H^1_{\text{ét}}(X, \mathbf{Z}/p\mathbf{Z})$. By Artin-Schreier, this group is identified with the cokernel of the endomorphism $f \mapsto f^p - f$ of R. Generally (and particularly if $R = k[T]$) this group is not invariant under base extension, and therefore the same can be said about $\pi_1(X)$. Similarly, the Künneth theorem fails for $R = k[T]$.

The following lemma states that under mild hypotheses, the Künneth theorem holds when permanence of π_1 under base extension is satisfied for one of the factors.

Lemma 16.1.2. Let X and Y be schemes over an algebraically closed field k, with Y qcqs and X connected. Assume that for all algebraically closed extensions k'/k, $Y_{k'}$ is connected, and the natural map $\pi_1(Y_{k'}) \to \pi_1(Y)$ is an isomorphism. Then $\pi_1(X \times Y) \to \pi_1(X) \times \pi_1(Y)$ is an isomorphism.

Proof. This can be proved by the same methods that we will use below for diamonds, and it has been written up by Kedlaya, [Ked19a, Corollary 4.1.23]. □

We also note that conversely, permanence of π_1 under algebraically closed base field extension k'/k is implied by the Künneth formula. Indeed, taking $X = \operatorname{Spec} k'$, we get $\pi_1(Y_{k'}) = \pi_1(\operatorname{Spec} k') \times \pi_1(Y) = \pi_1(Y)$. One might argue that the Künneth formula could maybe only hold for X of finite type, and fail for $X = \operatorname{Spec} k'$; however, this cannot happen, as the Künneth formula for X of finite type implies it in general by a standard approximation argument.

16.2 DRINFELD'S LEMMA FOR SCHEMES

Let us recall the notions of absolute and relative Frobenii. For a scheme $X/\mathbf{F}_p$, let $F_X \colon X \to X$ be the *absolute Frobenius map*: this is the identity on $|X|$ and the pth power map on the structure sheaf. For $f \colon Y \to X$ a morphism of schemes, we have the pullback $F_X^* Y = Y \times_{X, F_X} X$ (this is often denoted $Y^{(p)}$). The *relative Frobenius* $F_{Y/X} \colon Y \to F_X^* Y$ is the unique morphism making the following diagram commute:

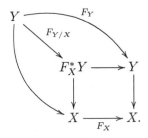

A crucial fact is that $F_{Y/X}$ is an isomorphism when $Y \to X$ is étale.

Definition 16.2.1. Let $X_1, \dots, X_n$ be connected qcqs schemes over $\mathbf{F}_p$ and

$X = X_1 \times \cdots \times X_n$. Consider the *ith partial Frobenius*

$$F_i = 1 \times \cdots \times F_{X_i} \times \cdots \times 1 \colon X_1 \times \cdots \times X_n \to X_1 \times \cdots \times X_n.$$

Let

$$(X_1 \times \cdots \times X_n/\mathrm{p.Fr.})_{\mathrm{f\acute{e}t}}$$

be the category of finite étale maps $Y \to X_1 \times \cdots \times X_n$ equipped with commuting isomorphisms $\beta_i \colon Y \xrightarrow{\sim} F_i^* Y$ such that $\beta_n \circ \cdots \circ \beta_1 = F_{Y/X} \colon Y \xrightarrow{\sim} F_X^* Y$.

Remark 16.2.2. Strictly speaking, the notation $\beta_n \circ \ldots \circ \beta_1$ is an abuse: the morphism β_2 should be the pullback of β_2 through F_1, and so forth.

Remark 16.2.3. To give an object of this category it suffices to produce all but one of the β_i, by the product relation. Thus if $n = 2$, the category $(X_1 \times X_2/\mathrm{p.Fr.})_{\mathrm{f\acute{e}t}}$ is the category of finite étale morphisms $Y \to X_1 \times X_2$ equipped with an isomorphism $\beta \colon Y \xrightarrow{\sim} F_1^* Y$.

This forms a Galois category in the sense of SGA1, so that (after choosing a geometric point s of $X_1 \times \cdots \times X_n$) one can define the fundamental group $\pi_1(X_1 \times \cdots \times X_n/\mathrm{p.Fr.})$; this is the automorphism group of the fiber functor on $(X_1 \times \cdots \times X_n/\mathrm{p.Fr.})_{\mathrm{f\acute{e}t}}$ determined by s.

Theorem 16.2.4 (Drinfeld's lemma for schemes, [Dri80, Theorem 2.1], [Laf97, IV.2, Theorem 4], [Lau, Theorem 8.1.4], [Ked19a, Theorem 4.2.12]). The natural map

$$\pi_1(X_1 \times \cdots \times X_n/\mathrm{p.Fr.}) \to \pi_1(X_1) \times \cdots \times \pi_1(X_n)$$

is an isomorphism.

Example 16.2.5. If $X_1 = X_2 = \mathrm{Spec}\,\mathbf{F}_p$, then $(X_1 \times X_2/\mathrm{p.Fr.})_{\mathrm{f\acute{e}t}}$ is the category of finite étale covers of $\mathrm{Spec}\,\mathbf{F}_p$ equipped with one partial Frobenius; these are indeed parametrized by $\widehat{\mathbf{Z}} \times \widehat{\mathbf{Z}}$.

The crucial step in the proof of Theorem 16.2.4 is to establish permanence of π_1 under extension of the base, once a relative Frobenius is added to the picture; let us only state the relevant permanence property of π_1.

Let $k/\mathbf{F}_p$ be algebraically closed. For a scheme $X/\mathbf{F}_p$, let $\overline{X} := X \otimes_{\mathbf{F}_p} k$; this has a relative Frobenius $F_{\overline{X}/k} \colon \overline{X} \to F_k^* \overline{X} = \overline{X}$. One can then define a category $(\overline{X}/F_{\overline{X}/k})_{\mathrm{f\acute{e}t}}$ and if X is connected a group $\pi_1(\overline{X}/F_{\overline{X}/k})$ (after choosing a geometric point).

Lemma 16.2.6 ([Lau, Lemma 8.12]). Let $X/\mathbf{F}_p$ be a connected scheme. Then $\pi_1(\overline{X}/F_{\overline{X}/k}) \to \pi_1(X)$ is an isomorphism. More generally, if X is not necessarily connected, there is an equivalence of categories between finite étale covers $Y_0 \to X$ and finite étale covers $Y \to \overline{X}$ equipped with an isomorphism $F_{\overline{X}/k}^* Y \xrightarrow{\sim} Y$.

Proof. (Sketch.)

1. The category of finite-dimensional φ-modules (V, φ_V) over k is equivalent to the category of finite-dimensional $\mathbf{F}_p$-vector spaces, via $(V, \varphi_V) \mapsto V^{\varphi=1}$ and its inverse $V_0 \mapsto (V_0 \otimes k, 1 \otimes \varphi_k)$.

2. Let X be projective over $\mathbf{F}_p$. Then there is an equivalence between pairs $(\mathcal{E}, \varphi_{\mathcal{E}})$, where $\mathcal{E}$ is a coherent sheaf on $\overline{X}$, and $\varphi_{\mathcal{E}} \colon F_k^* \mathcal{E} \xrightarrow{\sim} \mathcal{E}$, and coherent sheaves $\mathcal{E}_0/X$. (Describe everything in terms of graded modules, finite-dimensional over k (resp., $\mathbf{F}_p$) in each degree, then use (1). See [Lau, Lemma 8.1.1].)

3. Without loss of generality in the lemma, X is affine, and by noetherian approximation, X is of finite type. By cohomological descent, we can assume X is normal and connected. Choose an embedding $X \hookrightarrow X'$ into a normal projective $\mathbf{F}_p$-scheme. The following categories are equivalent:

a) $Y/\overline{X}$ finite étale with $F_{\overline{X}/k}^* Y \xrightarrow{\sim} Y$,

b) $Y/\overline{X}$ finite étale with $F_k^* Y \xrightarrow{\sim} Y$,

c) $Y'/\overline{X'}$ finite normal with $F_k^* Y' \xrightarrow{\sim} Y'$, such that Y' is étale over the open subset $\overline{X}$,

d) (using (2)) Y_0'/X' finite normal such that Y_0' is étale over X,

e) Y_0/X finite étale.

(The proof of the equivalence of (b) and (c) uses the normalization of $\overline{X'}$ in Y.)

$\square$

16.3   DRINFELD'S LEMMA FOR DIAMONDS

A diamond $\mathcal{D}$ is defined to be connected if it is not the disjoint union of two open subsheaves; equivalently, if $|\mathcal{D}|$ is connected. For a connected diamond $\mathcal{D}$, finite étale covers of $\mathcal{D}$ form a Galois category, so for a geometric point $x \in \mathcal{D}(C, \mathcal{O}_C)$ we can define a profinite group $\pi_1(\mathcal{D}, x)$, such that finite $\pi_1(\mathcal{D}, x)$-sets are equivalent to finite étale covers $\mathcal{E} \to \mathcal{D}$.

We would like to replace all the connected schemes X_i appearing in Drinfeld's lemma with $\operatorname{Spd} \mathbf{Q}_p$. Even though $\mathbf{Q}_p$ has characteristic 0, its diamond $\operatorname{Spd} \mathbf{Q}_p$ admits an absolute Frobenius $F \colon \operatorname{Spd} \mathbf{Q}_p \to \operatorname{Spd} \mathbf{Q}_p$, because after all it is a sheaf on the category Perf of perfectoid affinoids in characteristic p, and there is an absolute Frobenius defined on these.

Let

$$(\operatorname{Spd} \mathbf{Q}_p \times \cdots \times \operatorname{Spd} \mathbf{Q}_p / \mathrm{p.Fr.})_{\text{fét}}$$

be the category of finite étale covers $E \to (\operatorname{Spd} \mathbf{Q}_p)^n$ equipped with commuting isomorphisms $\beta_i \colon E \xrightarrow{\sim} F_i^* E$ (where F_i is the ith partial Frobenius), $i = 1, \ldots, n$

such that

$$\prod_i \beta_i = F_{E/(\mathrm{Spd}\,\mathbf{Q}_p)^n} : E \xrightarrow{\sim} F^* E .$$

As above, this is the same as the category of finite étale covers $E \to (\mathrm{Spd}\,\mathbf{Q}_p)^n$ equipped with commuting isomorphisms $\beta_1, \ldots, \beta_{n-1}$. A new feature of this story is that the action of $F_1^{\mathbf{Z}} \times \cdots \times F_{n-1}^{\mathbf{Z}}$ on $|(\mathrm{Spd}\,\mathbf{Q}_p)^n|$ is free and totally discontinuous. Thus the quotient $(\mathrm{Spd}\,\mathbf{Q}_p)^n/(F_1^{\mathbf{Z}} \times \cdots \times F_{n-1}^{\mathbf{Z}})$ is a diamond, and $((\mathrm{Spd}\,\mathbf{Q}_p)^n/\mathrm{p.Fr.})_{\mathrm{f\acute{e}t}}$ is simply the category of finite étale covers of it.

The version of Drinfeld's lemma one needs for the analysis of the cohomology of moduli spaces of shtukas is the following.

Theorem 16.3.1. $\pi_1((\mathrm{Spd}\,\mathbf{Q}_p)^n/\mathrm{p.Fr.}) \cong G_{\mathbf{Q}_p}^n$.

As before, this is related to a permanence property of π_1 that we will establish first. Define $X = \mathrm{Spd}\,\mathbf{Q}_p/F^{\mathbf{Z}}$. For any algebraically closed nonarchimedean field $C/\mathbf{F}_p$, we consider the base change X_C.

Lemma 16.3.2. For any algebraically closed nonarchimedean field $C/\mathbf{F}_p$, one has $\pi_1(X_C) \cong G_{\mathbf{Q}_p}$.

This will be equivalent to the fact that the Fargues-Fontaine curve is simply connected. Indeed, note that by Proposition 15.2.7

$$
\begin{aligned}
(X_C)_{\mathrm{f\acute{e}t}} &= ((\mathrm{Spd}\,\mathbf{Q}_p/F^{\mathbf{Z}}) \times \mathrm{Spd}\,C)_{\mathrm{f\acute{e}t}} \\
&= (\mathrm{Spd}\,\mathbf{Q}_p \times (\mathrm{Spd}\,C/F_C^{\mathbf{Z}}))_{\mathrm{f\acute{e}t}} \\
&= (\mathcal{X}_{\mathrm{FF}}^{\diamond})_{\mathrm{f\acute{e}t}}
\end{aligned}
$$

and so, using Theorem 10.4.2, we see that

$$(X_C)_{\mathrm{f\acute{e}t}} \cong (\mathcal{X}_{\mathrm{FF}}^{\diamond})_{\mathrm{f\acute{e}t}} \cong (\mathcal{X}_{\mathrm{FF}})_{\mathrm{f\acute{e}t}} .$$

Finally, Lemma 16.3.2 follows from the theorem of Fargues-Fontaine, Theorem 13.5.7, saying that $(\mathcal{X}_{\mathrm{FF}})_{\mathrm{f\acute{e}t}} \cong (X_{\mathrm{FF}})_{\mathrm{f\acute{e}t}} \cong (\mathbf{Q}_p)_{\mathrm{f\acute{e}t}}$. We remark that in this proof, we have used the formalism of diamonds rather heavily to transport finite étale maps between different presentations of a diamond as the diamond of an analytic adic space. The same argument is made much more explicit in [Wei17].

This establishes the desired permanence of π_1 under change of algebraically closed base field. Now we need to detail the argument that this implies the Künneth formula. For this, we will establish a sort of Stein factorization for certain morphisms of diamonds, Proposition 16.3.3. This implies the analogue of Lemma 16.1.2 for diamonds, Proposition 16.3.6. In the following, $\mathrm{Spd}\,k$ denotes the pro-étale sheaf sending any affinoid perfectoid $X = \mathrm{Spa}(R, R^+)$ to $\mathrm{Hom}(k, R)$; note that this is not a diamond as it is not analytic.

Proposition 16.3.3. Let k be a discrete algebraically closed field, and let

$D, X \to \operatorname{Spd} k$ be diamonds, with $D \to \operatorname{Spd} k$ qcqs.[1] Assume that for all algebraically closed nonarchimedean fields C/k, D_C is connected, and $\pi_1(D_C) \to \pi_1(D)$ is an isomorphism. Let $Y = D \times_k X$, and let $\widetilde{Y} \to Y$ be finite étale. There exists a finite étale morphism $\widetilde{X} \to X$ fitting into the diagram

$$\begin{array}{ccc} \widetilde{Y} & \longrightarrow & \widetilde{X} \\ \downarrow & & \downarrow \\ Y & \longrightarrow & X \end{array} \qquad\qquad (16.3.1)$$

such that $\widetilde{Y} \to \widetilde{X}$ has geometrically connected fibers. Furthermore, $\widetilde{X} \to X$ is unique up to unique isomorphism.

Remark 16.3.4. Proposition 16.3.3 can be interpreted as a Stein factorization of the morphism $\widetilde{Y} \to X$.

Proof. By the uniqueness claim and quasi-pro-étale descent of finite étale maps, Theorem 9.1.3, we can assume that X is an affinoid perfectoid space. First we establish the claim of uniqueness. This will follow from the following universal property of the diagram in Eq. (16.3.1): if such a diagram exists, it is the initial object in the category of diagrams

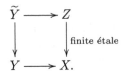

Suppose $\widetilde{X}$ and Z fit into diagrams as above, where $\widetilde{Y} \to \widetilde{X}$ has geometrically connected fibers; we will produce a unique morphism $\widetilde{X} \to Z$ over X and under $\widetilde{Y}$.

First, assume that $X = \operatorname{Spa}(C, C^+)$, where C is an algebraically closed nonarchimedean field, and $C^+ \subset C$ is an open and bounded valuation subring. In this case $\widetilde{X}$ and Z are finite disjoint unions of copies of X, where $\widetilde{X}$ is determined by the connected components of $\widetilde{Y}$. As the map $\widetilde{Y} \to Z$ contracts connected components, it factors uniquely over $\widetilde{X}$, as desired.

In general, a map $\widetilde{X} \to Z$ is determined by its behavior on geometric points (as the equalizer of two maps $\widetilde{X} \rightrightarrows Z$ over X is again finite étale over X, and so checking whether it equals $\widetilde{X}$ can be done on geometric points). This shows that the map is unique if it exists. To check existence of $\widetilde{X} \to Z$, we need the following analogue of Lemma 7.4.6 for diamonds:

---

[1]This is very different from asking that D be qcqs, as $\operatorname{Spd} k$ is not quasiseparated! For example, $\operatorname{Spd} \mathbf{Q}_p / F^{\mathbf{Z}} \times \operatorname{Spd} k$ is not quasiseparated, but $\operatorname{Spd} \mathbf{Q}_p / F^{\mathbf{Z}} \times \operatorname{Spd} k \to \operatorname{Spd} k$ is qcqs; cf. Example 17.3.3.

Lemma 16.3.5 ([Sch17, Proposition 11.23 (i)]). For any geometric point $\mathrm{Spa}(C, C^+) \to X$, the functors

$$2\text{-}\varinjlim_{U \ni x} (Y \times_X U)_{\mathrm{fét}} \to (Y \times_X \mathrm{Spa}(C, C^+))_{\mathrm{fét}}$$

and

$$2\text{-}\varinjlim_{U \ni x} U_{\mathrm{fét}} \to \mathrm{Spa}(C, C^+)_{\mathrm{fét}}$$

are equivalences, where U runs over étale neighborhoods of x in X.

As we have a factorization $\tilde{X} \to Z$ (over X and under $\tilde{Y}$) after pullback to any geometric point of X, this lemma allows us to spread it into a small neighborhood. Indeed, the condition that it is a map under $\tilde{Y}$ can be reformulated in terms of the commutativity of a corresponding diagram of finite étale covers of Y.

Thus, if $\tilde{X}$ exists, it is unique up to unique isomorphism. It remains to show that $\tilde{X}$ exists, and for this we can again localize. If $\mathrm{Spa}(C, C^+) \to X$ is a geometric point, then $\tilde{Y} \times_X \mathrm{Spa}(C, C^+) = \bigsqcup_{i=1}^n \tilde{Y}_i$ decomposes into a disjoint union of connected finite étale covers of $Y \times_X \mathrm{Spa}(C, C^+) = D \times_k \mathrm{Spa}(C, C^+)$. As

$$(D \times_k \mathrm{Spa}(C, C^+))_{\mathrm{fét}} \cong (D \times_k \mathrm{Spa}(C, \mathcal{O}_C)_{\mathrm{fét}} \cong D_{\mathrm{fét}}$$

by assumption, each $\tilde{Y}_i$ is of the form $\tilde{D}_i \times_D Y$ for some connected finite étale cover $\tilde{D}_i \to D$. Applying Lemma 16.3.5 again, we have an isomorphism $\tilde{Y} \times_X U \cong \bigsqcup_{i=1}^n \tilde{D}_i \times_k U$ on an étale neighborhood U of the given geometric point. Replacing X by U, we can assume to have such a decomposition globally. But then taking $\tilde{X} = \bigsqcup_{i=1}^n X$ has the desired property. □

Proposition 16.3.6. Let k be a discrete algebraically closed field, and let $D, X \to \mathrm{Spd}\, k$ be connected diamonds, with $D \to \mathrm{Spd}\, k$ qcqs. Assume that for all algebraically closed nonarchimedean fields C/k, D_C is connected, and $\pi_1(D_C) \to \pi_1(D)$ is an isomorphism. Then $D \times_k X$ is also connected, and the map $\pi_1(D \times_k X) \to \pi_1(D) \times \pi_1(X)$ is an isomorphism.

Proof. Let $Y = D \times_k X$.

1. We show that Y is connected. Assume $Y = Y_1 \sqcup Y_2$, with Y_i open and closed. We claim that the locus $\{x \in X | (Y_1)_x = \emptyset\}$ is open, and similarly for Y_2. If both of them are nonempty, we get a clopen decomposition of X, contradicting our assumption that X is connected. For the claim, we can assume that X is affinoid perfectoid by passing to a pro-étale cover. Suppose $\bar{x} = \mathrm{Spa}(C, C^+) \to X$ is a geometric point such that $(Y_1)_x = \emptyset$. Since the morphism $D \to \mathrm{Spd}\, k$ is quasicompact, so is the base change $Y = D \times_k X \to X$, and therefore Y and its closed subset Y_1 are also quasicompact. Choose

an affinoid perfectoid space Z surjecting onto Y_1. As

$$\overline{x} = \varprojlim_U U,$$

where U runs over étale neighborhoods of $\overline{x}$ in X, we also get

$$\emptyset = Z_{\overline{x}} = \varprojlim_U Z \times_X U.$$

On the level of topological spaces, we have here an empty inverse limit of spectral spaces along spectral maps, which implies that $Z \times_X U = \emptyset$ for some U (cf. the proof of Lemma 8.2.4), and therefore $Y_1 \times_X U$ is empty.

2. We show that $\pi_1(D \times_k X) \to \pi_1(X)$ is surjective. By (1), for all finite étale covers $\widetilde{X} \to X$ with $\widetilde{X}$ connected, we have that $D \times_k \widetilde{X}$ is connected. Now, a connected étale cover of X corresponds to a continuous transitive action of $\pi_1(X)$ on a finite set. So the claim is equivalent to saying that every such action restricts to a transitive action of $\pi_1(D \times_k X)$. It is a simple exercise to see that this is equivalent to the surjectivity of $\pi_1(D \times_k X) \to \pi_1(X)$.

3. Let $\overline{x} = \mathrm{Spa}(C, \mathcal{O}_C) \to X$ be a geometric point of rank 1. We claim that the sequence

$$\pi_1(D_C) \to \pi_1(D \times_k X) \to \pi_1(X)$$

is exact in the middle.

Given a finite quotient G of $\pi_1(D \times_k X)$, corresponding to a Galois cover $\widetilde{Y}$ of $Y = D \times_k X$, there exists a quotient $G \to H$, corresponding to a Galois cover $\widetilde{X} \to X$ as in Proposition 16.3.3. Consider the homomorphism $\pi_1(D_C) \to \ker(G \to H)$: the cosets of its image correspond to connected components in the fiber of $\widetilde{Y} \to \widetilde{X}$ over a geometric point. But the fibers of $\widetilde{Y} \to \widetilde{X}$ are geometrically connected, so $\pi_1(D_C) \to \ker(G \to H)$ is surjective. This suffices to prove the claim.

Putting together (1), (2), and (3), we have the following diagram of groups, where the top row is an exact sequence:

$$\pi_1(D_C) \longrightarrow \pi_1(D \times_k X) \longrightarrow \pi_1(X) \longrightarrow 1$$
$$\searrow_{\sim} \qquad \downarrow$$
$$\pi_1(D)$$

This shows that $\pi_1(D \times_k X) \to \pi_1(D) \times \pi_1(X)$ is an isomorphism. $\qquad \square$

Proof of Theorem 16.3.1. We know by Lemma 16.3.2 that

$$D = \mathrm{Spd}\, \mathbf{Q}_p / F^{\mathbf{Z}} \times \mathrm{Spd}\, k$$

satisfies the hypothesis of Proposition 16.3.6. By induction, this implies that

$$\pi_1(D\times_k D\times_k\cdots\times_k D) = \pi_1(D)\times\pi_1(D\times_k\cdots\times_k D) = \ldots = \pi_1(D)\times\cdots\times\pi_1(D)\,.$$

Now $\pi_1(D) = G_{\mathbf{Q}_p}$, and taking $k = \overline{\mathbf{F}}_p$, one one can identify $\pi_1(D\times_k\cdots\times_k D)$ with $\pi_1((\operatorname{Spd}\mathbf{Q}_p)^n/(\text{p.Fr.}))$. $\qquad\square$

Lecture 17

The v-topology

In this lecture, we develop a powerful technique for proving results about diamonds. There is a topology even finer than the pro-étale topology, the *v-topology*, which is reminiscent of the fpqc topology on schemes but which is more "topological" in nature. The class of v-covers is extremely general, which will reduce many proofs to very simple base cases. We will give a sample application of this philosophy in the appendix to this lecture by establishing a general classification of p-divisible groups over integral perfectoid rings in terms of Breuil-Kisin-Fargues modules, generalizing the theory of Lecture 14 for p-divisible groups over $\mathcal{O}_C$.

Another use of the v-topology is to prove that certain pro-étale sheaves on Perf are diamonds without finding an explicit pro-étale cover. This will be established as Theorem 17.3.9.

17.1 THE v-TOPOLOGY ON Perfd

We consider the following big topology on the category Perfd of all perfectoid spaces (not necessarily of characteristic p).

Definition 17.1.1. The *v-topology* on Perfd is the topology generated by open covers and *all* surjective maps of affinoids. In other words, $\{f_i\colon X_i \to Y\}_{i\in I}$ is a cover if and only if for all quasicompact open subsets $V \subset Y$ there is some finite subset $I_V \subset I$ and quasicompact open $U_i \subset X_i$ for $i \in I_U$ such that $V = \bigcup_{i\in I_U} f_i(U_i)$.

Remark 17.1.2. We ignore set-theoretic issues; they are adressed in [Sch17] by choosing suitable cut-off cardinals (and without the use of Grothendieck universes).

It may appear at first sight that the v-topology admits far too many covers to be a workable notion. By Proposition 4.3.3, we still see that v-covers induce quotient maps on topological spaces. Moreover, we have the following surprising theorem, which shows that the structure sheaf is a sheaf for the v-topology on Perfd, just as it is for the fpqc topology on schemes.

Theorem 17.1.3 ([Sch17, Theorem 8.7, Proposition 8.8]). The functors $X \mapsto H^0(X, \mathcal{O}_X)$ and $X \mapsto H^0(X, \mathcal{O}_X^+)$ are sheaves on the v-site. Moreover if X is affinoid then $H_v^i(X, \mathcal{O}_X) = 0$ for $i > 0$, and $H_v^i(X, \mathcal{O}_X^+)$ is almost zero for $i > 0$.

Proof. (Sketch.) By pro-étale descent, it suffices to check these assertions for totally disconnected spaces. But then one gets an automatic flatness result: □

Proposition 17.1.4 ([Sch17, Proposition 7.23]). Let $X = \mathrm{Spa}(R, R^+)$ be a totally disconnected perfectoid space, and let $(R, R^+) \to (S, S^+)$ be a map to any Huber pair, and let $\varpi \in R$ be a pseudo-uniformizer. Then S^+/ϖ is flat over R^+/ϖ, and faithfully flat if $\mathrm{Spa}(S, S^+) \to \mathrm{Spa}(R, R^+)$ is surjective.

Proof. (Sketch.) The idea is to check flatness on stalks of $\mathrm{Spec}\, R^+/\varpi$. After passing to stalks, we are reduced to the case that X is connected, in which case $(R, R^+) = (K, K^+)$, where K^+ is a valuation ring. Thus, $K^+ \to S^+$ is flat (as it is equivalent to torsion-freeness), and so $K^+/\varpi \to S^+/\varpi$ is flat. □

Corollary 17.1.5. Representable presheaves are sheaves on the v-site.

Proof. As in the pro-étale case—cf. Proposition 8.2.8—this follows from Theorem 17.1.3. □

In fact, more generally, diamonds are v-sheaves. This is an analogue of a result of Gabber saying that algebraic spaces are fpqc sheaves.

Proposition 17.1.6 ([Sch17, Proposition 11.9]). Let Y be a diamond. Then Y is a v-sheaf.

Moreover, one has the following descent result, already mentioned (in the pro-étale case) in Theorem 9.1.3 (i).

Corollary 17.1.7 ([Sch17, Proposition 9.3]). The functor which assigns to a totally disconnected affinoid perfectoid X the category $\{Y/X$ affinoid perfectoid$\}$ is a stack for the v-topology.

Proof. (Sketch.) This follows from Proposition 17.1.4 and faithfully flat descent. The actual argument is rather subtle because of rings of integral elements to which faithfully flat descent cannot be applied directly. □

Another important descent result is the following result on vector bundles.

Lemma 17.1.8. The fibered category sending any $X \in \mathrm{Perfd}$ to the category of locally finite free $\mathcal{O}_X$-modules is a stack on the v-site on Perfd.

Proof. Suppose $\widetilde{X} \to X$ is a surjective morphism of perfectoid affinoids, with $X = \mathrm{Spa}(R, R^+)$ and $\widetilde{X} = \mathrm{Spa}(\widetilde{R}, \widetilde{R}^+)$. We will show that the base change functor from finite projective R-modules to finite projective $\widetilde{R}$-modules equipped with a descent datum is an equivalence of categories; this suffices by Theorem 5.2.8. Full faithfulness follows from the sheaf property of the structure presheaf on the v-site, Theorem 17.1.3.

Essential surjectivity can be checked locally, as vector bundles glue over open covers by Theorem 5.2.8. Consider first the case where $R = K$ is a perfectoid field. Then $\widetilde{R}$ is a nonzero Banach K-algebra, and we need to prove that finite projective $\widetilde{R}$-modules with descent data descend to K. We can write $\widetilde{R}$ as a ω_1-filtered colimit of topologically countably generated Banach K-algebras; thus, any descent datum of finite projective $\widetilde{R}$-modules is defined over a topologically countably generated Banach K-algebra, and we can assume that $\widetilde{R}$ itself is topologically countably generated. In particular, by [BGR84, §2.7, Theorem 4], $\widetilde{R}$ is topologically free as a K-Banach space. This implies that a complex C of K-Banach spaces is exact if and only if $C \widehat{\otimes}_K \widetilde{R}$ is exact.

Recall the proof of faithfully flat descent for modules (for instance, [Sta, Tag 023F]). The same proof now carries over to $\widetilde{R}/K$ (where now all occuring sums are ϖ-adically convergent instead of finite).

Thus we have established Lemma 17.1.8 over a point. Returning to the general case, suppose that $\widetilde{M}/\widetilde{R}$ is a finite projective module equipped with a descent datum

$$\widetilde{M}\widehat{\otimes}_{\widetilde{R},i_1}(\widetilde{R}\widehat{\otimes}_R\widetilde{R}) \cong \widetilde{M}\widehat{\otimes}_{\widetilde{R},i_2}(\widetilde{R}\widehat{\otimes}_R\widetilde{R}),$$

where $i_1, i_2 \colon \widetilde{R} \rightrightarrows \widetilde{R}\widehat{\otimes}_R\widetilde{R}$ are the two obvious homomorphisms. We wish to descend $\widetilde{M}$ to M/R.

After replacing $\widetilde{X}$ with an open cover, we may assume that $\widetilde{M} = \widetilde{R}^r$ is free. The descent datum is given by a matrix $B \in \mathrm{GL}_r(\widetilde{R}\widehat{\otimes}_R\widetilde{R})$ that satisfies a cocycle condition. Pick any $x \in X$ with completed residue field $K(x)$. We can descend the fiber of $\widetilde{M}$ over x, so there exists $A_x \in GL_r(\widetilde{R}\widehat{\otimes}_R K(x))$ such that $B(x) = \mathrm{pr}_1^*(A_x)\,\mathrm{pr}_2^*(A_x)^{-1} \in \mathrm{GL}_r(\widetilde{R}\widehat{\otimes}_R\widetilde{R}\widehat{\otimes}_R K(x))$. Approximate A_x by some $A_U \in \mathrm{GL}_r(\widetilde{R}\widehat{\otimes}_R \mathcal{O}_X(U))$ for a rational neighborhood $U \subset X$ of x in X. After conjugating by A_U, we may assume $B|_U \in \mathrm{GL}_r(\widetilde{R}^+\widehat{\otimes}_{R^+}\widetilde{R}^+\widehat{\otimes}_{R^+}\mathcal{O}_X^+(U))$, and even that $B|_U \equiv 1 \pmod{\varpi}$ for a pseudo-uniformizer $\varpi \in R$. Replacing X by U, we may assume that $B \equiv 1 \pmod{\varpi}$ to begin with.

Now $(B-1)/\varpi$ modulo ϖ satisfies the *additive* cocycle condition, so it lives in $\check{H}^1(\widetilde{X}/X, M_r(\mathcal{O}_X^+/\varpi))$, but this group is almost zero by Theorem 17.1.3. Thus we can conjugate B by a matrix in $1 + \varpi^{1-\epsilon}M_r(\widetilde{R}^+)$ so as to assume that $B \equiv 1 \pmod{\varpi^{2-\epsilon}}$ for some $\varepsilon > 0$. Continuing, we find that we can conjugate B to 1, as desired. $\qquad\square$

Corollary 17.1.9. For any $n \geq 1$, including $n = \infty$, the functor sending an affinoid perfectoid space $S = \mathrm{Spa}(R, R^+)$ to the category of vector bundles on $B_{\mathrm{dR}}^+(R^\sharp)/\xi^n$ is a v-stack.

Proof. The case $n = 1$ is the previous result. The case $n > 1$ follows by a simple induction using the vanishing of $H_v^1(S, \mathcal{E})$ for any vector bundle $\mathcal{E}$ on S, and the case $n = \infty$ by passage to the limit. $\qquad\square$

17.2   SMALL v-SHEAVES

Definition 17.2.1. A v-sheaf $\mathcal{F}$ on Perf is small if there is a surjective map of v-sheaves $X \to \mathcal{F}$ from (the sheaf represented by) a perfectoid space X.

A slightly surprising statement is that any small v-sheaf admits something like a geometric structure.

Proposition 17.2.2 ([Sch17, Proposition 12.3]). *Let $\mathcal{F}$ be a small v-sheaf, and let $X \to \mathcal{F}$ be a surjective map of v-sheaves from a diamond X (e.g., a perfectoid space). Then $R = X \times_{\mathcal{F}} X$ is a diamond, and $\mathcal{F} = X/R$ as v-sheaves.*

Proof. (Sketch.) Note that $R \subset X \times X$ is a sub-v-sheaf, where $X \times X$ is a diamond. The key statement now is that any sub-v-sheaf of a diamond is again a diamond; cf. [Sch17, Proposition 11.10]. $\qquad\square$

This makes small v-sheaves accessible in a two-step procedure: First analyze diamonds as quotients of perfectoid spaces by representable equivalence relations, and then small v-sheaves as quotients of perfectoid spaces by diamond equivalence relations.[1]

For example, we can define the underlying topological space of a small v-sheaf.

Definition 17.2.3. Let $\mathcal{F}$ be a small v-sheaf, and let $X \to \mathcal{F}$ be a surjective map of v-sheaves from a diamond X, with $R = X \times_{\mathcal{F}} X$. Then the underlying topological space of $\mathcal{F}$ is $|\mathcal{F}| = |X| \, / \, |R|$.

To ensure that this is well-defined and functorial, one uses Proposition 4.3.3; cf. [Sch17, Proposition 12.7].

In the rest of this lecture, we will present a criterion for when a small v-sheaf is a diamond. The first condition is that it is spatial, which is already an interesting condition for diamonds.

17.3   SPATIAL v-SHEAVES

We want to single out those diamonds for which $|\mathcal{F}|$ is well-behaved.

Definition 17.3.1. A v-sheaf $\mathcal{F}$ is *spatial* if

1. $\mathcal{F}$ is qcqs (in particular, small), and
2. $|\mathcal{F}|$ admits a neighborhood basis consisting of $|\mathcal{G}|$, where $\mathcal{G} \subset \mathcal{F}$ is quasicompact open.

---

[1] In the original lectures, a new term of "v-diamonds" was introduced, until we noticed that, if formulated correctly, this is essentially no condition!

More generally, $\mathcal{F}$ is *locally spatial* if it admits a covering by spatial open sub-sheaves.

Remark 17.3.2. 1. For algebraic spaces, (1) implies (2); however (1) does not imply (2) in the context of small v-sheaves, or even diamonds. See Example 17.3.6 below.
2. If $\mathcal{F}$ is quasicompact, then so is $|\mathcal{F}|$. Indeed, any open cover of $|\mathcal{F}|$ pulls back to a cover of $\mathcal{F}$. However, the converse need not hold true, but it does when $\mathcal{F}$ is locally spatial; cf. [Sch17, Proposition 12.14 (iii)].
3. If $\mathcal{F}$ is quasiseparated, then so is any subsheaf of $\mathcal{F}$. Thus if $\mathcal{F}$ is spatial, then so is any quasicompact open subsheaf.

Example 17.3.3. Let K be a perfectoid field in characteristic p, and let $\mathcal{F} = \operatorname{Spa} K/\operatorname{Frob}^{\mathbf{Z}}$, so that $|\mathcal{F}|$ is one point. Then $\mathcal{F}$ is not quasiseparated. Indeed if $X = Y = \operatorname{Spa} K$ (which are quasicompact), then $X \times_{\mathcal{F}} Y$ is a disjoint union of $\mathbf{Z}$ copies of $\operatorname{Spa} K$, and so is not quasicompact. In particular $\mathcal{F}$ is not spatial. However, $\mathcal{F} \times \operatorname{Spa} \mathbf{F}_p((t^{1/p^{\infty}})) = (\mathbf{D}_K^*/\operatorname{Frob}^{\mathbf{Z}})^{\diamond}$ *is spatial.*

Proposition 17.3.4 ([Sch17, Proposition 12.13]). Let $\mathcal{F}$ be a spatial v-sheaf. Then $|\mathcal{F}|$ is a spectral space, and for any perfectoid space X with a map $X \to \mathcal{F}$, the map $|X| \to |\mathcal{F}|$ is a spectral map.

Proof. (Sketch that $|\mathcal{F}|$ is spectral.) Choose a surjection $X \to \mathcal{F}$ from an affinoid perfectoid space and let $R = X \times_{\mathcal{F}} X$, which is a qcqs diamond (in fact, itself spatial; cf. [Sch17, Proposition 11.20]). By [Sch17, Lemma 2.9], it is enough to construct many quasicompact open subsets $U \subset |X|$ that are stable under the equivalence relation $|R|$. By (2) in Definition 17.3.1, we can just take the preimages of $|\mathcal{G}|$ for $\mathcal{G} \subset \mathcal{F}$ quasicompact open. Since $\mathcal{G}$ is quasicompact and $\mathcal{F}$ is quasiseparated, $\mathcal{G} \times_{\mathcal{F}} X \subset X$ is still quasicompact, and so $|\mathcal{G} \times_{\mathcal{F}} X| \subset |X|$ is a quasicompact open subset. $\qquad\square$

To check whether a small v-sheaf is spatial, we can use the following proposition.

Proposition 17.3.5 ([Sch17, Lemma 2.10]). Let X be a spectral space, and $R \subset X \times X$ a spectral equivalence relation such that each $R \to X$ is open and spectral. Then X/R is a spectral space, and $X \to X/R$ is spectral.

Proof. (Sketch.) We need to produce many $U \subset X$ which are qc open and R-stable. Let $s, t \colon R \to X$ be the maps to X. Let $V \subset X$ be any qc open. Then $s^{-1}(V) \subset R$ is qc open (since $R \to X$ is spectral), so $t(s^{-1}(V)) \subset X$ is qc open (since $R \to X$ is open) and R-stable. $\qquad\square$

Remark 17.3.6. There are counterexamples to Proposition 17.3.5 if $R \to X$ is generalizing but not open. For example, even when X and R are profinite sets, one can produce any compact Hausdorff space as X/R; indeed, if T is any compact Hausdorff space, one can find a surjection $X \to T$ from a profinite set

X (e.g., the Stone-Cech compactification of T considered as a discrete set), and then $R \subset X \times X$ is closed, and thus profinite itself. Repeating this construction in the world of diamonds, i.e. taking $\operatorname{Spa} K \times \underline{X} / \operatorname{Spa} K \times \underline{R}$, produces a qcqs diamond $\mathcal{D}$ with $|\mathcal{D}| = T$.

Corollary 17.3.7. Let $\mathcal{F}$ be a small v-sheaf. Assume there exists a presentation $R \rightrightarrows X \to \mathcal{F}$, where R and X are spatial v-sheaves (e.g., qcqs perfectoid spaces), and each $R \to X$ is open. Then $\mathcal{F}$ is spatial.

Proof. Since X is quasicompact, $\mathcal{F}$ is quasicompact. Since R is quasicompact, $\mathcal{F}$ is quasiseparated. Then Proposition 17.3.5 shows that $|\mathcal{F}| = |X| / |R|$ is spectral, and $|X| \to |\mathcal{F}|$ is spectral. Any quasicompact open $U \subset |\mathcal{F}|$ defines an open subdiamond $\mathcal{G} \subset \mathcal{F}$ covered by $\mathcal{G} \times_{\mathcal{F}} X \subset X$, which is quasicompact. Thus $\mathcal{G}$ itself is quasicompact. $\qquad\square$

Proposition 17.3.8. If X is a qcqs analytic adic space over $\operatorname{Spa} \mathbf{Z}_p$, then $X^\diamond$ is spatial.

Proof. By finding a finite cover of X by affinoid perfectoid spaces, we see that $X^\diamond$ is also quasicompact. By Proposition 10.3.7, $|X^\diamond| \cong |X|$; this implies that $|X^\diamond|$ has a basis of opens $|U|$, where $U \subset X$ is quasicompact open. By Proposition 10.3.6, these correspond to open subdiamonds $U^\diamond \subset X^\diamond$. $\qquad\square$

We can now state the main theorem of today's lecture, which says that a spatial v-sheaf is a diamond as soon as its points are sufficiently nice.

Theorem 17.3.9 ([Sch17, Theorem 12.18]). Let $\mathcal{F}$ be a spatial v-sheaf. Assume that for all $x \in |\mathcal{F}|$, there is a quasi-pro-étale map $X_x \to \mathcal{F}$ from a perfectoid space X_x such that x lies in the image of $|X_x| \to |\mathcal{F}|$. Then $\mathcal{F}$ is a diamond.

Proof. (Sketch.) It is enough to find a quasi-pro-étale surjection $X \to \mathcal{F}$ from an affinoid perfectoid space X. For this, we "simplify" the space $\mathcal{F}$ in several steps. More specifically, we replace $\mathcal{F}$ by the inverse limit over "all" étale covers; after such a reduction, one can assume that $\mathcal{F}$ has no nonsplit étale covers. This step is actually the hardest, as one needs to ensure that the "spatial" condition is preserved; we refer to [Sch17, Lemmas 12.16, 12.17]. In this step, one needs to ensure that all étale covers can be taken to be quasicompact, which is a consequence of the "spatial" condition; this is clear for open covers, but in fact extends to étale covers. The goal is now to show that after this reduction, $\mathcal{F}$ is actually representable by a perfectoid space; cf. [Sch17, Proposition 12.20].

First we check that all connected components are representable. Let $K \subset \mathcal{F}$ be a connected component. Then K has a unique closed point x. (This is because, after our reduction, every cover of $\mathcal{F}$ by quasicompact opens is split.) Let $X_x = \operatorname{Spa}(C, C^+) \to \mathcal{F}$ be a quasi-pro-étale morphism such that x lies in the image of $|X_x|$. Here we can assume that C is an algebraically closed nonarchimedean field and $C^+ \subset C$ is an open and bounded valuation subring. Then the image of $|X_x|$ is exactly K.

We claim that the map $X_x \to \mathcal{F}$ is an isomorphism. For simplicity, we assume that $C^+ = \mathcal{O}_C$ in the argument. Then the product $X_x \times_K X_x$ is of the form $\mathrm{Spa}(C, \mathcal{O}_C) \times \underline{G}$ for some profinite set G, which automatically gets a group structure through the equivalence relation structure. Then $K = X_x/\underline{G}$. For any open subgroup $H \subset G$, $X_x/\underline{H} \to K$ is finite étale. By a variant of Theorem 7.4.8 for small v-sheaves (cf. [Sch17, Lemma 12.17]), this finite étale cover spreads to a small neighborhood of x, which together with a complementary open subset not containing x forms an étale cover of $\mathcal{F}$. Using our assumption about no nonsplit étale covers, this implies that the cover was trivial to start with, so that in fact G is trivial, and so $K = \mathrm{Spd}(C, C^+)$.

Thus every connected component of $\mathcal{F}$ really is a geometric "point" of the form $\mathrm{Spd}(C, C^+)$. Finally, one shows that this implies that $\mathcal{F}$ itself is representable, cf. [Sch17, Lemma 12.21]. $\qquad\qquad\qquad\qquad\qquad\qquad\qquad\qquad\qquad\qquad\qquad\qquad\qquad\qquad\square$

17.4 MORPHISMS OF V-SHEAVES

There are several useful lemmas about morphisms of v-sheaves. First, we have an analogue of Proposition 8.3.3 characterizing injective maps.

Lemma 17.4.1 ([Sch17, Proposition 12.15]). Let $f : \mathcal{G} \to \mathcal{F}$ be a map of small v-sheaves, and assume that f is qcqs, or that $\mathcal{G}$ and $\mathcal{F}$ are locally spatial. The following conditions are equivalent.

1. The map f is injective.
2. For all affinoid fields (C, C^+) where C is algebraically closed, the map $\mathcal{G}(C, C^+) \to \mathcal{F}(C, C^+)$ is injective.
3. The map $|\mathcal{G}| \to |\mathcal{F}|$ is injective and for all perfectoid spaces T, the map

$$\mathcal{G}(T) \to \mathcal{F}(T) \times_{C^0(|T|,|\mathcal{F}|)} C^0(|T|, |\mathcal{G}|)$$

is bijective, i.e. $\mathcal{G} = \mathcal{F} \times_{|\mathcal{F}|} |\mathcal{G}|$.

We can thus define closed immersions.

Definition 17.4.2. A map $f : \mathcal{G} \to \mathcal{F}$ of small v-sheaves is a closed immersion if it is quasicompact, injective, and $|\mathcal{G}| \to |\mathcal{F}|$ is a closed immersion.

Using [Sch17, Corollary 10.6], this is equivalent to [Sch17, Definition 10.7 (ii)]. For a map of affinoid perfectoid spaces $\mathrm{Spa}(S, S^+) \to \mathrm{Spa}(R, R^+)$, it does not imply that $R \to S$ is surjective, i.e. that the map is *Zariski closed*; we refer to [Sch17, Section 5] for a discussion of the relation. Now we can also define separated maps.

Definition 17.4.3 ([Sch17, Definition 10.7 (iii)]). A map $f : \mathcal{G} \to \mathcal{F}$ of small v-sheaves is separated if $\Delta_f : \mathcal{G} \to \mathcal{G} \times_{\mathcal{F}} \mathcal{G}$ is a closed immersion.

There is a valuative criterion.

Proposition 17.4.4 ([Sch17, Proposition 10.9]). A map $f : \mathcal{G} \to \mathcal{F}$ of small v-sheaves is separated if and only if it is quasiseparated and for all perfectoid affinoid fields (K, K^+) and any diagram

$$
\begin{array}{ccc}
\mathrm{Spa}(K, \mathcal{O}_K) & \longrightarrow & \mathcal{G} \\
\Big\downarrow & \nearrow & \Big\downarrow f \\
\mathrm{Spa}(K, K^+) & \longrightarrow & \mathcal{F}
\end{array}
$$

there exists at most one dotted arrow making the diagram commute.

There is also a notion of proper maps.

Definition 17.4.5 ([Sch17, Definition 18.1]). A map $f : \mathcal{G} \to \mathcal{F}$ of small v-sheaves is proper if it is quasicompact, separated, and universally closed, i.e. after any pullback the map $|\mathcal{G}| \to |\mathcal{F}|$ is closed.

Again, there is a valuative criterion.

Proposition 17.4.6 ([Sch17, Proposition 18.3]). A map $f : \mathcal{G} \to \mathcal{F}$ of small v-sheaves is proper if and only if it is qcqs and for all perfectoid affinoid fields (K, K^+) and any diagram

$$
\begin{array}{ccc}
\mathrm{Spa}(K, \mathcal{O}_K) & \longrightarrow & \mathcal{G} \\
\Big\downarrow & \nearrow & \Big\downarrow f \\
\mathrm{Spa}(K, K^+) & \longrightarrow & \mathcal{F}
\end{array}
$$

there exists a unique dotted arrow making the diagram commute.

In practice, there are many examples where the functors are independent of R^+ and hence seem to satisfy the valuative criteria, except for the condition that f is quasicompact. Such maps are called partially proper.

Definition 17.4.7 ([Sch17, Definition 18.4]). A map $f : \mathcal{G} \to \mathcal{F}$ of small v-sheaves is partially proper if it is quasiseparated and for every affinoid perfectoid space $\mathrm{Spa}(R, R^+)$ and any diagram

$$
\begin{array}{ccc}
\mathrm{Spa}(R, R^\circ) & \longrightarrow & \mathcal{G} \\
\Big\downarrow & \nearrow & \Big\downarrow f \\
\mathrm{Spa}(R, R^+) & \longrightarrow & \mathcal{F}
\end{array}
$$

there exists a unique dotted arrow making the diagram commute.

We note that uniqueness is equivalent to separatedness by [Sch17, Proposition 10.10]. In particular, we get the following corollary.

Corollary 17.4.8. Let $f : \mathcal{G} \to \mathcal{F}$ be a map of small v-sheaves.

1. The map f is proper if and only if it is partially proper and quasicompact.
2. The map f is a closed immersion if and only if it is proper and for all algebraically closed nonarchimedean fields C of characteristic p, the map $\mathcal{G}(C, \mathcal{O}_C) \to \mathcal{F}(C, \mathcal{O}_C)$ is injective.

Proof. Part (1) follows from Proposition 17.4.6. In part (2), we have to prove that if f is proper and satisfies the injectivity conditon, then f is a closed immersion. By definition of closed immersions and proper maps, it suffices to see that f is injective. Using Lemma 17.4.1, it suffices to check that $\mathcal{G}(C, C^+) \to \mathcal{F}(C, C^+)$ is bijective for all affinoid fields (C, C^+) with C algebraically closed. But the valuative criterion reduces this to the case of $(C, \mathcal{O}_C)$-points, as desired. □

Moreover, we have a simple characterization of surjective maps.

Lemma 17.4.9 ([Sch17, Lemma 12.11]). Let $f : \mathcal{G} \to \mathcal{F}$ be a map of small v-sheaves. If f is surjective as a map of v-sheaves, then $|\mathcal{G}| \to |\mathcal{F}|$ is a quotient map. Conversely, if f is quasicompact and $|\mathcal{G}| \to |\mathcal{F}|$ is surjective, then f is surjective as a map of v-sheaves.

Note that the condition that $|\mathcal{G}| \to |\mathcal{F}|$ is surjective is satisfied for example if for all affinoid fields (C, C^+) with C algebraically closed, the map $\mathcal{G}(C, C^+) \to \mathcal{F}(C, C^+)$ is surjective.

Combining this with Lemma 17.4.1, we get the following characterization of isomorphisms.

Corollary 17.4.10 ([Sch17, Lemma 12.5]). Let $f : \mathcal{G} \to \mathcal{F}$ be a qcqs map of small v-sheaves. Then f is an isomorphism if and only if for all affinoid fields (C, C^+) with C algebraically closed, the map $\mathcal{G}(C, C^+) \to \mathcal{F}(C, C^+)$ is a bijection.

These results show that a surprising amount can be deduced from quasicompactness/quasiseparatedness together with the behavior on geometric points.

Appendix to Lecture 17:
Dieudonné theory over perfectoid rings

In Theorem 14.4.1, we established a classification result for p-divisible groups over the ring of integers $\mathcal{O}_C$ in an algebraically closed nonarchimedean extension $C/\mathbf{Q}_p$. Here, we wish to generalize this to a classification result for p-divisible groups over general integral perfectoid rings, by using v-descent. In fact, we will use v-descent in the setting of perfect schemes as established in [BS17].

We recall the definition of integral perfectoid rings from [BMS18, Section 3].

Definition 17.5.1. An integral perfectoid ring is a p-complete $\mathbf{Z}_p$-algebra R such that Frobenius is surjective on R/p, there is some element $\pi \in R$ such that $\pi^p = pu$ for a unit $u \in R^\times$, and the kernel of $\theta : A_{\inf}(R) = W(R^\flat) \to R$ is principal, where $R^\flat = \varprojlim_{x \mapsto x^p} R$.

Thus, for any integral perfectoid ring R, there is some element $\xi \in A_{\inf}(R)$ such that $R = A_{\inf}(R)/\xi$; such an element is necessarily a nonzerodivisor by [BMS18, Lemma 3.10]. We let $A_{\mathrm{crys}}(R)$ be the p-adic completion of the PD thickening of $A_{\inf}(R) \to R$. The map $A_{\inf}(R) \to A_{\mathrm{crys}}(R)$ is injective. Note that if $pR = 0$ so that $A_{\inf}(R) = W(R)$ and one can take $\xi = p$, then $A_{\mathrm{crys}}(R)$ is a slightly pathological ring which has extra divided powers of p; in particular, it is not p-torsion free. However, in that case there is a section $A_{\mathrm{crys}}(R) \to A_{\inf}(R)$ sending the extra divided powers of p to the usual divided powers.

Theorem 17.5.2. The category of p-divisible groups G over R is equivalent to the category of finite projective $A_{\inf}(R)$-modules M together with a φ-linear isomorphism $\varphi_M : M[\frac{1}{\xi}] \cong M[\frac{1}{\varphi(\xi)}]$ such that $M \subset \varphi_M(M) \subset \frac{1}{\varphi(\xi)}M$. The equivalence is functorial in R, and agrees with usual Dieudonné theory if $pR = 0$, and with the classification from Theorem 14.4.1 if $R = \mathcal{O}_C$; these requirements specify the equivalence uniquely.

Moreover, under this equivalence $M \otimes_{A_{\inf}(R)} A_{\mathrm{crys}}(R)$ is the Dieudonné module of G evaluated on the PD thickening $A_{\mathrm{crys}}(R) \to R$, compatibly with φ_M.

Note that if R is of characteristic p, our normalization of φ_M differs from the usual normalization by a factor of p^{-1}; this is in fact our normalization of covariant Dieudonné theory. If $p \neq 2$, Theorem 17.5.2 has also been obtained by E. Lau, [Lau18], using the theory of displays.

Our proof of this theorem involves some rather nonstandard arguments, which however are rather versatile.

Proof. If $pR = 0$, the result is due to Gabber; cf. [Lau13].[2] Next, if $R = V$ is a valuation ring and $pR \neq 0$, then $K = V[\frac{1}{p}]$ is a perfectoid field. Assume that K is algebraically closed. Let $\mathcal{O}_K \subset K$ be its ring of integers, so that $V \subset \mathcal{O}_K$. Moreover, let k be the residue field of $\mathcal{O}_K$, and $\overline{V} \subset k$ the image of V, which is a valuation ring of k. Then p-divisible groups over V are equivalent to pairs of p-divisible groups over $\mathcal{O}_K$ and $\overline{V}$ with common base change to k, and the same is true on the side of Breuil-Kisin-Fargues modules. Thus, the result follows from Theorem 14.4.1 in case K is algebraically closed.

In general, we can now give a description of the functor $G \mapsto M(G)$. Indeed, if $M_{\mathrm{crys}}(G)$ is the Dieudonné module of G over $A_{\mathrm{crys}}(R)$, then $M(G) \subset M_{\mathrm{crys}}(G)$ is the largest submodule mapping into $M(G_V) \subset M_{\mathrm{crys}}(G_V)$ for all maps $R \to V$ where V is an integral perfectoid valuation ring with algebraically closed fraction field. This association is clearly functorial in R and G, but it is not clear that it has the desired properties. We will establish this in increasing generality.

Assume first that $R = \prod_{i \in I} V_i$ is a product of (infinitely many) valuation rings V_i with algebraically closed fraction field. In this case, the category of finite projective R-modules M of fixed rank r is equivalent to the product over all $i \in I$ of the category of finite projective V_i-modules M_i of rank r, via $M = \prod_{i \in I} M_i$. This implies that the category of p-divisible groups G over R of fixed height h is equivalent to the product over all $i \in I$ of the category of p-divisible groups G_i over V_i of height h, and similarly for Breuil-Kisin-Fargues modules. Thus, the equivalence of categories follows (and it is uniquely determined). However, it is not clear that it agrees with the general description. To check this, we can decompose I into two disjoint parts according to whether V_i is of characteristic p or mixed characteristic. The characteristic p part reduces to the usual Dieudonné theory, so we can assume that all V_i are of mixed characteristic. Fix a compatible system of primitive p-power roots $\epsilon \in R$, and let $\mu = [\epsilon] - 1 \in A_{\mathrm{inf}}$. Let $T = \mathrm{Hom}_R(\mathbf{Q}_p/\mathbf{Z}_p, G) = \prod_{i \in I} T_i$, where $T_i = \mathrm{Hom}_{V_i}(\mathbf{Q}_p/\mathbf{Z}_p, G_i)$ is the p-adic Tate module of G_i. Then T is a finite projective module over $C^0(\mathrm{Spec}\, R, \mathbf{Z}_p) = \prod_{i \in I} \mathbf{Z}_p$. As in Proposition 14.8.1, we have a natural isomorphism

$$T \otimes_{C^0(\mathrm{Spec}\, R, \mathbf{Z}_p)} B_{\mathrm{crys}}(R) \cong M_{\mathrm{crys}}(G) \otimes_{A_{\mathrm{crys}}(R)} B_{\mathrm{crys}}(R) \ .$$

On the other hand, for each i, we have

$$T_i \otimes_{\mathbf{Z}_p} A_{\mathrm{inf}}(V_i) \subset M(G_i) \subset T_i \otimes_{\mathbf{Z}_p} \tfrac{1}{\mu} A_{\mathrm{inf}}(V_i) \ .$$

---

[2]Incidentally, Gabber's first proof of this result was along very similar lines as the proof we will give. In fact, he reduces to the case of perfect valuation rings which is due to Berthelot, [Ber80]. Our proof actually only needs the case of perfect valuation rings as input.

By taking a product over all $i \in I$, this implies that

$$T \otimes_{C^0(\operatorname{Spec} R, \mathbf{Z}_p)} A_{\inf}(R) \subset M \subset T \otimes_{C^0(\operatorname{Spec} R, \mathbf{Z}_p)} \tfrac{1}{\mu} A_{\inf}(R) .$$

In particular, we see that

$$M \otimes_{A_{\inf}(R)} B_{\mathrm{crys}}(R) \cong M_{\mathrm{crys}}(G) \otimes_{A_{\mathrm{crys}}(R)} B_{\mathrm{crys}}(R) .$$

We claim that under this isomorphism, we have an equality

$$M \otimes_{A_{\inf}(R)} A_{\mathrm{crys}}(R) = M_{\mathrm{crys}}(G) .$$

To check this, note that checking whether an element of $B_{\mathrm{crys}}(R) = A_{\mathrm{crys}}(R)[\frac{1}{\mu}]$ lies in $A_{\mathrm{crys}}(R)$ can be done by looking at the image in $B_{\mathrm{crys}}(V_i)$ for all $i \in I$. This way, one sees that one gets a map $M \otimes_{A_{\inf}(R)} A_{\mathrm{crys}}(R) \to M_{\mathrm{crys}}(G)$. As both sides are finite projective modules of the same rank, checking whether it is an isomorphism amounts to checking whether the determinant is an invertible element of $A_{\mathrm{crys}}(R)$. Again, this can be checked after mapping to $A_{\mathrm{crys}}(V_i)$ for all $i \in I$.

Now, for a general map $R \to V$ with V of mixed characteristic, we see that the base extension of M to $A_{\inf}(V)$ is correct after inverting μ and after base extension to $A_{\mathrm{crys}}(V)$, in a compatible way. This determines $M(G_V)$, so indeed $M \otimes_{A_{\inf}(R)} A_{\inf}(V) = M(G_V)$, as desired. For maps $R \to V$ with $pV = 0$, the problem reduces to the usual Dieudonné theory.

Finally, we handle the general case. Fix some integral perfectoid ring R with tilt $R^\flat$. For any perfect $R^\flat$-algebra S, we get a corresponding integral perfectoid R-algebra $W(S)/\xi$, where ξ is a generator of the kernel of $\theta : A_{\inf}(R) = W(R^\flat) \to R$. Recall that in [BS17], a map of qcqs schemes $Y \to X$ is a v-cover if $Y^{\mathrm{ad}} \to X^{\mathrm{ad}}$ is surjective. By the v-descent results of [BS17], it follows that both p-divisible groups over R and Breuil-Kisin-Fargues modules satisfy v-descent along v-covers $\operatorname{Spec} S \to \operatorname{Spec} R^\flat$. In other words, we are free to replace $R^\flat$ by S if $\operatorname{Spec} S \to \operatorname{Spec} R^\flat$ is a v-cover. But then we can take for S a product $S = \prod_{i \in I} V_i$ of valuation rings with algebraically closed fraction field, where I ranges over a set of representatives of all maps from $R^\flat$ to such V_i. In this case $W(S)/\xi = \prod_{i \in I} W(V_i)/\xi$, and each $W(V_i)/\xi$ is again a valuation ring with algebraically closed fraction field.   □

Lecture 18

---

v-sheaves associated with perfect and formal schemes

The more general formalism of v-sheaves makes it possible to consider not only analytic adic spaces as diamonds, but also certain non-analytic objects as v-sheaves.

18.1 DEFINITION

Let X be any pre-adic space over $\mathbf{Z}_p$. Consider the presheaf $X^\diamond$ on Perf whose S-valued points for $S \in$ Perf are given by untilts $S^\sharp$ of S together with a map $S^\sharp \to X$. If $X = \operatorname{Spa} \mathbf{F}_p$, this is the trivial functor, sending any S to a point. This is not a diamond, but we will immediately show that it is a v-sheaf. In this lecture, we analyze some properties of this construction.

Lemma 18.1.1. For any pre-adic space X over $\mathbf{Z}_p$, the presheaf $X^\diamond$ is a v-sheaf.

Proof. First, $\operatorname{Spd} \mathbf{Z}_p = \operatorname{Untilt}$ is a v-sheaf; this is [Sch17, Lemma 15.1 (i)], and can be proved by following the proof of Lemma 9.4.5. It remains to see that for any pre-adic space X over $\mathbf{Z}_p$, the presheaf $\operatorname{Hom}(-, X)$ on Perfd is a v-sheaf. This follows from Theorem 17.1.3 if X is affinoid, to which the general case reduces easily. $\qquad\square$

We will sometimes need the following lemma.

Lemma 18.1.2. Let K be a complete nonarchimedean field in which p is topologically nilpotent, with completed algebraic closure C and absolute Galois group G_K. Let $\mathcal{O}_K \subset K$ and $\mathcal{O}_C \subset C$ be the rings of integers. Then the map $\operatorname{Spd} \mathcal{O}_C \to \operatorname{Spd} \mathcal{O}_K$ is a proper v-cover, and induces an isomorphism of v-sheaves

$$\operatorname{Spd} \mathcal{O}_C / \underline{G_K} \xrightarrow{\cong} \operatorname{Spd} \mathcal{O}_K .$$

Proof. It is enough to prove the result v-locally, so we can take as test object $X = \operatorname{Spa}(R, R^+)$ given as follows: Pick any set (C_i, C_i^+) of complete algebraically closed fields C_i of characteristic p with open and bounded valuation subrings $C_i^+ \subset C_i$, and pseudouniformizers $\varpi_i \in C_i^+$, and let R^+ be the $\varpi = (\varpi_i)_{i \in I}$-adic completion of $\prod_i C_i^+$, and $R = R^+[1/\varpi]$.

Note that both sides live over $\operatorname{Spd}\mathcal{O}_K$, and so we may pick an untilt $(R^\sharp, R^{\sharp+})$ of (R, R^+) over $\mathcal{O}_K$, which is equivalent to a collection $(C_i^\sharp, C_i^{\sharp+})$ of untilts of (C_i, C_i^+) over $\mathcal{O}_K$, subject to the condition that $\varpi^\sharp$ divides some power of ϖ_K, where ϖ_K is a pseudouniformizer of K. (This condition is necessary as the image of ϖ_K under $\mathcal{O}_K \to R^{\sharp+}$ is topologically nilpotent and $\varpi^\sharp$ is a pseudouniformizer.) Now as C_i is algebraically closed, we can extend $\mathcal{O}_K \to C_i^{\sharp+}$ to $\mathcal{O}_C \to C_i^{\sharp+}$, and by passing to the product we get a map $\mathcal{O}_C \to R^{\sharp+}$, proving surjectivity of $\operatorname{Spd}\mathcal{O}_C \to \operatorname{Spd}\mathcal{O}_K$. On the other hand, any two lifts to maps $\mathcal{O}_C \to R^{\sharp+} = \prod C_i^{\sharp+}$ differ at each $i \in I$ by the action of some $\gamma_i \in G_K$, and so passing to the quotient modulo $\underline{G_K}(X) = \prod_{i \in I} G_K$, we get

$$\operatorname{Spd}\mathcal{O}_C(X)/\underline{G_K}(X) \cong \operatorname{Spd}\mathcal{O}_K(X) \,,$$

and in particular $\operatorname{Spd}\mathcal{O}_C/\underline{G_K} \cong \operatorname{Spd}\mathcal{O}_K$. This also implies that $\operatorname{Spd}\mathcal{O}_C \to \operatorname{Spd}\mathcal{O}_K$ is proper: As source and target are separated, it is separated; and in general a map of the form $X \to X/\underline{G}$ for a profinite group G acting on a v-sheaf X is quasicompact and universally closed. $\qquad\square$

18.2   TOPOLOGICAL SPACES

First, we analyze the behaviour on topological spaces. We start with an instructive example, showing that in general $|X|$ and $|X^\diamond|$ are different.

Example 18.2.1. Let $X = \operatorname{Spa}(\mathbb{F}_p[t], \mathbb{F}_p[t])$. Then there is an open subfunctor $U \subset |X^\diamond|$, which on $\operatorname{Spa}(R, R^+)$-valued points is given by $R^{\circ\circ} \subset R^+$. After base change to any nonarchimedean field, this is the open unit disc inside the closed unit disc. However, U does not arise from an open subscheme of X.

We do not know how to describe $|X^\diamond|$ in general, except for recalling that as for any v-sheaf, its points are in bijection with equivalence classes of maps $\operatorname{Spa}(L, L^+) \to X^{\mathrm{ad}}$ from perfectoid fields L with an open and bounded valuation subring $L^+ \subset L$. We can say the following.

Proposition 18.2.2. Let X be a pre-adic space over $\operatorname{Spa}\mathbf{Z}_p$ with associated v-sheaf $X^\diamond$. There is a natural continuous map $|X^\diamond| \to |X|$ that is surjective.

Proof. To construct the continuous map $|X^\diamond| \to |X|$, note that for any $S \in \operatorname{Perf}$ with a map $S \to X^\diamond$, one has a continuous natural map $|S| \cong |S^\sharp| \to |X|$ induced by the map $S^\sharp \to X$. If X is analytic, this is a homeomorphism by Proposition 10.3.7. In general, the map is surjective: This reduces to the case that $X = \operatorname{Spa}(K, K^+)$ is an affinoid field. If K is nonarchimedean, the result follows from the analytic case already handled. Otherwise, K is discrete, and $\operatorname{Spa}(K((t)), K^+[\![t]\!] + tK[\![t]\!]) \to \operatorname{Spa}(K, K^+)$ is surjective on topological spaces, where the source is analytic. $\qquad\square$

18.3   PERFECT SCHEMES

Note that if X is any scheme of characteristic p (considered as an adic space via the functor taking $\operatorname{Spec} R$ to $\operatorname{Spa}(R, R)$), one has $X^\diamond = (X_{\text{perf}})^\diamond$. In particular, in characteristic p, the best one can hope for is the following proposition.

Proposition 18.3.1. The functor $X \mapsto X^\diamond$ from perfect schemes of characteristic p to small v-sheaves on Perf is fully faithful.

Proof. First, we check that if $X = \operatorname{Spec} R$ is any affine scheme, then for any perfect scheme Y of characteristic p, we have $\operatorname{Hom}(Y, X) = \operatorname{Hom}(Y^\diamond, X^\diamond)$. Both constructions take colimits to limits, so we can reduce to the case that $Y = \operatorname{Spec} S$ is affine. In this case $Y^\diamond$ has a v-cover by $\widetilde{Y} = \operatorname{Spa}(S((t^{1/p^\infty})), S[[t^{1/p^\infty}]])$. In particular, it follows that $\operatorname{Hom}(Y^\diamond, X^\diamond)$ injects into $\operatorname{Hom}(R, S[[t^{1/p^\infty}]])$. It remains to see that the image is contained in the maps taking R into $S \subset S[[t^{1/p^\infty}]]$. If $f : Y^\diamond \to X^\diamond$ is any map of v-sheaves and $r \in R$ any element, the image $f^*(r)$ of r in $S[[t^{1/p^\infty}]]$ has the property that the image under the two maps

$$S[[t^{1/p^\infty}]] \to \mathcal{O}^+(\widetilde{Y} \times_{Y^\diamond} \widetilde{Y})$$

agree. We claim that this implies that $f^*(r) \in S$. This can be checked after mapping S to fields, so we can assume that $S = k$ is a field. But then

$$\mathcal{O}^+(\widetilde{Y} \times_{Y^\diamond} \widetilde{Y}) = \mathcal{O}^+(\mathbf{D}^*_{k((t_1^{1/p^\infty}))}) = k[[t_1^{1/p^\infty}, t_2^{1/p^\infty}]] ,$$

so the result is clear.

Now assume that X and Y are general. If two maps $f, g : Y \to X$ induce the same map $f^\diamond = g^\diamond$, then in particular we see that $|f| = |g| : |Y| \to |X|$ (as by Proposition 18.2.2, the map $|Y^\diamond| \to |Y^{\text{ad}}| \to |Y|$ is surjective). But then $f = g$ by the affine case already handled.

Finally, if $\psi : Y^\diamond \to X^\diamond$ is any map of v-sheaves, we have to see that it is induced by a map $f : Y \to X$. We do this in several steps, successively generalizing which schemes Y are allowed.[1]

Case 1: $Y = \operatorname{Spec} K$, K a field. In this case $Y^\diamond$ is covered by $\operatorname{Spa}(K((t^{1/p^\infty})))$, which is itself a point, and in particular the image of $|Y^\diamond| \to |X^\diamond| \to |X|$ is a point $x \in X$. Thus, the map factors over $U^\diamond$ for some open affine $U \subset X$, any neighborhood of $x \in X$. By the affine case already handled, we get the result.

Case 2: $Y = \operatorname{Spec} V$, V a valuation ring. This is the key case. Let k be the residue field of V; by Case 1, the map $(\operatorname{Spec} k)^\diamond \to Y^\diamond \to X^\diamond$ comes from a map $\operatorname{Spec} k \to X$; let $x \in X$ be its image. Let U be any open affine neighborhood of x. Our goal is to show that $Y^\diamond \to X^\diamond$ factors over $U^\diamond \subset X^\diamond$, which implies the result by the affine case.

To prove that $Y^\diamond \to X^\diamond$ factors over $U^\diamond \subset X^\diamond$, consider the closed subset

---

[1] The argument seems unnecessarily complicated; we apologize to the reader, but we have not found a more simple argument.

$Z \subset |Y^\diamond|$ of all $y \in |Y^\diamond|$ that do not map into $|U^\diamond|$. We want to show that Z is empty; assume otherwise. We claim that the image of $Z \to |\operatorname{Spec} V|$ has a minimal point (recalling that $|\operatorname{Spec} V|$ is a totally ordered chain of points), where the map $Z \to |\operatorname{Spec} V|$ sends a point $z \in Z$ represented by a map $(V, V) \to (L, L^+)$ to the prime ideal $\ker(V \to L)$. This follows from quasicompacity of Z and Lemma 18.3.2 below. Let $w \in |\operatorname{Spec} V|$ be this minimal point, and let $V \to W$ be the valuation ring quotient of V such that $|\operatorname{Spec} W| \subset |\operatorname{Spec} V|$ is the closure of w. Then replacing V by W, we can assume that w is the generic point of $\operatorname{Spec} V$. In other words, $Z \subset |Y^\diamond|$ sits over the generic point $\operatorname{Spec} K \subset \operatorname{Spec} V$, i.e. it lies in $|\operatorname{Spa}(K, V)^\diamond| \subset |\operatorname{Spa}(V, V)^\diamond|$.

Pick a point $z \in Z \subset |Y^\diamond|$, corresponding to a map $(K, V) \to (L, L^+)$ for some perfectoid field L with open and bounded valuation subring $L^+ \subset L$. Assume that the induced norm $|\cdot| : K \to \mathbb{R}_{\geq 0}$ is nontrivial. Then inside $\operatorname{Spa}(V((t^{1/p^\infty})), V[[t^{1/p^\infty}]])$, any preimage of z will have the property that its orbit under the map $t \mapsto t^p$ will converge out of $\operatorname{Spa}(K, V)^\diamond$. As Z is closed and contained in $|\operatorname{Spa}(K, V)^\diamond|$, it follows that this cannot happen. In other words, the induced norm $|\cdot| : K \to \mathbb{R}_{\geq 0}$ must be trivial.

The preceding implies that Z is contained in the image of the map

$$W := \operatorname{Spa}(K((t^{1/p^\infty})), V[[t^{1/p^\infty}]] + t^{1/p^\infty} K[[t^{1/p^\infty}]]) \to \operatorname{Spa}(K, V)^\diamond \to Y^\diamond .$$

Note that $|W| \cong |\operatorname{Spec} V|$ is local. In particular, the map $W \to Y^\diamond \to X^\diamond$ factors over $(U')^\diamond$ for some open affine $U' = \operatorname{Spec} B \subset X$, inducing a map $B \to V[[t^{1/p^\infty}]] + t^{1/p^\infty} K[[t^{1/p^\infty}]]$. On the other hand, this map is compatible with the map $\operatorname{Spec} K \to X$ from Case 1, which implies that the composite

$$B \to V[[t^{1/p^\infty}]] + t^{1/p^\infty} K[[t^{1/p^\infty}]] \to K[[t^{1/p^\infty}]]$$

actually lands in K; in other words, B maps into V. This gives a map $g : \operatorname{Spec} V \to \operatorname{Spec} B \subset X$.

At this point, we would like to say that $\psi = g^\diamond$ (and we remark that we could have constructed g directly without the previous reductions), but we will not be able to see this directly, and we will only use g to prove that Z is empty, reaching the desired contradiction. For this, it is enough to prove that g factors over $U \subset X$.

The subset $W \subset \operatorname{Spa}(V((t^{1/p^\infty})), V[[t^{1/p^\infty}]])$ is pro-quasicompact open, being the intersection of the subsets $|a| \geq |t|$ over all $a \in V$. On the other hand, the preimage of $U' \subset X$ in $\operatorname{Spa}(V((t^{1/p^\infty})), V[[t^{1/p^\infty}]])$ is some open subspace containing W. It follows that this preimage contains an open subset $\widetilde{W} \subset \operatorname{Spa}(V((t^{1/p^\infty})), V[[t^{1/p^\infty}]])$ of the form $|a| \geq |t|$ for some $a \in V$. Thus, we have a map $\widetilde{W} \to Y^\diamond \to X^\diamond$ that factors over $(\operatorname{Spec} B)^\diamond \subset X^\diamond$, inducing a map $B \to \mathcal{O}(\widetilde{W})$. We claim that this agrees with the composite $B \xrightarrow{g} V \to \mathcal{O}(\widetilde{W})$.

This follows from injectivity of $\mathcal{O}(\widetilde{W}) \to K((t^{1/p^\infty}))$, using the fact that

$$\mathcal{O}(\widetilde{W}) = V[[t^{1/p^\infty}]]\langle (\frac{t}{a})^{1/p^\infty}\rangle[t^{-1}] \subset K((t^{1/p^\infty})) \ .$$

In other words, the map $\widetilde{W} \to Y^\diamond \to X^\diamond$ agrees with the composite $\widetilde{W} \to (\operatorname{Spec} V)^\diamond \xrightarrow{g^\diamond} X^\diamond$.

We note that for all $w \in \widetilde{W} \setminus Z$, the image of w in X lies inside $U \subset X$, while it is also given as the image of $|\widetilde{W}| \to |(\operatorname{Spec} V)^\diamond| \to |\operatorname{Spec} V| \xrightarrow{g} |X|$. Thus, if we can find some $w \in \widetilde{W} \setminus Z$ that projects to the closed point of $\operatorname{Spec} V$, then it follows that g maps the closed point of $\operatorname{Spec} V$ into U, and thus g factors over U, as desired.

In other words, we need to find a map $h : (V, V) \to (L, L^+)$ to some perfectoid field L with an open and bounded valuation subring $L^+ \subset L$, a topologically nilpotent element $t \in L$ such that $th(a)^{-1} \in L^+$, the map $\operatorname{Spec} L^+ \to \operatorname{Spec} V$ is surjective, and the norm $|\cdot| : V \to L \to \mathbb{R}_{\geq 0}$ is not the discrete norm of K. Replacing V by a quotient, we can assume that any element of V divides a^n for n large enough. In that case $K = V[a^{-1}]$ admits itself the structure of a perfectoid field, with a norm normalized by $|a| = \frac{1}{2}$. We can then take $(L, L^+) = (K, V)$ and $t = a$ to get the desired point.

Case 3: $Y = \operatorname{Spec} \prod V_i$, V_i *valuation rings.* Let $Y' = \operatorname{Spec} \prod_i K_i \subset Y$, where K_i is the fraction field of V_i. Note that $|Y'|$ is an extremally disconnected profinite set (the Stone-Čech compactification of the index set I), and that the map $|(Y')^\diamond| \to |Y'|$ is a homeomorphism, as $(Y')^\diamond$ is covered by $Z' = \operatorname{Spa}((\prod_i K_i[[t^{1/p^\infty}]])[t^{-1}], \prod_i K_i[[t^{1/p^\infty}]])$, which is a totally disconnected space with spectrum again the Stone-Čech compactification of I. It follows that the map $|(Y')^\diamond| \to |X^\diamond| \to |X|$ factors over an affine open cover of X, so by the affine case already handled, the map $(Y')^\diamond \to X^\diamond$ is induced by a map $f' : Y' \to X$.

Similarly,

$$Z = \operatorname{Spa}((\prod_i K_i[[t^{1/p^\infty}]])[t^{-1}], \prod_i V_i[[t^{1/p^\infty}]] + t^{1/p^\infty} \prod_i K_i[[t^{1/p^\infty}]]) \to Y$$

is a totally disconnected affinoid perfectoid space with $|Z| \to |Y|$ being a homeomorphism. The map $Z \to Y^\diamond \to X^\diamond$ will then factor over the pullback of an affine cover of X. Any open cover of Z is refined by one induced from a decomposition $I = I_1 \sqcup \ldots \sqcup I_r$, so up to such a refinement we can assume that $Z \to X^\diamond$ factors over $U^\diamond$ for some $U = \operatorname{Spec} A \subset X$ open affine. Then $Z \to X^\diamond$ is induced by a map

$$A \to \prod_i V_i[[t^{1/p^\infty}]] + t^{1/p^\infty} \prod_i K_i[[t^{1/p^\infty}]] \ .$$

On the other hand, we know that the map $Z' \to X^\diamond$ comes from a map $A \to$

$\prod_i K_i$. This means that the composite

$$A \to \prod_i V_i[[t^{1/p^\infty}]] + t^{1/p^\infty} \prod_i K_i[[t^{1/p^\infty}]] \subset \prod_i K_i[[t^{1/p^\infty}]]$$

lands in elements that are independent of t. In particular, the map

$$A \to \prod_i V_i[[t^{1/p^\infty}]] + t^{1/p^\infty} \prod_i K_i[[t^{1/p^\infty}]]$$

factors over a map $A \to \prod_i V_i$. This gives a map

$$f : Y = \mathrm{Spec} \prod_i V_i \to U = \mathrm{Spec}\, A \subset X \ .$$

It remains to see that $\psi = f^\diamond$. For this, it suffices to show that the locus inside

$$W = \mathrm{Spa}((\prod_i V_i[[t^{1/p^\infty}]])[t^{-1}], \prod_i V_i[[t^{1/p^\infty}]])$$

(which is a v-cover of $Y^\diamond$) where ψ and $f^\diamond$ agree is all of W. This reduces to the case of a single valuation ring (possibly an ultraproduct of the V_i): This implies that on each connected component of W, the maps ψ and $f^\diamond$ agree. In particular, ψ factors over $U^\diamond \subset X^\diamond$, and so we reduce to the case of affine X. But for a single valuation ring, the result follows from Case 2.

Case 4: Y general. By [BS17], we can v-localize on Y and in particular assume that $Y = \mathrm{Spec} \prod_i V_i$, where the V_i are valuation rings.　　　□

The following lemma was used in the proof.

Lemma 18.3.2. Let V be a valuation ring, Z a quasicompact topological space, and $Z \to \mathrm{Spec}\, V$ a continuous map. Then the image of Z in $\mathrm{Spec}\, V$ has a (unique) minimal point.

Proof. Write V as a filtered colimit of valuation rings $V_i \subset V$ of finite rank. Then $\mathrm{Spec}\, V$ is the limit of $\mathrm{Spec}\, V_i$, where each $\mathrm{Spec}\, V_i$ is a finite totally ordered chain of points. The image of Z in $\mathrm{Spec}\, V_i$ has a unique minimal point $z_i \in \mathrm{Spec}\, V_i$, and the points z_i are compatible under the transition maps, and so define a point $z \in \mathrm{Spec}\, V$. We need to see that z lies in the image of Z. Let $U_i \subset \mathrm{Spec}\, V_i$ be the open subset of all proper generalizations of z_i. If z is not in the image of Z, then the preimages of the U_i cover Z. By quasicompacity of Z, this implies that finitely many preimages cover it, but as the limit is cofiltered, this implies that the preimage of some U_i is all of Z. This contradicts that z_i is in the image of Z.　　　□

18.4 FORMAL SCHEMES

For applications to local models and integral models of Rapoport-Zink spaces, we are interested in the case where $\mathfrak{X}$ is a formal scheme over the ring of integers $\mathcal{O}_E$ in some complete discretely valued $E/\mathbf{Q}_p$ with perfect residue field k. Moreover, we assume $\mathfrak{X}$ is locally formally of finite type, flat, and normal, as is known in many cases of Rapoport-Zink spaces.

Proposition 18.4.1. The functor $\mathfrak{X} \mapsto \mathfrak{X}^\diamond$ from flat and normal formal schemes locally formally of finite type over $\operatorname{Spf} \mathcal{O}_E$ towards the category of small v-sheaves over $\operatorname{Spd} \mathcal{O}_E$ is fully faithful.

Proof. By passage to the maximal unramified extension and Galois descent, we can assume that k is algebraically closed. We will deduce the proposition from the combination of Proposition 10.2.3 concerning the generic fiber, Proposition 18.3.1 concerning the perfection of the special fiber, and the following result proved by Lourenço.

Theorem 18.4.2 ([Lou17]). Assume that k is algebraically closed. Consider the category $C_{\mathcal{O}_E}$ of triples $(X_\eta, X_s, \mathrm{sp})$ where X_η is a rigid space over E, X_s is a perfect scheme over k, and $\mathrm{sp} : |X_\eta| \to X_s(k)$ is a map of sets, where $|X_\eta|$ denotes the set of classical points of X_η.

The functor $\mathfrak{X} \mapsto (\mathfrak{X}_\eta, (\mathfrak{X}_{\mathrm{red}})^{\mathrm{perf}}, \mathrm{sp})$ from the category of flat and normal formal schemes locally formally of finite type over $\operatorname{Spf} \mathcal{O}_E$ towards $C_{\mathcal{O}_E}$ is fully faithful.

Remark 18.4.3. In the theorem, as $|\mathfrak{X}| = |(\mathfrak{X}_{\mathrm{red}})^{\mathrm{perf}}|$, one immediately recovers the underlying topological space. To recover the structure sheaf, one uses the fact that if $\operatorname{Spf} R \subset \mathfrak{X}$ is any open subset, then R agrees with the ring of powerbounded functions on the generic fiber $(\operatorname{Spf} R)_\eta$. Lourenço observes that this is analogous to Riemann's First Hebbarkeitssatz, and proves it this way. An alternative argument is due to de Jong, [dJ95].

To prove Proposition 18.4.1, we first define a functor from small v-sheaves Y in the essential image of $\mathfrak{X} \mapsto \mathfrak{X}^\diamond$ towards $C_{\mathcal{O}_E}$. The small v-sheaf $Y_\eta = Y \times_{\operatorname{Spd} \mathcal{O}_E} \operatorname{Spd} E$ lies in the essential image of the category of rigid spaces over E by Proposition 10.2.3, giving the first part of the triple. There is a maximal open subfunctor $Y_a \subset Y$ that is a diamond, given as $\mathfrak{X}_a^\diamond$, where $\mathfrak{X}_a \subset \mathfrak{X}^{\mathrm{ad}}$ is the analytic locus. The complementary closed subfunctor $Y_{\mathrm{red}} \subset Y$ is given by $(\mathfrak{X}_{\mathrm{red}}^{\mathrm{perf}})^\diamond$; thus, Proposition 18.3.1 shows that we may recover $\mathfrak{X}_{\mathrm{red}}^{\mathrm{perf}}$. Finally, if $x \in |Y_\eta|$ is any classical point, its closure in $|Y| \cong |\mathfrak{X}^\diamond|$ meets the special fiber in a point of $\mathfrak{X}(k) \subset |Y_{\mathrm{red}}|$. This defines the specialization mapping.

Now Theorem 18.4.2 implies faithfulness of the functor $\mathfrak{X} \mapsto \mathfrak{X}^\diamond$, and it also implies that if $\mathfrak{X}$ and $\mathfrak{Y}$ are flat normal formal schemes locally formally of finite type over $\operatorname{Spf} \mathcal{O}_E$ and $\psi : \mathfrak{X}^\diamond \to \mathfrak{Y}^\diamond$ is a map of the associated v-sheaves over $\operatorname{Spd} \mathcal{O}_E$, then there is a map $f : \mathfrak{X} \to \mathfrak{Y}$ over $\operatorname{Spf} \mathcal{O}_E$ such that the map on the generic fiber is given by the map determined by ψ. We need to see that this implies $\psi = f^\diamond$. It suffices to see that the graphs

$$\Gamma_\psi, \Gamma_{f^\diamond} : \mathfrak{X}^\diamond \hookrightarrow (\mathfrak{X} \times_{\mathrm{Spf}\, \mathcal{O}_E} \mathfrak{Y})^\diamond = \mathfrak{X}^\diamond \times_{\mathrm{Spd}\, \mathcal{O}_E} \mathfrak{Y}^\diamond$$

agree. As both are closed immersions, this is equivalent to checking that the closed subspaces

$$\Gamma_\psi(|\mathfrak{X}^\diamond|), \Gamma_f(|\mathfrak{X}^\diamond|) \subset |(\mathfrak{X} \times_{\mathrm{Spf}\, \mathcal{O}_E} \mathfrak{Y})^\diamond|$$

agree. This follows from the observation that the subspaces agree in the generic fiber, and $|\mathfrak{X}_\eta^\diamond| \subset |\mathfrak{X}^\diamond|$ is dense. $\qquad\square$

Lecture 19

The B_{dR}^+-affine Grassmannian

In this lecture, we will define an object that was one of the big motivations to develop a theory of diamonds.

Recall that the affine Grassmannian over the complex numbers $\mathbb{C}$ of a reductive group G is the quotient

$$\mathrm{Gr}_G = G(\mathbb{C}((t)))/G(\mathbb{C}[[t]]) \, ,$$

endowed with a suitable structure of ind-complex analytic space. More precisely, one has the Cartan decomposition

$$G(\mathbb{C}((t))) = \bigsqcup_{\mu: \mathbf{G}_m \to G} G(\mathbb{C}[[t]])\mu(t)G(\mathbb{C}[[t]])$$

where μ runs over conjugacy classes of cocharacters. The $G(\mathbb{C}[[t]])$-orbit of (the image of) $\mu(t)$ in Gr_G is a finite-dimensional complex manifold, and its closure in Gr_G is a complex-analytic space $\mathrm{Gr}_{G,\mu}$, called a Schubert variety.

In our situation, we want to replace $\mathbb{C}$ with an algebraically closed nonarchimedean field $C/\mathbf{Q}_p$, and the field $C((t))$ of Laurent series with Fontaine's field $B_{\mathrm{dR}}(C)$ of p-adic periods. Thus, we want to give a geometric structure to the quotient

$$\mathrm{Gr}_G^{B_{\mathrm{dR}}^+}(C) = G(B_{\mathrm{dR}}(C))/G(B_{\mathrm{dR}}^+(C)) \, .$$

The idea is to consider it as a functor on all perfectoid algebras, and then prove that the analogue of $\mathrm{Gr}_{G,\mu}$ is a (spatial) diamond.

For some basic results, our arguments follow closely Zhu's survey [Zhu17b].

19.1 DEFINITION OF THE B_{dR}^+-AFFINE GRASSMANNIAN

Definition 19.1.1. Let $C/\mathbf{Q}_p$ be an algebraically closed nonarchimedean field and G/C a reductive group. The B_{dR}^+-affine Grassmannian $\mathrm{Gr}_G = \mathrm{Gr}_G^{B_{\mathrm{dR}}^+}$ is the étale sheafification of the functor taking $S = \mathrm{Spa}(R, R^+) \in \mathrm{Perf}$ with a map $S \to \mathrm{Spd}\, C$ corresponding to an untilt $(R^\sharp, R^{\sharp+})$ over $(C, \mathcal{O}_C)$ to the coset space $G(B_{\mathrm{dR}}(R^\sharp))/G(B_{\mathrm{dR}}^+(R^\sharp))$.

Moreover, we define the loop group LG and the positive loop group L^+G by sending $S = \operatorname{Spa}(R, R^+)$ over $\operatorname{Spd} C$ to $G(B_{\mathrm{dR}}(R^\sharp))$ resp. $G(B_{\mathrm{dR}}^+(R^\sharp))$, so $\operatorname{Gr}_G = LG/L^+G$.

One can give an equivalent definition that does not involve a sheafification.

Proposition 19.1.2. The B_{dR}^+-affine Grassmannian Gr_G is the functor taking any affinoid perfectoid $S = \operatorname{Spa}(R, R^+) \in \operatorname{Perf}$ with an untilt $S^\sharp = \operatorname{Spa}(R^\sharp, R^{\sharp+})$ over $\operatorname{Spa} C$ to the set of G-torsors $\mathcal{E}$ on $\operatorname{Spec} B_{\mathrm{dR}}^+(R^\sharp)$ together with a trivialization of $\mathcal{E}|_{\operatorname{Spec} B_{\mathrm{dR}}(R^\sharp)}$. Moreover, for any open subset $U \subset \operatorname{Spa} C^\flat \dot\times \operatorname{Spa} \mathbf{Z}_p$ containing the point $\operatorname{Spa} C$ and on which p is invertible, it is also equivalent to the functor taking any $S \in \operatorname{Perf}$ with untilt $S^\sharp$ over $\operatorname{Spa} C$ to the set of G-torsors $\mathcal{E}_U$ on

$$S \dot\times \operatorname{Spa} \mathbf{Z}_p \times_{\operatorname{Spa} C^\flat \dot\times \operatorname{Spa} \mathbf{Z}_p} U$$

together with a trivialization of

$$\mathcal{E}|_{S \dot\times \operatorname{Spa} \mathbf{Z}_p \times_{\operatorname{Spa} C^\flat \dot\times \operatorname{Spa} \mathbf{Z}_p} (U \setminus \operatorname{Spa} C)}$$

that is meromorphic along $S^\sharp = S \dot\times \operatorname{Spa} \mathbf{Z}_p \times_{\operatorname{Spa} C^\flat \dot\times \operatorname{Spa} \mathbf{Z}_p} \operatorname{Spa} C$. Finally, Gr_G is a small v-sheaf.

Remark 19.1.3. For our purposes, it is easiest to define G-torsors in terms of the Tannakian formalism, i.e., as exact $\otimes$-functors from the category $\operatorname{Rep}_C G$ of algebraic representations of G towards the category of vector bundles; cf. Theorem 19.5.2 in the appendix. In this language, a trivialization of

$$\mathcal{E}|_{S \dot\times \operatorname{Spa} \mathbf{Z}_p \times_{\operatorname{Spa} C^\flat \dot\times \operatorname{Spa} \mathbf{Z}_p} (U \setminus \operatorname{Spa} C)}$$

is meromorphic along $S^\sharp$ if and only if this holds true for the corresponding vector bundles associated to all algebraic representations of G.

Proof. Any G-torsor $\mathcal{E}$ on $\operatorname{Spec} B_{\mathrm{dR}}^+(R^\sharp)$ is locally trivial for the étale topology on $\operatorname{Spa}(R^\sharp, R^{\sharp+})$. Indeed, its reduction to $\operatorname{Spec} R^\sharp$ is étale locally trivial, so we can assume this reduction is trivial. We claim that then $\mathcal{E}$ is already trivial. To check this, it suffices to find compatible trivializations over $\operatorname{Spec} B_{\mathrm{dR}}^+(R^\sharp)/\xi^n$ for all $n \geq 1$. At each step, lifting a trivialization amounts to trivializing a torsor under the Lie algebra of G; as $H_{\text{ét}}^1(S^\sharp, \mathcal{O}_{S^\sharp}) = 0$, we see that it is trivial.

Recall that $S^\sharp \hookrightarrow S \dot\times \operatorname{Spa} \mathbf{Z}_p \times_{\operatorname{Spa} C^\flat \dot\times \operatorname{Spa} \mathbf{Z}_p} U$ is a closed Cartier divisor by Proposition 11.3.1. Thus, the identification with G-torsors over this locus follows from the Tannakian formalism and the Beauville-Laszlo lemma, Lemma 5.2.9. Therefore Gr_G is a v-sheaf by Proposition 19.5.3; alternatively, this follows more directly from Corollary 17.1.9. As the datum of a G-torsor over these schemes or adic spaces with a trivialization over some open subset is a set-theoretically bounded amount of data, it is clear that Gr_G is small. □

Lemma 19.1.4. The v-sheaf Gr_G is separated in the sense of [Sch17, Definition 10.7]. In particular, the map $\mathrm{Gr}_G \to \mathrm{Spd}\, C$ is separated, and therefore also quasiseparated. Moreover, Gr_G is partially proper.

Proof. Once we have proved that Gr_G is separated, it is clear that Gr_G is partially proper as its (R, R^+)-valued points depend only on R and not on R^+.

Given $X = \mathrm{Spa}(R, R^+)$ with untilt $R^\sharp$, and two $B_{\mathrm{dR}}(R^\sharp)^+$-lattices $M_1, M_2 \subset B_{\mathrm{dR}}(R^\sharp)^r$, we want to show that the locus where $\{M_1 = M_2\}$ is representable by a closed subdiamond of $X^\diamond$; we will in fact show that it is representable by an affinoid perfectoid space $X_0 \subset X$ that is closed in X. It is enough to show this for the locus where $M_1 \subset M_2$ (as then by symmetry the same applies to the locus where $M_2 \subset M_1$).

Let $\xi \in B_{\mathrm{dR}}^{+}(R^\sharp)$ generate Fil^1. We have the loci $\{M_1 \subset \xi^{-i} M_2\}$ for $i \in \mathbf{Z}$. For $i \gg 0$, this is all of X. By induction we may assume $M_1 \subset \xi^{-1} M_2$. Then

$$\{M_1 \subset M_2\} = \bigcap_{m \in M_1} \{m \mapsto 0 \in \xi^{-1} M_2 / M_2\}$$

So it suffices to show that $\{m \mapsto 0 \in \xi^{-1} M_2 / M_2\}$ is closed and representable by an affinoid perfectoid space.

The quotient $\xi^{-1} M_2 / M_2$ is a finite projective $R^\sharp$-module. Writing it as a direct summand of $(R^\sharp)^r$, we see that $\{m \mapsto 0 \in \xi^{-1} M_2 / M_2\}$ is the vanishing locus of an r-tuple of elements of $R^\sharp$. Finally we are reduced to showing that the vanishing locus $\{f = 0\}$ of a single $f \in R^\sharp$ is closed and representable. But $\{f = 0\}$ is the intersection of the $\{|f| \le |\varpi|^n\}$ for $n \ge 1$ (with $\varpi \in R$ a uniformizer), and each of these is rational, hence affinoid perfectoid. Thus the limit $\{f = 0\}$ is also affinoid perfectoid. The complement $\{f \neq 0\}$ is clearly open. $\qquad\square$

Lemma 19.1.5. Let $\rho : G \hookrightarrow H$ be a closed embedding of reductive groups over C. Then the induced map $\mathrm{Gr}_G \to \mathrm{Gr}_H$ is a closed embedding.

Proof. The proof is identical to [PR08, Theorem 1.4]: Recall that the quotient H/G is affine. Given a map $S = \mathrm{Spa}(R, R^+) \to \mathrm{Gr}_H$, it comes étale locally from an element in $H(B_{\mathrm{dR}}(R^\sharp))$. We get an induced point of $(H/G)(B_{\mathrm{dR}}(R^\sharp))$. The locus where this lies in $(H/G)(B_{\mathrm{dR}}^{+}(R^\sharp))$ is closed, by reducing to the same question for $\mathbb{A}^1$ in place of H/G, where it follows from the arguments in the previous lemma. Now the map $\mathrm{Gr}_G \to \mathrm{Gr}_H$ is clearly injective as $G(B_{\mathrm{dR}}^{+}(R^\sharp)) = H(B_{\mathrm{dR}}^{+}(R^\sharp)) \cap G(B_{\mathrm{dR}}(R^\sharp))$, and the image is identified with the previously identified closed sublocus, noting that the map

$$H(B_{\mathrm{dR}}^{+}(R^\sharp))/G(B_{\mathrm{dR}}^{+}(R^\sharp)) \to (H/G)(B_{\mathrm{dR}}^{+}(R^\sharp))$$

becomes an isomorphism after étale sheafification (as G-torsors on $\mathrm{Spec}\, B_{\mathrm{dR}}^{+}(R^\sharp)$ are trivial étale locally on $\mathrm{Spa}(R, R^+)$). $\qquad\square$

19.2   SCHUBERT VARIETIES

Next, we define the Schubert varieties inside Gr_G. For this, we fix a maximal torus and a Borel $T \subset B \subset G$.

Proposition 19.2.1. The action of $L^+G(C) = G(B_{\mathrm{dR}}^+(C))$ on $\mathrm{Gr}_G(C)$ gives a disjoint decomposition

$$\mathrm{Gr}_G(C) = \bigsqcup_{\mu \in X_*(T)^+} G(B_{\mathrm{dR}}^+(C)) \cdot \xi^\mu \ ,$$

where $\xi^\mu \in \mathrm{Gr}_G(C)$ is the image of $\mu(\xi) \in G(B_{\mathrm{dR}}(C))$.

Proof. This is the usual Cartan decomposition

$$G(B_{\mathrm{dR}}(C)) = \bigsqcup_{\mu \in X_*(T)^+} G(B_{\mathrm{dR}}^+(C)) \cdot \xi^\mu \cdot G(B_{\mathrm{dR}}^+(C)) \ ,$$

as follows for example by choosing an isomorphism $B_{\mathrm{dR}}(C) \cong C(\!(\xi)\!)$. $\qquad\square$

Definition 19.2.2. For a dominant cocharacter $\mu \in X_*(T)^+$, consider the subfunctors

$$\mathrm{Gr}_\mu \subset \mathrm{Gr}_{\leq\mu} \subset \mathrm{Gr}_G$$

defined by the condition that a map $S \to \mathrm{Gr}_G$ with $S \in \mathrm{Perf}$ factors over Gr_μ resp. $\mathrm{Gr}_{\leq\mu}$ if and only if for all geometric points $x = \mathrm{Spa}(C(x), C(x)^+) \to S$, the corresponding $C(x)$-valued point of Gr_G lies in

$$L^+G(C(x)) \cdot \xi^\mu$$

resp.

$$\bigsqcup_{\mu' \leq \mu} L^+G(C(x)) \cdot \xi^{\mu'} \ .$$

Here, $\leq$ denotes the Bruhat order on $X_*(T)^+$, i.e., $\mu' \leq \mu$ if and only if $\mu - \mu'$ can be written as a sum of positive coroots with nonnegative integral coefficients.

Proposition 19.2.3. For all $\mu \in X_*(T)^+$, the subfunctor $\mathrm{Gr}_{\leq\mu} \subset \mathrm{Gr}_G$ is a closed subfunctor that is proper over $\mathrm{Spd}\,C$, and $\mathrm{Gr}_\mu \subset \mathrm{Gr}_{\leq\mu}$ is an open subfunctor.

Proof. We follow the argument of [Zhu17b, Proposition 2.1.4]. For the first part, it is enough to prove that $\mathrm{Gr}_{\leq\mu} \subset \mathrm{Gr}_G$ is closed and proper over $\mathrm{Spd}\,C$, as this formally implies that $\mathrm{Gr}_\mu = \mathrm{Gr}_{\leq\mu} \setminus \bigcup_{\mu' < \mu} \mathrm{Gr}_{\leq\mu'}$ is open in $\mathrm{Gr}_{\leq\mu}$. First, we handle the case $G = \mathrm{GL}_n$. Note that a $B_{\mathrm{dR}}^+(R^\sharp)$-lattice $\Xi \subset B_{\mathrm{dR}}(R^\sharp)^n$ has relative position $\mu = (\mu_1, \ldots, \mu_n)$ with $\mu_1 \geq \ldots \geq \mu_n$ if and only if for all $i = 1, \ldots, n$,

$$\Lambda^i(\Xi) \subset \xi^{\mu_n + \ldots + \mu_{n-i+1}} B_{\mathrm{dR}}^+(R^\sharp)^{\binom{n}{i}}$$

with equality for $i = n$. Such conditions determine closed subsets by the proof of Lemma 19.1.4. Therefore to see that $\mathrm{Gr}_{\leq \mu}$ is proper, it suffices to treat the case $\mu = (N, 0, \ldots, 0)$ for each integer $N \geq 0$, as each $\mathrm{Gr}_{\leq \mu}$ is a closed subset of a space of this form. It is clear from the definitions that $\mathrm{Gr}_{\leq \mu}$ is partially proper, so it suffices to show that it is quasicompact. Consider the functor $\widetilde{\mathrm{Gr}}_N$ parametrizing chains of $B_{\mathrm{dR}}^+(R^\sharp)$-lattices $\Xi_N \subset \Xi_{N-1} \subset \cdots \subset \Xi_0 = B_{\mathrm{dR}}^+(R^\sharp)^n$ such that each Ξ_i/Ξ_{i+1} is an invertible $R^\sharp$-module. Then $\widetilde{\mathrm{Gr}}_N$ defines a successive $(\mathbb{P}^{n-1})^\diamond$-bundle over $\mathrm{Spd}\, C$, and therefore is proper. We have a natural map of small v-sheaves $\pi \colon \widetilde{\mathrm{Gr}}_N \to \mathrm{Gr}_{\leq \mu}$ which sends a chain to Ξ_N. This map is surjective on (C, C^+)-points, for example by choosing an isomorphism $B_{\mathrm{dR}}(C) \cong C(\!(\xi)\!)$ and using the similar assertion for the usual affine Grassmannian. The map is also quasicompact, because $\widetilde{\mathrm{Gr}}_N$ is quasicompact. Therefore by [Sch17, Lemma 12.11], the map is a v-cover. This implies that $\mathrm{Gr}_{\leq \mu}$ is also quasicompact, and hence proper.

Next, we treat the case that the derived group of G is simply connected. In that case—cf. [Zhu17b, Proposition 2.1.4]—the subfunctor $\mathrm{Gr}_{G, \leq \mu} \subset \mathrm{Gr}_G$ is the intersection of the preimages of $\mathrm{Gr}_{\mathrm{GL}_n, \leq \rho(\mu)} \subset \mathrm{Gr}_{\mathrm{GL}_n}$ over all representations $\rho \colon G \to \mathrm{GL}_n$, so the result follows from the case of GL_n and Lemma 19.1.5.

In general, we choose a z-extension $1 \to D \to \widetilde{G} \to G \to 1$ whose kernel is a central torus D, such that the derived group of $\widetilde{G}$ is simply connected. Moreover, lift μ to some dominant cocharacter $\widetilde{\mu}$ of $\widetilde{G}$. Then $\mathrm{Gr}_{\widetilde{G}, \leq \widetilde{\mu}} \to \mathrm{Gr}_{G, \leq \mu}$ is surjective on (C, C^+)-valued points. As the source is quasicompact, [Sch17, Lemma 12.5] implies that it is surjective as a map of v-sheaves, so in particular $\mathrm{Gr}_{G, \leq \mu}$ is quasicompact, and thus proper, and in particular closed in Gr_G. $\quad\square$

The following theorem is one of the main results of the lectures.

Theorem 19.2.4. For any μ, the small v-sheaf $\mathrm{Gr}_{\leq \mu}$ is a spatial diamond.

Proof. Using a faithful representation $\rho \colon G \hookrightarrow \mathrm{GL}_n$, we get a closed embedding $\mathrm{Gr}_{G, \leq \mu} \hookrightarrow \mathrm{Gr}_{\mathrm{GL}_n, \leq \rho(\mu)}$ by Lemma 19.1.5, reducing us to the case of $G = \mathrm{GL}_n$. Moreover, we can assume that μ is of the form $(N, 0, \ldots, 0)$ for some $N \geq 0$. This case will be handled in the next subsection. $\quad\square$

19.3 THE DEMAZURE RESOLUTION

In the study of the usual Grassmannian variety G/B attached to a reductive group G, one defines a Schubert variety to be the closure of a B-orbit in G/B. Generally, Schubert varieties are singular varieties. Desingularizations of Schubert varieties are constructed by Demazure, [Dem74]. We will make use of an analogue of this construction in the context of the B_{dR}^+-Grassmannian.

In the following, $G = \mathrm{GL}_n$.

Definition 19.3.1. Suppose μ corresponds to $(k_1 \geq \cdots \geq k_n)$, with $k_n \geq 0$. The *Demazure resolution* $\widetilde{\mathrm{Gr}}_\mu / \operatorname{Spd} C$ sends a characteristic p perfectoid Huber pair (R, R^+) with untilt $R^\sharp$ over C to the set of chains of $B_{\mathrm{dR}}^+(R^\sharp)$-lattices

$$M_{k_1} \subset M_{k_1-1} \subset \cdots \subset M_0 = B_{\mathrm{dR}}^+(R^\sharp)^n$$

where for all $i = 0, \ldots, k_1 - 1$, $\xi M_i \subset M_{i+1} \subset M_i$, and M_i / M_{i+1} is a finite projective $R^\sharp$-module of rank j_i, where $k_{j_i} > i \geq k_{j_i+1}$ (with convention $k_{n+1} = 0$).

The idea behind this definition is to write M_{k_1} as a series of successive minuscule modifications. Analyzing $\widetilde{\mathrm{Gr}}_\mu$ is easier than analyzing $\mathrm{Gr}_{\mathrm{GL}_n}$ directly. It is a succession of Grassmannian bundles.

Lemma 19.3.2. The v-sheaf $\widetilde{\mathrm{Gr}}_\mu$ is a spatial diamond.

Proof. First note that $\widetilde{\mathrm{Gr}}_\mu$ is indeed a v-sheaf by Corollary 17.1.9. By induction it is enough to prove that if $X / \operatorname{Spd} \mathbf{Q}_p$ is a spatial diamond, and $\mathcal{E} / \mathcal{O}_X^\sharp$ is locally free of finite rank, then $\mathrm{Grass}(d, \mathcal{E}) \to X$ is a spatial diamond. Here $\mathrm{Grass}(d, \mathcal{E}) \to X$ associates to a morphism $\operatorname{Spa}(R, R^+) \to X$ the set of projective rank d quotients of $\mathcal{E}|_{\operatorname{Spa}(R^\sharp, R^{\sharp+})}$.

Since X is a spatial diamond, we can choose a universally open quasi-proétale surjection $\widetilde{X} \to X$ from an affinoid perfectoid space $\widetilde{X}$ by [Sch17, Proposition 11.24], and it is enough to prove the result after pullback to $\widetilde{X}$; thus, we can assume that X is an affinoid perfectoid space. Moreover, passing to an open cover, we can assume that $\mathcal{E}$ is trivial. Then $\mathrm{Grass}(d, \mathcal{E})$ decomposes into a product $X \times_{\operatorname{Spd} \mathbf{Q}_p} \mathrm{Grass}(d, n)^\diamond$, and the result follows from Proposition 17.3.8. $\square$

Lemma 19.3.3. The map $\widetilde{\mathrm{Gr}}_\mu \to \mathrm{Gr}_{\mathrm{GL}_n}$ sending $\{M_{k_1} \subset \ldots \subset M_0\}$ to $M_{k_1} \subset M_0$ factors over $\mathrm{Gr}_{\leq\mu}$, and induces a surjective map of v-sheaves $\widetilde{\mathrm{Gr}}_\mu \to \mathrm{Gr}_{\leq\mu}$. When restricted to $\mathrm{Gr}_\mu \subset \mathrm{Gr}_{\leq\mu}$, this map is an isomorphism.

Proof. These assertions are well-known for the usual affine Grassmannian, and therefore follow on geometric points by fixing an isomorphism $B_{\mathrm{dR}}(C) \cong C((\xi))$. Now, to check that the map factors over $\mathrm{Gr}_{\leq\mu}$, it is by definition enough to check on geometric points. By Lemma 17.4.9, the surjectivity as a map of v-sheaves follows from surjectivity on geometric points. By Corollary 17.4.10, the final assertion can also be checked on geometric points. $\square$

Corollary 19.3.4. The v-sheaf Gr_μ is a locally spatial diamond. $\square$

In order to prove Theorem 19.2.4 in the case $G = \mathrm{GL}_n$, $\mu = (N, 0, \ldots, 0)$, we want to apply Theorem 17.3.9. We already know that $\mathrm{Gr}_{\leq\mu}$ is proper over $\operatorname{Spd} C$, and in particular qcqs. Moreover, any point $x \in |\mathrm{Gr}_{\leq\mu}|$ lies in $|\mathrm{Gr}_{\mu'}|$ for some $\mu' \leq \mu$, and so the previous corollary ensures that there is a quasi-pro-étale map $X_x \to \mathrm{Gr}_{\mu'} \to \mathrm{Gr}_{\leq\mu}$ having x in its image, as (locally) closed immersions are quasi-pro-étale. It remains to see that $\mathrm{Gr}_{\leq\mu}$ is spatial. For this, we apply the criterion of Corollary 17.3.7. Thus, it is enough to find a surjective map of v-sheaves $X \to \mathrm{Gr}_{\leq\mu}$ such that for the equivalence relation $R = X \times_{\mathrm{Gr}_{\leq\mu}} X$, the maps $s, t \colon |R| \to |X|$ are open. For this, we consider the functor $X/\mathrm{Spd}\, C$ given by

$$ X(R, R^+) = \left\{ A \in M_r(W(R^+)) \mid \det A \in \xi^N W(R^+)^\times \right\}. $$

Lemma 19.3.5. The functor X is represented by an affinoid perfectoid space.

Proof. First we observe that the functor $(R, R^+) \mapsto M_r(W(R^+))$ is representable by an infinite-dimensional closed unit ball $B_{\mathbf{C}_p}^\infty$ and is thus affinoid perfectoid. For an element $f \in W(R^+)$, the condition that $f \equiv 0 \pmod{\xi}$ is closed and relatively representable, as it is equivalent to the condition that $f^\sharp = 0 \in R^{\sharp+} = W(R^+)/\xi$. The condition that f is invertible is a rational subset, as it is equivalent to $\{|\overline{f}| = 1\}$, where $\overline{f} = f \pmod{p} \in R^+$. These assertions easily imply the claim. $\qquad\square$

Let $\lambda \colon X \to \mathrm{Gr}_{\mathrm{GL}_n}$ be the map which sends A to $A \cdot (B_{\mathrm{dR}}^+)^r$.

Lemma 19.3.6. The map λ factors over $\mathrm{Gr}_{\leq\mu}$ for $\mu = (N, 0, \ldots, 0)$, and $\lambda \colon X \to \mathrm{Gr}_{\leq\mu}$ is a surjective map of v-sheaves.

Proof. As above, this can be checked on (C, C^+)-valued points. It is clear that the image is contained in $\mathrm{Gr}_{\leq\mu}$ by using its description as in the proof of Proposition 19.2.3. Conversely, given $\Xi \subset (B_{\mathrm{dR}}^+)^r$, which is finite projective with $\det \Xi = \xi^N B_{\mathrm{dR}}^+$, let $M = \Xi \cap W(C^+)^r \subset (B_{\mathrm{dR}}^+)^r$. Then M is a ξ-torsion free $W(C^+)$-module such that $M[\xi^{-1}] = W(C^+)[\xi^{-1}]^r$, and $M[1/p]_\xi^\wedge \cong M$. By the Beauville-Laszlo lemma, Lemma 5.2.9, M is finite and projective away from the locus $\{p = \xi = 0\}$ in $\mathrm{Spa}\, W(C^+)$. By Proposition 14.2.6, we see that M is actually a finite projective $W(C^+)$-module (as it is the space of global sections of restriction to the punctured spectrum). Fixing an isomorphism $M \cong W(C^+)^r$, we get a matrix $A \in M_r(W(C^+))$ as desired. $\qquad\square$

Consider the equivalence relation $R = X \times_{\mathrm{Gr}_{\leq\mu}} X$. Let $L_W^+\, \mathrm{GL}_r / \mathrm{Spd}\, C$ be the functor sending (R, R^+) to $\mathrm{GL}_r(W(R^+))$; this is also represented by an affinoid perfectoid space, as in Lemma 19.3.5.

Lemma 19.3.7. The map $(A, B) \mapsto (A, BA)$ defines an isomorphism

$$ X \times_{\mathrm{Spd}\, C} L_W^+\, \mathrm{GL}_r \overset{\sim}{\to} R = X \times_{\mathrm{Gr}_{\leq\mu}} X. $$

Proof. The map $W(R^+) \to B_{\mathrm{dR}}^+(R^\sharp)$ is injective, as can be checked on geometric points. We need to show that if $A_1, A_2 \in X(R, R^+)$ give the same point of $\mathrm{Gr}_{\leq\mu}$, then their ratio $A_2 A_1^{-1}$ lies in $\mathrm{GL}_r(W(R^+))$. But we know that it lies in

$GL_r(W(R^+)[\xi^{-1}])$ and in $GL_r(B_{\mathrm{dR}}^+(R^\sharp))$, so this follows from

$$W(R^+) = W(R^+)[\xi^{-1}] \cap B_{\mathrm{dR}}^+(R^\sharp) .$$

$\square$

To finish the proof that $\mathrm{Gr}_{\leq\mu}$ is spatial, it remains to show that $|R| \to |X|$ is open. But $R = X \times_{\mathrm{Spd}\, C} L_W^+\, GL_r$, and the map $L_W^+\, GL_r \to \mathrm{Spd}\, C$ is universally open, as follows by writing it as an inverse limit with surjective transition maps of affinoid perfectoid spaces corresponding to smooth rigid spaces over $\mathrm{Spd}\, C$ (representing the functors $(R, R^+) \to GL_r(W(R^+)/p^n)$).

19.4   MINUSCULE SCHUBERT VARIETIES

Coming back to the case of a general reductive group G and conjugacy class μ of cocharacters, we note that if μ is minuscule, one has $\mathrm{Gr}_\mu = Gr_{\leq\mu}$. In this case, one can even identify this space explicitly.

Definition 19.4.1. The parabolic $P_\mu \subset G$ associated with μ is the parabolic containing the Borel B^- opposite to B and with Levi component given by the centralizer of μ when μ is realized as a cocharacter $\mathbf{G}_{\mathrm{m}} \to T$ which is dominant with respect to B. Equivalently,

$$P_\mu = \{g \in G \mid \lim_{t\to\infty} \mu(t)g\mu(t)^{-1} \text{ exists}\} .$$

The flag variety $\mathscr{F}\ell_{G,\mu} = G/P_\mu$ parametrizes parabolics in G in the conjugacy class of P_μ.

The following result has appeared in [CS17], using some results from [Sch13] on modules with integrable connection satisfying Griffiths transversality to show that π_μ is an isomorphism if μ is minuscule. Here, we observe that this can be deduced as a simple application of Corollary 17.4.10.

Proposition 19.4.2 ([CS17, Proposition 3.4.3, Theorem 3.4.5]). For any μ, there is a natural Bialynicki-Birula map

$$\pi_\mu \colon \mathrm{Gr}_\mu \to \mathscr{F}\ell_{G,\mu}^\Diamond .$$

If μ is minuscule, the Bialynicki-Birula map is an isomorphism.

Proof. By the Tannakian formalism, it is enough to define the Bialynicki-Birula map for GL_n. In this case, if $\mu = (m_1, \ldots, m_n)$, with $m_1 \geq \ldots \geq m_n$, then Gr_μ parametrizes lattices $\Xi \subset B_{\mathrm{dR}}(R^\sharp)^n$ of relative position $(m_1, \ldots, m_n)$; this includes the lattice

$$\Xi_0 = \bigoplus_{i=1}^{n} \xi^{m_i} B_{\mathrm{dR}}^+(R^\sharp) e_i \ ,$$

where $(e_1, \ldots, e_n)$ is the basis of $B_{\mathrm{dR}}(R^\sharp)^n$. Now for any such lattice, we can define a descending filtration $\mathrm{Fil}_{\Xi}^\bullet$ on $(R^\sharp)^n$ with

$$\mathrm{Fil}_{\Xi}^i = (\xi^i \Xi \cap B_{\mathrm{dR}}^+(R^\sharp)^n) / (\xi^i \Xi \cap \xi B_{\mathrm{dR}}^+(R^\sharp)^n) \ .$$

Now, if M is a finitely generated $R^\sharp$-module for which the dimension of $M \otimes_{R^\sharp} K$ is constant for all maps $R^\sharp \to K$ to nonarchimedean fields K, then M is finite projective; cf. [KL15, Proposition 2.8.4]. Using this fact, one checks that all Fil_{Ξ}^i and $(R^\sharp)^n / \mathrm{Fil}_{\Xi}^i$ are finite projective $R^\sharp$-modules, by induction. More precisely, we argue by descending induction on i. For i very large, one has $\mathrm{Fil}_{\Xi}^i = 0$ and the claim is clear. If the claim is true for $i+1$, then the short exact sequence

$$0 \to \xi^i \Xi \cap B_{\mathrm{dR}}^+(R^\sharp)^n \xrightarrow{\xi} \xi^{i+1} \Xi \cap B_{\mathrm{dR}}^+(R^\sharp) \to \mathrm{Fil}_{\Xi}^{i+1} \to 0$$

shows that $\xi^i \Xi \cap B_{\mathrm{dR}}^+(R^\sharp)^n$ is a finite projective $B_{\mathrm{dR}}^+(R^\sharp)$-module, as the middle term is finite projective and the right term is of projective dimension ≤ 1 over $B_{\mathrm{dR}}^+(R^\sharp)$. Moreover, the formation of $\xi^i \Xi \cap B_{\mathrm{dR}}^+(R^\sharp)$ commutes with any base change in R (again, by the displayed short exact sequence and the same result for $i+1$). Now we can identify Fil_{Ξ}^i with the image of the map

$$(\xi^i \Xi \cap B_{\mathrm{dR}}^+(R^\sharp)^n)/\xi \to (B_{\mathrm{dR}}^+(R^\sharp)^n)/\xi = (R^\sharp)^n$$

of finite projective $R^\sharp$-modules. The formation of the cokernel of this map commutes with any base change, and is of the same dimension at any point, and thus the cokernel is finite projective by [KL15, Proposition 2.8.4]. This implies that the image Fil_{Ξ}^i of this map is also finite projective, as desired.

The stabilizer of the filtration $\mathrm{Fil}_{\Xi}^\bullet$ defines a parabolic in the conjugacy class of P_μ, as desired.

To see that if μ is minuscule, the map is an isomorphism, note that source $\mathrm{Gr}_\mu = \mathrm{Gr}_{\leq \mu}$ and target are qcqs over $\mathrm{Spd}\,C$; thus by Corollary 17.4.10 it is enough to check bijectivity on (C, C^+)-valued points. In this case, fixing an isomorphism $B_{\mathrm{dR}}(C) \cong C(\!(\xi)\!)$, the result follows from the classical case. $\square$

Appendix to Lecture 19:
$\mathcal{G}$-torsors

Let $\mathcal{G}$ be a flat linear algebraic group over $\mathbf{Z}_p$, and let $\operatorname{Rep}\mathcal{G}$ denote its exact $\otimes$-category of representations, i.e., the category of algebraic representations on finite free $\mathbf{Z}_p$-modules. In this section, we briefly recall a few alternative characterizations of $\mathcal{G}$-torsors, first in the case of schemes, and then in the case of adic spaces.

Let X be a scheme over $\mathbf{Z}_p$. There are three possible definitions of $\mathcal{G}$-torsors.

- (Geometric) A geometric $\mathcal{G}$-torsor is a scheme $\mathcal{P} \to X$ over X with an action of $\mathcal{G}$ over X such that fppf locally on X, there is a $\mathcal{G}$-equivariant isomorphism $\mathcal{P} \cong \mathcal{G} \times X$.
- (Cohomological) A cohomological $\mathcal{G}$-torsor is an fppf sheaf $\mathcal{Q}$ on X with an action of $\mathcal{G}$ such that fppf locally on X, there is a $\mathcal{G}$-equivariant isomorphism $\mathcal{Q} \cong \mathcal{G}$.
- (Tannakian) A Tannakian $\mathcal{G}$-torsor is an exact $\otimes$-functor $P : \operatorname{Rep}\mathcal{G} \to \operatorname{Bun}(X)$, where $\operatorname{Bun}(X)$ is the category of vector bundles on X.

Theorem 19.5.1 ([SR72], [Bro13][1]). The categories of geometric, cohomological, and Tannakian $\mathcal{G}$-torsors are canonically equivalent. More precisely:

1. If $\mathcal{P}$ is a geometric $\mathcal{G}$-torsor, then the fppf sheaf of sections of $\mathcal{P}$ is a cohomological $\mathcal{G}$-torsor.
2. If $\mathcal{Q}$ is a cohomological $\mathcal{G}$-torsor, then for any representation $V \in \operatorname{Rep}\mathcal{G}$, the pushout $\mathcal{Q} \times^{\mathcal{G}} (V \otimes_{\mathbf{Z}_p} \mathcal{O}_X)$ of fppf sheaves is a vector bundle on X, and this defines an exact $\otimes$-functor $\operatorname{Rep}\mathcal{G} \to \operatorname{Bun}(X)$.
3. If $P : \operatorname{Rep}\mathcal{G} \to \operatorname{Bun}(X)$ is an exact $\otimes$-functor, then, by writing $\mathcal{O}_{\mathcal{G}}(\mathcal{G})$ as a filtered colimit of objects of $\operatorname{Rep}\mathcal{G}$ and extending P in a colimit-preserving way, the object $P(\mathcal{O}_{\mathcal{G}}(\mathcal{G}))$ is a faithfully flat quasicoherent $\mathcal{O}_X$-algebra whose relative spectrum defines a geometric $\mathcal{G}$-torsor.

The composite of any three composable functors is equivalent to the identity.

If $\mathcal{G}$ is smooth, these categories are also equivalent to the categories obtained by replacing the fppf site with the étale site in the definitions of geometric and

---

[1]In [Bro13], the assumption that X is (faithfully) flat over $\mathbf{Z}_p$ is unnecessary. Also, any exact $\otimes$-functor is faithful, as the trivial representation embeds into $V \otimes V^*$ for any $V \in \operatorname{Rep}\mathcal{G}$, and so $\mathcal{O}_X$ embeds into $P(V) \otimes P(V^*)$, thus $P(V) \neq 0$.

cohomological $\mathcal{G}$-torsors.

Proof. This is standard, so let us briefly recall the arguments. Part (1) is clear. If $\mathcal{G}$ is smooth, then (by descent of smoothness) also $\mathcal{P} \to X$ is smooth, and hence admits étale local sections; this shows that in this case, one even gets a geometric $\mathcal{G}$-torsor on the étale site. Part (2) follows from fppf descent of vector bundles. Finally, in part (3), first note that one can indeed write $\mathcal{O}_{\mathcal{G}}(\mathcal{G})$ as such a filtered colimit; simply take any finitely generated $\mathbf{Z}_p$-submodule, and saturate it under the action of $\mathcal{G}$. As $\mathcal{O}_{\mathcal{G}}(\mathcal{G})$ is flat, this gives the result. It is then easy to see that $P(\mathcal{O}_{\mathcal{G}}(\mathcal{G}))$ defines a faithfully flat quasicoherent $\mathcal{O}_X$-algebra whose spectrum $\mathcal{P}$ has an action of $\mathcal{G}$ (by the remaining action on the $\mathcal{G}$-representation $\mathcal{O}_{\mathcal{G}}(\mathcal{G})$). After pullback along $\mathcal{P} \to X$, one checks that there is an isomorphism $\mathcal{P} \times_X \mathcal{P} \cong \mathcal{G} \times \mathcal{P}$, which one obtains from a similar identity of $\mathcal{G}$-representations upon taking P and the relative spectrum. $\square$

Now we wish to extend the results to adic spaces. There is the problem that in general, if X is an adic space over $\mathbf{Z}_p$, it is not clear whether $\mathcal{G} \times X$ is also an adic space. For this reason, we restrict to one class of spaces where this happens, at least when $\mathcal{G}$ is smooth. Many variants of the results are possible.

So, assume from now on that $\mathcal{G}$ is smooth, and that X is an analytic adic space over $\mathbf{Z}_p$ that is sousperfectoid, i.e., locally of the form $\mathrm{Spa}(R, R^+)$ with R sousperfectoid; cf. Definition 6.3.1. By Proposition 6.3.3, any $Y \in X_{\text{ét}}$ is itself sousperfectoid, and in particular an adic space. This implies that $\mathcal{O}_{X_{\text{ét}}}$ is an étale sheaf by [KL15, Theorem 8.2.22 (c)] (that is acyclic on affinoid subsets), and the same holds for vector bundles. We consider the following kinds of $\mathcal{G}$-torsors. Here, and in the following, $\mathcal{G}$ denotes the adic space $\mathrm{Spa}(R, R^+)$ if $\mathcal{G} = \mathrm{Spec}\, R$ and $R^+ \subset R$ is the integral closure of $\mathbf{Z}_p$. In other words, for any adic space S over $\mathbf{Z}_p$, one has $\mathcal{G}(S) = \mathcal{G}(\mathcal{O}_S(S))$.

- (Geometric) A geometric $\mathcal{G}$-torsor is a an adic space $\mathcal{P} \to X$ over X with an action of $\mathcal{G}$ over X such that étale locally on X, there is a $\mathcal{G}$-equivariant isomorphism $\mathcal{P} \cong \mathcal{G} \times X$.
- (Cohomological) A cohomological $\mathcal{G}$-torsor is an étale sheaf $\mathcal{Q}$ on X with an action of $\mathcal{G}$ such that étale locally on X, there is a $\mathcal{G}$-equivariant isomorphism $\mathcal{Q} \cong \mathcal{G}$.
- (Tannakian) A Tannakian $\mathcal{G}$-torsor is an exact $\otimes$-functor $P : \mathrm{Rep}\,\mathcal{G} \to \mathrm{Bun}(X)$, where $\mathrm{Bun}(X)$ is the category of vector bundles on X.

Again, the three points of view are equivalent.

Theorem 19.5.2. The categories of geometric, cohomological, and Tannakian $\mathcal{G}$-torsors on X are canonically equivalent. More precisely:

1. If $\mathcal{P}$ is a geometric $\mathcal{G}$-torsor, then the étale sheaf of sections of $\mathcal{P}$ is a cohomological $\mathcal{G}$-torsor.
2. If $\mathcal{Q}$ is a cohomological $\mathcal{G}$-torsor, then for any representation $V \in \mathrm{Rep}\,\mathcal{G}$, the pushout $\mathcal{Q} \times^{\mathcal{G}} (V \otimes_{\mathbf{Z}_p} \mathcal{O}_X)$ of étale sheaves is a vector bundle on X, and this

defines an exact $\otimes$-functor $\operatorname{Rep} \mathcal{G} \to \operatorname{Bun}(X)$.

3. If $P : \operatorname{Rep} \mathcal{G} \to \operatorname{Bun}(X)$ is an exact $\otimes$-functor, then, by writing $\mathcal{O}_{\mathcal{G}}(\mathcal{G})$ as a filtered colimit of objects of $\operatorname{Rep} \mathcal{G}$ and extending P in a colimit-preserving way and assuming $X = \operatorname{Spa}(R, R^+)$ is affinoid, the object $P(\mathcal{O}_{\mathcal{G}}(\mathcal{G}))$ is a faithfully flat R-algebra such that the analytification of its spectrum defines a geometric $\mathcal{G}$-torsor.

The composite of any three composable functors is equivalent to the identity. Moreover, any geometric $\mathcal{G}$-torsor $\mathcal{P} \to X$ is itself a sousperfectoid adic space that is locally étale over a ball over X.

Proof. Part (1) is clear, and part (2) follows from étale descent of vector bundles, [KL15, Theorem 8.2.22 (d)]. Finally, for part (3) note that on any open affinoid subset $U = \operatorname{Spa}(R, R^+) \subset X$ of X, the datum is equivalent to a Tannakian $\mathcal{G}$-torsor on $\operatorname{Spec} R$. Thus, this is equivalent to a geometric $\mathcal{G}$-torsor over $\operatorname{Spec} R$, which is in particular smooth over $\operatorname{Spec} R$, and so locally étale over an affine space. This implies that its analytification is locally étale over a ball over $\operatorname{Spa}(R, R^+)$, and so by Proposition 6.3.3 defines a sousperfectoid adic space itself if R is sousperfectoid. This also verifies the final assertion. $\square$

In the following we will be lax about the distinction between the three kinds of $\mathcal{G}$-torsors, and simply talk about $\mathcal{G}$-torsors unless the distinction is important. Finally, we will need the following result about the descent of $\mathcal{G}$-torsors on (open subsets of) $S \dot{\times} \operatorname{Spa} \mathbf{Z}_p$.

Proposition 19.5.3. Let $S \in \operatorname{Perf}$ be a perfectoid space of characteristic p and let $U \subset S \dot{\times} \operatorname{Spa} \mathbf{Z}_p$ be an open subset. The functor on Perf_S sending any $S' \to S$ to the groupoid of $\mathcal{G}$-torsors on

$$U \times_{S \dot{\times} \operatorname{Spa} \mathbf{Z}_p} S' \dot{\times} \operatorname{Spa} \mathbf{Z}_p$$

is a v-stack.

Proof. We use the Tannakian interpretation of $\mathcal{G}$-torsors. This directly reduces us to the case of $\mathcal{G} = \operatorname{GL}_n$, i.e., vector bundles. We may assume that $S = \operatorname{Spa}(R, R^+)$ is affinoid, and that $\widetilde{S} = \operatorname{Spa}(\widetilde{R}, \widetilde{R}^+) \to S$ is a v-cover, along which we want to descend vector bundles.

Recall that U is sousperfectoid; in fact, more precisely, the product $U' = U \times_{\operatorname{Spa} \mathbf{Z}_p} \operatorname{Spa} \mathbf{Z}_p[p^{1/p^\infty}]_p^\wedge$ is a perfectoid space. Moreover,

$$\widetilde{U}' = \widetilde{U} \times_{\operatorname{Spa} \mathbf{Z}_p} \operatorname{Spa} \mathbf{Z}_p[p^{1/p^\infty}]_p^\wedge = \widetilde{U} \times_{S \dot{\times} \operatorname{Spa} \mathbf{Z}_p} S' \dot{\times} \operatorname{Spa} \mathbf{Z}_p$$

is a perfectoid space that is a v-cover of $\widetilde{U}'$, and the construction is compatible with fiber products. By Proposition 17.1.8, this implies that the functor sending $S' \to S$ to the category of vector bundles on $\widetilde{U}'$ is a v-stack.

Now if $\widetilde{\mathcal{E}}$ is a vector bundle on $\widetilde{U}$ with a descent datum, then its pullback

$\widetilde{\mathcal{E}}'$ to $\widetilde{U}'$ descends to a vector bundle $\mathcal{E}'$ on U'. Assume that $U = \mathrm{Spa}(A, A^+)$ is affinoid perfectoid, where A is sousperfectoid. Then also $U' = \mathrm{Spa}(A', A'^+)$, $\widetilde{U} = \mathrm{Spa}(\widetilde{A}, \widetilde{A}^+)$ and $\widetilde{U}' = \mathrm{Spa}(\widetilde{A}', \widetilde{A}'^+)$ are affinoid adic spaces. We have a finite projective A'-module M' and a finite projective $\widetilde{A}$-module $\widetilde{M}$ with the same base extension $\widetilde{M}'$ to $\widetilde{A}'$; moreover, $\widetilde{M}$ comes with a descent datum. To descend M' to A, it suffices to obtain a descent datum $M' \widehat{\otimes}_A A' \cong A' \widehat{\otimes}_A M'$ as the map $A \to A'$ splits as topological A-modules (and so the usual proof of descent works; note that being finite projective descends, e.g. by [Sta, Tag 08XD]). To define this descent datum, it suffices to define a descent datum $\widetilde{M}' \widehat{\otimes}_{\widetilde{A}} \widetilde{A}' \cong \widetilde{A}' \widehat{\otimes}_{\widetilde{A}} \widetilde{M}'$ satisfying a cocycle condition over $\widetilde{A} \widehat{\otimes}_A \widetilde{A}$. But we have such an isomorphism as defined by $\widetilde{M}$. $\qquad\square$

Lecture 20

Families of affine Grassmannians

In the geometric case, if X is a smooth curve over a field k, Beilinson-Drinfeld, [BD96], defined a family of affine Grassmannians

$$\mathrm{Gr}_{G,X} \to X$$

whose fiber over $x \in X$ parametrizes G-torsors on X with a trivialization over $X \setminus \{x\}$. If one fixes a coordinate at x, this gets identified with the affine Grassmannian considered previously.

In fact, more generally Beilinson-Drinfeld considered such families over many copies of X,

$$\mathrm{Gr}_{G,X^n} \to X^n,$$

whose fiber over $(x_1, \ldots, x_n) \in X^n$ parametrizes G-torsors on X with a trivialization over $X \setminus \{x_1, \ldots, x_n\}$. Something interesting happens when the points x_i collide: Over fibers with distinct points x_i, one gets a product of n copies of the affine Grassmannian, while over fibers with all points $x_i = x$ equal, one gets just one copy of the affine Grassmannian: This is possible as the affine Grassmannian is infinite-dimensional. However, sometimes it is useful to remember more information when the points collide; then one can consider the variant

$$\widetilde{\mathrm{Gr}}_{G,X^n} \to X^n$$

whose fiber over $(x_1, \ldots, x_n) \in X^n$ parametrizes G-torsors $P_1, \ldots, P_n$ on X together with a trivialization of P_1 on $X \setminus \{x_1\}$ and isomorphisms between P_i and P_{i+1} on $X \setminus \{x_{i+1}\}$ for $i = 1, \ldots, n-1$. There is a natural forgetful map $\widetilde{\mathrm{Gr}}_{G,X^n} \to \mathrm{Gr}_{G,X^n}$ which is an isomorphism over the locus where all x_i are distinct, but the fibers of $\widetilde{\mathrm{Gr}}_{G,X^n}$ over points with all $x_i = x$ equal is the convolution affine Grassmannian parametrizing successive modifications of the trivial G-torsor.

Our aim in this lecture is to study analogues of this picture if X is replaced by $X = \mathrm{Spd}\, \mathbf{Q}_p$ or also $X = \mathrm{Spd}\, \mathbf{Z}_p$.

20.1 THE CONVOLUTION AFFINE GRASSMANNIAN

As a preparation, we discuss the convolution affine Grassmannian in the setting of the last lecture.

Definition 20.1.1. For any integer $m \geq 1$, let $\widetilde{\mathrm{Gr}}_G^m / \mathrm{Spd}\, C$ be the presheaf sending $S = \mathrm{Spa}(R, R^+)$ with an untilt $S^\sharp = \mathrm{Spa}(R^\sharp, R^{\sharp+})$ over $\mathrm{Spa}\, C$ to the set of G-torsors $P_1, \ldots, P_m$ over $\mathrm{Spec}\, B_{\mathrm{dR}}^+(R^\sharp)$ together with a trivialization of P_1 over $\mathrm{Spec}\, B_{\mathrm{dR}}(R^\sharp)$ and isomorphisms between P_i and P_{i-1} over $\mathrm{Spec}\, B_{\mathrm{dR}}(R^\sharp)$ for $i = 2, \ldots, m$.

Remark 20.1.2. By composing isomorphisms, this is the same as parametrizing G-torsors $P_1, \ldots, P_m$ over $\mathrm{Spec}\, B_{\mathrm{dR}}^+(R^\sharp)$ together with a trivialization of each P_i over $\mathrm{Spec}\, B_{\mathrm{dR}}(R^\sharp)$; thus

$$\widetilde{\mathrm{Gr}}_G^m \cong \prod_{i=1}^m \mathrm{Gr}_G \, ,$$

where the product is taken over $\mathrm{Spd}\, C$. In particular, $\widetilde{\mathrm{Gr}}_G^m$ is a partially proper small v-sheaf.

The reason we defined $\widetilde{\mathrm{Gr}}_G^m$ not simply as a product is that the following definition of convolution Schubert varieties is more natural in this perspective.

Definition 20.1.3. For $\mu_1, \ldots, \mu_m \in X_*(T)^+$, let

$$\widetilde{\mathrm{Gr}}_{G, \leq \mu_\bullet} = \widetilde{\mathrm{Gr}}_{G, \leq (\mu_1, \ldots, \mu_m)} \subset \widetilde{\mathrm{Gr}}_G^m$$

be the closed subfunctor given by the condition that at all geometric rank 1-points, the relative position of P_i and P_{i-1} is bounded by μ_i in the Bruhat order (where P_0 is the trivial G-torsor).

This is indeed a closed subfunctor by Proposition 19.2.3.

Proposition 20.1.4. The v-sheaf $\widetilde{\mathrm{Gr}}_{G, \leq \mu_\bullet}$ is a spatial diamond that is proper over $\mathrm{Spd}\, C$. Under the map $\widetilde{\mathrm{Gr}}_{G, \leq \mu_\bullet} \to \mathrm{Gr}_G$ sending $(P_1, \ldots, P_m)$ to P_m with the composite trivialization, the image is contained in the Schubert variety $\mathrm{Gr}_{G, \leq |\mu_\bullet|}$, where $|\mu_\bullet| = \mu_1 + \ldots + \mu_m$, and the induced map

$$\widetilde{\mathrm{Gr}}_{G, \leq \mu_\bullet} \to \mathrm{Gr}_{G, \leq |\mu_\bullet|}$$

is a surjective map of v-sheaves.

Proof. We prove the first point by induction on m, the case $m = 1$ being known. The map

$$\widetilde{\mathrm{Gr}}_{G, \leq (\mu_1, \ldots, \mu_m)} \to \widetilde{\mathrm{Gr}}_{G, \leq (\mu_1, \ldots, \mu_{m-1})}$$

forgetting P_m is locally on the target isomorphic to a product over $\operatorname{Spd} C$ with $\operatorname{Gr}_{G,\leq\mu_m}$, by taking an étale local trivialization of P_{m-1}. By [Sch17, Proposition 13.4, Proposition 18.3], it follows that the map is representable in spatial diamonds and proper. By induction, we see that $\widetilde{\operatorname{Gr}}_{G,\leq\mu_\bullet}$ is a spatial diamond, proper over $\operatorname{Spd} C$.

The factorization can be checked on geometric points, as can surjectivity as v-sheaves by Lemma 17.4.9, using the fact that the source is quasicompact (and the target quasiseparated). But on geometric points, the situation agrees with the classical affine Grassmannian by choosing an isomorphism $B_{\mathrm{dR}}(C) \cong C((\xi))$. □

20.2   OVER $\operatorname{Spd} \mathbf{Q}_p$

First, we consider the case of a single copy of $\operatorname{Spd} \mathbf{Q}_p$. This is a minor variation on the previous lecture.

Definition 20.2.1. Let G be a reductive group over $\mathbf{Q}_p$. The Beilinson-Drinfeld Grassmannian over $\operatorname{Spd} \mathbf{Q}_p$ is the presheaf $\operatorname{Gr}_{G,\operatorname{Spd}\mathbf{Q}_p} / \operatorname{Spd} \mathbf{Q}_p$ whose $S = \operatorname{Spa}(R, R^+)$-points parametrize the data of an untilt $S^\sharp = \operatorname{Spa}(R^\sharp, R^{\sharp+})$ of S of characteristic 0 together with a G-torsor $\mathcal{P}$ on $S \dot\times \operatorname{Spa} \mathbf{Q}_p$ and a trivialization of $\mathcal{P}|_{S \dot\times \operatorname{Spa} \mathbf{Q}_p \setminus S^\sharp}$ that is meromorphic along $S^\sharp$.

Proposition 20.2.2. The Beilinson-Drinfeld Grassmannian $\operatorname{Gr}_{G,\operatorname{Spd}\mathbf{Q}_p}$ is a small v-sheaf, and is the étale sheafification of the functor taking $S = \operatorname{Spa}(R, R^+)$ with untilt $S^\sharp = \operatorname{Spa}(R^\sharp, R^{\sharp+})$ of characteristic 0 to $G(B_{\mathrm{dR}}(R^\sharp))/G(B_{\mathrm{dR}}^+(R^\sharp))$. In particular, for any algebraically closed nonarchimedean field $C/\mathbf{Q}_p$, the base change of $\operatorname{Gr}_{G,\operatorname{Spd}\mathbf{Q}_p} \times_{\operatorname{Spd}\mathbf{Q}_p} \operatorname{Spd} C$ is the affine Grassmannian $\operatorname{Gr}_G = \operatorname{Gr}_{G,\operatorname{Spd}C}$ considered in the previous lecture.

Proof. This is identical to the case of Proposition 19.1.2.   □

In particular, many statements about $\operatorname{Gr}_{G,\operatorname{Spd}\mathbf{Q}_p}$ can be obtained by descent from $\mathbf{C}_p$. There is one subtlety however, namely that Schubert varieties are not themselves defined over $\mathbf{Q}_p$. More precisely, if one fixes a maximal torus and a Borel $T \subset B \subset G_{\overline{\mathbf{Q}}_p}$, one can identify the set of dominant cocharacters $X_*(T)^+$ with the set of conjugacy classes of cocharacters $\mathbf{G}_m \to G_{\overline{\mathbf{Q}}_p}$; in particular, there is a natural continuous action of $\Gamma = \operatorname{Gal}(\overline{\mathbf{Q}}_p/\mathbf{Q}_p)$ on this set. Thus, for $\mu \in X_*(T)^+$ corresponding to a conjugacy class of cocharacters, its field of definition E is a finite extension of $\mathbf{Q}_p$, and the Schubert variety $\operatorname{Gr}_{G,\operatorname{Spd}C_p,\leq\mu}$ descends to E to give a closed subfunctor

$$\operatorname{Gr}_{G,\operatorname{Spd}E,\leq\mu} \subset \operatorname{Gr}_{G,\operatorname{Spd}E} = \operatorname{Gr}_{G,\operatorname{Spd}\mathbf{Q}_p} \times_{\operatorname{Spd}\mathbf{Q}_p} \operatorname{Spd}E \ .$$

From the previous lecture, one immediately obtains the following proposition. Here $\mathscr{Fl}_{G,\mu}$ is the flag variety over E that parametrizes parabolics in the conjugacy class of P_μ.

Proposition 20.2.3. The v-sheaf $\mathrm{Gr}_{G,\mathrm{Spd}\,E,\leq\mu}$ is a spatial diamond, and the map $\mathrm{Gr}_{G,\mathrm{Spd}\,E,\leq\mu} \to \mathrm{Spd}\,E$ is proper. If μ is minuscule, then

$$\mathrm{Gr}_{G,\mathrm{Spd}\,E,\leq\mu} = \mathrm{Gr}_{G,\mathrm{Spd}\,E,\mu} = \mathscr{Fl}^{\diamond}_{G,\mu} \ .$$

20.3   OVER $\mathrm{Spd}\,\mathbf{Z}_p$

Next, we consider the case of $\mathrm{Spd}\,\mathbf{Z}_p$. This is more interesting, as there are now two distinct geometric fibers. Now we have to fix a smooth model $\mathcal{G}$ of G over $\mathbf{Z}_p$; the easiest case is when $\mathcal{G}$ itself is reductive, but we do not assume this for the moment.

Definition 20.3.1. The Beilinson-Drinfeld Grassmannian $\mathrm{Gr}_{\mathcal{G},\mathrm{Spd}\,\mathbf{Z}_p} / \mathrm{Spd}\,\mathbf{Z}_p$ is the presheaf whose $S = \mathrm{Spa}(R, R^+)$-points parametrize the data of an untilt $S^\sharp = \mathrm{Spa}(R^\sharp, R^{\sharp+})$ together with a $\mathcal{G}$-torsor $\mathcal{P}$ on $S \dot{\times} \mathrm{Spa}\,\mathbf{Z}_p$ and a trivialization of $\mathcal{P}|_{S\dot{\times}\,\mathrm{Spa}\,\mathbf{Z}_p\backslash S^\sharp}$ that is meromorphic along $S^\sharp$.

Again, we have the following proposition.

Proposition 20.3.2. The Beilinson-Drinfeld Grassmannian $\mathrm{Gr}_{\mathcal{G},\mathrm{Spd}\,\mathbf{Z}_p}$ is a small v-sheaf, and is the étale sheafification of the functor taking $S = \mathrm{Spa}(R, R^+)$ with untilt $S^\sharp = \mathrm{Spa}(R^\sharp, R^{\sharp+})$ to $\mathcal{G}(B_{\mathrm{dR}}(R^\sharp))/\mathcal{G}(B_{\mathrm{dR}}^+(R^\sharp))$.   □

Recall that if $R^\sharp = R$ is of characteristic p, then $B_{\mathrm{dR}}^+(R^\sharp) = W(R)$ is the ring of Witt vectors. Thus, on the special fiber, we look at the functor sending $S = \mathrm{Spa}(R, R^+)$ to $\mathcal{G}(W(R)[\frac{1}{p}])/\mathcal{G}(W(R))$, which is precisely the Witt vector affine Grassmannian considered in [Zhu17a] and [BS17].

Definition 20.3.3. The Witt vector affine Grassmannian $\mathrm{Gr}_{\mathcal{G}}^W$ is the functor on affine perfect schemes $S = \mathrm{Spec}\,R$ of characteristic p taking $S = \mathrm{Spec}\,R$ to the set of $\mathcal{G}$-torsors on $\mathrm{Spec}\,W(R)$ together with a trivialization over $\mathrm{Spec}\,W(R)[\frac{1}{p}]$; it is also the étale sheafification of

$$R \mapsto \mathcal{G}(W(R)[\tfrac{1}{p}])/\mathcal{G}(W(R)) \ .$$

Theorem 20.3.4 ([BS17, Corollary 9.6]). The Witt vector affine Grassmannian $\mathrm{Gr}_{\mathcal{G}}^W$ can be written as an increasing union of perfections of quasiprojective varieties over $\mathbf{F}_p$ along closed immersions.

Thus, $\mathrm{Gr}_{\mathcal{G},\mathrm{Spd}\,\mathbf{Z}_p} \to \mathrm{Spd}\,\mathbf{Z}_p$ defines a deformation from the Witt vector affine Grassmannian to the B_{dR}^+-affine Grassmannian.

If $\mathcal{G}$ itself is reductive, then for any conjugacy class of cocharacters μ, the field of definition E is unramified over $\mathbf{Q}_p$ with residue field $\mathbf{F}_q$, and the Schubert varieties

$$\mathrm{Gr}^W_{\mathcal{G},\mathbf{F}_q,\leq\mu} \subset \mathrm{Gr}^W_{\mathcal{G},\mathbf{F}_q} = \mathrm{Gr}^W_{\mathcal{G}} \times_{\mathrm{Spec}\,\mathbf{F}_p} \mathrm{Spec}\,\mathbf{F}_q$$

are perfections of projective varieties over $\mathbf{F}_q$. One can then also define a family of Schubert varieties over $\mathrm{Spd}\,\mathcal{O}_E$:

Definition 20.3.5. Assume that $\mathcal{G}$ is reductive, and let $\mu \in X_*(T)^+$ be a conjugacy class of cocharacters defined over E. The Schubert variety

$$\mathrm{Gr}_{\mathcal{G},\mathrm{Spd}\,\mathcal{O}_E,\leq\mu} \subset \mathrm{Gr}_{\mathcal{G},\mathrm{Spd}\,\mathcal{O}_E} = \mathrm{Gr}_{\mathcal{G},\mathrm{Spd}\,\mathbf{Z}_p} \times_{\mathrm{Spd}\,\mathbf{Z}_p} \mathrm{Spd}\,\mathcal{O}_E$$

is the subfunctor defined by the condition that at all geometric rank 1-points, the relative position of $\mathcal{P}$ and the trivial $\mathcal{G}$-torsor is bounded by μ in the Bruhat order.

Proposition 20.3.6. The map $\mathrm{Gr}_{\mathcal{G},\mathrm{Spd}\,\mathcal{O}_E,\leq\mu} \to \mathrm{Spd}\,\mathcal{O}_E$ is proper and representable in spatial diamonds, and $\mathrm{Gr}_{\mathcal{G},\mathrm{Spd}\,\mathcal{O}_E,\leq\mu} \to \mathrm{Gr}_{\mathcal{G},\mathrm{Spd}\,\mathcal{O}_E}$ is a closed immersion.

Proof. The proof works as in the last lecture. To understand points on the special fiber, one uses Theorem 20.3.4. □

Proposition 20.3.7. For a closed immersion $\rho\colon \mathcal{G} \to \mathcal{H}$ of reductive groups over $\mathbf{Z}_p$, the functor $\mathrm{Gr}_{\mathcal{G},\mathrm{Spd}\,\mathbf{Z}_p} \to \mathrm{Gr}_{\mathcal{H},\mathrm{Spd}\,\mathbf{Z}_p}$ is a closed immersion.

Proof. Again, this can be proved as in the last lecture. □

In the next lecture, we will generalize some of these results to the case of parahoric groups $\mathcal{G}$.

20.4   OVER $\mathrm{Spd}\,\mathbf{Q}_p \times \ldots \times \mathrm{Spd}\,\mathbf{Q}_p$

Next, we want to understand a Beilinson-Drinfeld Grassmannian over $(\mathrm{Spd}\,\mathbf{Q}_p)^m = \mathrm{Spd}\,\mathbf{Q}_p \times \ldots \times \mathrm{Spd}\,\mathbf{Q}_p$.

Definition 20.4.1. The Beilinson-Drinfeld Grassmannian

$$\mathrm{Gr}_{G,(\mathrm{Spd}\,\mathbf{Q}_p)^m} \to (\mathrm{Spd}\,\mathbf{Q}_p)^m$$

is the presheaf sending $S = \mathrm{Spa}(R, R^+)$ with m untilts $S_i^\sharp = \mathrm{Spa}(R_i^\sharp, R_i^{\sharp+})$ of characteristic 0 to the set of G-torsors $\mathcal{P}$ on $S \dot\times \mathrm{Spa}\,\mathbf{Q}_p$ together with a trivialization of

$$\mathcal{P}\big|_{S\dot\times\,\mathrm{Spa}\,\mathbf{Q}_p \backslash \bigcup_{i=1}^m S_i^\sharp}$$

that is meromorphic along the closed Cartier divisor $\bigcup_{i=1}^m S_i^\sharp \subset S \dot\times \mathrm{Spa}\,\mathbf{Q}_p$.

As explained in the beginning of this lecture, it is sometimes useful to consider the following variant.

Definition 20.4.2. The convolution Beilinson-Drinfeld Grassmannian

$$\widetilde{\mathrm{Gr}}_{G,(\mathrm{Spd}\,\mathbf{Q}_p)^m} \to (\mathrm{Spd}\,\mathbf{Q}_p)^m$$

is the presheaf sending $S = \mathrm{Spa}(R, R^+)$ with m untilts $S_i^\sharp = \mathrm{Spa}(R_i^\sharp, R_i^{\sharp+})$ of characteristic 0 to the set of G-torsors $\mathcal{P}_1, \ldots, \mathcal{P}_m$ on $S \dot\times \mathrm{Spa}\,\mathbf{Q}_p$, together with a trivialization of

$$\mathcal{P}_1\big|_{S \dot\times \,\mathrm{Spa}\,\mathbf{Q}_p \setminus S_1^\sharp}$$

meromorphic along $S_1^\sharp$ and isomorphisms

$$\mathcal{P}_i\big|_{S \dot\times \,\mathrm{Spa}\,\mathbf{Q}_p \setminus S_i^\sharp} \cong \mathcal{P}_{i-1}\big|_{S \dot\times \,\mathrm{Spa}\,\mathbf{Q}_p \setminus S_i^\sharp}$$

for $i = 2, \ldots, n$ that are meromorphic along $S_i^\sharp$.

As before, Proposition 19.5.3 implies the following result.

Proposition 20.4.3. The presheaves $\mathrm{Gr}_{G,(\mathrm{Spd}\,\mathbf{Q}_p)^m}$ and $\widetilde{\mathrm{Gr}}_{G,(\mathrm{Spd}\,\mathbf{Q}_p)^m}$ are partially proper small v-sheaves.

We want to understand the analogue of Schubert varieties in this setting. For this, fix conjugacy classes of cocharacters $\mu_1, \ldots, \mu_m$, defined over the finite extensions $E_1, \ldots, E_m/\mathbf{Q}_p$.

Definition 20.4.4. Define a subfunctor $\mathrm{Gr}_{G,\mathrm{Spd}\,E_1 \times \ldots \times \mathrm{Spd}\,E_m, \leq \mu_\bullet}$ of

$$\mathrm{Gr}_{G,\mathrm{Spd}\,E_1 \times \ldots \times \mathrm{Spd}\,E_m} = \mathrm{Gr}_{G,(\mathrm{Spd}\,\mathbf{Q}_p)^m} \times_{(\mathrm{Spd}\,\mathbf{Q}_p)^m} (\mathrm{Spd}\,E_1 \times \ldots \times \mathrm{Spd}\,E_m)$$

and a subfunctor $\widetilde{\mathrm{Gr}}_{G,\mathrm{Spd}\,E_1 \times \ldots \times \mathrm{Spd}\,E_m, \leq \mu_\bullet}$ of

$$\widetilde{\mathrm{Gr}}_{G,\mathrm{Spd}\,E_1 \times \ldots \times \mathrm{Spd}\,E_m} = \widetilde{\mathrm{Gr}}_{G,(\mathrm{Spd}\,\mathbf{Q}_p)^m} \times_{(\mathrm{Spd}\,\mathbf{Q}_p)^m} (\mathrm{Spd}\,E_1 \times \ldots \times \mathrm{Spd}\,E_m)$$

by asking that the following condition be satisfied at all geometric rank 1-points $S = \mathrm{Spa}(C, \mathcal{O}_C)$. In the first case, the relative position of $\mathcal{P}$ and the trivial G-torsor at $S_i^\sharp$ is bounded by $\sum_{j|S_j^\sharp = S_i^\sharp} \mu_j$ in the Bruhat order for all $i = 1, \ldots, n$. In the second case, the relative position of $\mathcal{P}_i$ and $\mathcal{P}_{i-1}$ at $S_i^\sharp$ is bounded by μ_i in the Bruhat order (where $\mathcal{P}_0$ is the trivial G-torsor) for all $i = 1, \ldots, n$.

Proposition 20.4.5. The Beilinson-Drinfeld Schubert varieties have the following properties.

1. The v-sheaf $\widetilde{\mathrm{Gr}}_{G,\mathrm{Spd}\,E_1 \times \ldots \times \mathrm{Spd}\,E_m, \leq \mu_\bullet}$ is a locally spatial diamond that is proper over $\mathrm{Spd}\,E_1 \times \ldots \times \mathrm{Spd}\,E_m$.

2. The map

$$\widetilde{\mathrm{Gr}}_{G,\mathrm{Spd}\,E_1\times...\times\mathrm{Spd}\,E_m,\leq\mu_\bullet} \to \mathrm{Gr}_{G,\mathrm{Spd}\,E_1\times...\times\mathrm{Spd}\,E_m}$$

factors over $\mathrm{Gr}_{G,\mathrm{Spd}\,E_1\times...\times\mathrm{Spd}\,E_m,\leq\mu_\bullet}$, and induces a surjective map of v-sheaves

$$\widetilde{\mathrm{Gr}}_{G,\mathrm{Spd}\,E_1\times...\times\mathrm{Spd}\,E_m,\leq\mu_\bullet} \to \mathrm{Gr}_{G,\mathrm{Spd}\,E_1\times...\times\mathrm{Spd}\,E_m,\leq\mu_\bullet}\ .$$

3. The v-sheaf $\mathrm{Gr}_{G,\mathrm{Spd}\,E_1\times...\times\mathrm{Spd}\,E_m,\leq\mu_\bullet}$ is proper over $\mathrm{Spd}\,E_1\times\ldots\times\mathrm{Spd}\,E_m$, and

$$\mathrm{Gr}_{G,\mathrm{Spd}\,E_1\times...\times\mathrm{Spd}\,E_m,\leq\mu_\bullet} \subset \mathrm{Gr}_{G,\mathrm{Spd}\,E_1\times...\times\mathrm{Spd}\,E_m}$$

   is a closed embedding.
4. Finally, $\mathrm{Gr}_{G,\mathrm{Spd}\,E_1\times...\times\mathrm{Spd}\,E_m,\leq\mu_\bullet}$ is a locally spatial diamond.

Proof. The v-sheaf $\widetilde{\mathrm{Gr}}_{G,\mathrm{Spd}\,E_1\times...\times\mathrm{Spd}\,E_m,\leq\mu_\bullet}$ is a successive extension of copies of $\mathrm{Gr}_{G,\mathrm{Spd}\,E_i,\leq\mu_i}$. This implies that it is a locally spatial diamond, proper over $\mathrm{Spd}\,E_1\times\ldots\times\mathrm{Spd}\,E_m$, giving part (1).

   To check that the map to $\mathrm{Gr}_{G,\mathrm{Spd}\,E_1\times...\times\mathrm{Spd}\,E_m}$ factors over the Schubert variety $\mathrm{Gr}_{G,\mathrm{Spd}\,E_1\times...\times\mathrm{Spd}\,E_m,\leq\mu_\bullet}$, we can check on geometric rank 1-points, by definition; then, by [Sch17, Lemma 12.11], surjectivity as a map of v-sheaves can also be checked on such points. But over a geometric point of $\mathrm{Spd}\,E_1\times\ldots\times\mathrm{Spd}\,E_m$, the situation decomposes into a finite product of copies of the situation of Proposition 20.1.4, giving part (ii).

   We already know that $\mathrm{Gr}_{G,\mathrm{Spd}\,E_1\times...\times\mathrm{Spd}\,E_m}$, and thus the Schubert variety $\mathrm{Gr}_{G,\mathrm{Spd}\,E_1\times...\times\mathrm{Spd}\,E_m,\leq\mu_\bullet}$, is partially proper over $\mathrm{Spd}\,E_1\times\ldots\times\mathrm{Spd}\,E_m$. Thus, to see that it is proper, it remains to see that it is quasicompact, which follows from part (ii). As it is proper, the inclusion into $\mathrm{Gr}_{G,\mathrm{Spd}\,E_1\times...\times\mathrm{Spd}\,E_m}$ must be a closed embedding.

   For the final part, by applying Theorem 17.3.9 to fibers over quasicompact open subsets of the locally spatial diamond $\mathrm{Spd}\,E_1\times\ldots\times\mathrm{Spd}\,E_m$, it is enough to show that the v-sheaf $\mathrm{Gr}_{G,\mathrm{Spd}\,E_1\times...\times\mathrm{Spd}\,E_m,\leq\mu_\bullet}$ is locally spatial; indeed, any point lies in a fiber over $\mathrm{Spd}\,\mathbf{Q}_p\times\ldots\times\mathrm{Spd}\,\mathbf{Q}_p$, and all fibers are locally spatial (as they are finite products of Schubert varieties). To check that it is locally spatial, this can be reduced to the case of $G=\mathrm{GL}_n$ and $\mu_i=(N_i,0,\ldots,0)$ for some $N_i\geq 0$, and it is enough to check after pullback to $(\mathrm{Spd}\,\mathbf{C}_p)^m$. In this case, an argument similar to the previous lecture with the space $X/(\mathrm{Spd}\,\mathbf{C}_p)^m$ whose (R,R^+)-valued points with untilts given by primitive elements $\xi_i\in W(R^+)$ is given by

$$X(R,R^+) = \{A\in\mathrm{GL}_n(W(R^+)) \mid \det A\in\prod_{i=1}^{m}\xi_i^{N_i}W(R^+)^\times\}\ .$$

Then X is representable by a perfectoid space, the map $X\to(\mathrm{Spd}\,\mathbf{C}_p)^m$ is qcqs, and the map $X\to\mathrm{Gr}_{\mathrm{GL}_n,(\mathrm{Spd}\,\mathbf{C}_p)^m,\leq\mu_\bullet}$ is a surjective map of v-sheaves, and the induced equivalence relation $R = X \times_{\mathrm{Gr}_{\mathrm{GL}_n, (\mathrm{Spd}\,\mathbf{C}_p)^m, \leq \mu_\bullet}} X$ is isomorphic to $X \times_{\mathrm{Spd}\,\mathbf{C}_p} L_W^+ \mathrm{GL}_n$, and so as before $s, t \colon |R| \to |X|$ are open. $\qquad\square$

20.5 OVER $\mathrm{Spd}\,\mathbf{Z}_p \times \ldots \times \mathrm{Spd}\,\mathbf{Z}_p$

Finally, we extend the previous discussion to $\mathrm{Spd}\,\mathbf{Z}_p \times \ldots \times \mathrm{Spd}\,\mathbf{Z}_p$. Let $\mathcal{G}/\mathbf{Z}_p$ be a smooth group scheme with generic fiber G.

Definition 20.5.1. The Beilinson-Drinfeld Grassmannian

$$\mathrm{Gr}_{\mathcal{G}, (\mathrm{Spd}\,\mathbf{Z}_p)^m} \to (\mathrm{Spd}\,\mathbf{Z}_p)^m$$

is the presheaf sending $S = \mathrm{Spa}(R, R^+)$ with m untilts $S_i^\sharp = \mathrm{Spa}(R_i^\sharp, R_i^{\sharp+})$ to the set of $\mathcal{G}$-torsors $\mathcal{P}$ on $S \dot{\times} \mathrm{Spa}\,\mathbf{Z}_p$ together with a trivialization of

$$\mathcal{P}|_{S \dot{\times} \mathrm{Spa}\,\mathbf{Z}_p \setminus \bigcup_{i=1}^m S_i^\sharp}$$

that is meromorphic along the closed Cartier divisor $\bigcup_{i=1}^m S_i^\sharp \subset S \dot{\times} \mathrm{Spa}\,\mathbf{Z}_p$.
   Similarly, the convolution Beilinson-Drinfeld Grassmannian

$$\widetilde{\mathrm{Gr}}_{\mathcal{G}, (\mathrm{Spd}\,\mathbf{Z}_p)^m} / (\mathrm{Spd}\,\mathbf{Z}_p)^m$$

is the presheaf sending $S = \mathrm{Spa}(R, R^+)$ with m untilts $S_i^\sharp = \mathrm{Spa}(R_i^\sharp, R_i^{\sharp+})$ to the set of $\mathcal{G}$-torsors $\mathcal{P}_1, \ldots, \mathcal{P}_m$ on $S \dot{\times} \mathrm{Spa}\,\mathbf{Z}_p$ together with a trivialization of

$$\mathcal{P}_1|_{S \dot{\times} \mathrm{Spa}\,\mathbf{Z}_p \setminus S_1^\sharp}$$

meromorphic along $S_1^\sharp$ and isomorphisms

$$\mathcal{P}_i|_{S \dot{\times} \mathrm{Spa}\,\mathbf{Z}_p \setminus S_i^\sharp} \cong \mathcal{P}_{i-1}|_{S \dot{\times} \mathrm{Spa}\,\mathbf{Z}_p \setminus S_i^\sharp}$$

for $i = 2, \ldots, m$ that are meromorphic along $S_i^\sharp$.

Proposition 20.5.2. The presheaves $\mathrm{Gr}_{\mathcal{G}, (\mathrm{Spd}\,\mathbf{Z}_p)^m}$ and $\widetilde{\mathrm{Gr}}_{\mathcal{G}, (\mathrm{Spd}\,\mathbf{Z}_p)^m}$ are partially proper small v-sheaves.

   To discuss Schubert varieties, we make the assumption that $\mathcal{G}$ is reductive again. Fix conjugacy classes of cocharacters $\mu_1, \ldots, \mu_m$, defined over the finite extensions $E_1, \ldots, E_m / \mathbf{Q}_p$, unramified over $\mathbf{Z}_p$, and let $\mathcal{O}_1, \ldots, \mathcal{O}_m$ be their rings of integers.

Definition 20.5.3. Define a subfunctor $\mathrm{Gr}_{\mathcal{G}, \mathrm{Spd}\,\mathcal{O}_1 \times \ldots \times \mathrm{Spd}\,\mathcal{O}_m, \leq \mu_\bullet}$ of

$$\mathrm{Gr}_{\mathcal{G}, \mathrm{Spd}\,\mathcal{O}_1 \times \ldots \times \mathrm{Spd}\,\mathcal{O}_m} = \mathrm{Gr}_{\mathcal{G}, (\mathrm{Spd}\,\mathbf{Z}_p)^m} \times_{(\mathrm{Spd}\,\mathbf{Z}_p)^m} (\mathrm{Spd}\,\mathcal{O}_1 \times \ldots \times \mathrm{Spd}\,\mathcal{O}_m)$$

and a subfunctor $\widetilde{\mathrm{Gr}}_{\mathcal{G},\mathrm{Spd}\,\mathcal{O}_1\times\ldots\times\mathrm{Spd}\,\mathcal{O}_m,\leq\mu_\bullet}$ of

$$\widetilde{\mathrm{Gr}}_{\mathcal{G},\mathrm{Spd}\,\mathcal{O}_1\times\ldots\times\mathrm{Spd}\,\mathcal{O}_m} = \widetilde{\mathrm{Gr}}_{\mathcal{G},(\mathrm{Spd}\,\mathbf{Z}_p)^m} \times_{(\mathrm{Spd}\,\mathbf{Z}_p)^m} (\mathrm{Spd}\,\mathcal{O}_1\times\ldots\times\mathrm{Spd}\,\mathcal{O}_m)$$

by asking that the following condition be satisfied at all geometric rank 1-points $S = \mathrm{Spa}(C,\mathcal{O}_C)$. In the first case, the relative position of $\mathcal{P}$ and the trivial G-torsor at $S_i^\sharp$ is bounded by $\sum_{j|S_j^\sharp=S_i^\sharp} \mu_j$ in the Bruhat order for all $i = 1,\ldots,n$.

In the second case, the relative position of $\mathcal{P}_i$ and $\mathcal{P}_{i-1}$ at $S_i^\sharp$ is bounded by μ_i in the Bruhat order (where $\mathcal{P}_0$ is the trivial G-torsor) for all $i = 1,\ldots,n$.

Proposition 20.5.4. The Beilinson-Drinfeld Schubert varieties have the following properties.

1. The v-sheaf $\widetilde{\mathrm{Gr}}_{\mathcal{G},\mathrm{Spd}\,\mathcal{O}_1\times\ldots\times\mathrm{Spd}\,\mathcal{O}_m,\leq\mu_\bullet}$ is representable in spatial diamonds and proper over $\mathrm{Spd}\,\mathcal{O}_1 \times \ldots \times \mathrm{Spd}\,\mathcal{O}_m$.
2. The map

$$\widetilde{\mathrm{Gr}}_{\mathcal{G},\mathrm{Spd}\,\mathcal{O}_1\times\ldots\times\mathrm{Spd}\,\mathcal{O}_m,\leq\mu_\bullet} \to \mathrm{Gr}_{\mathcal{G},\mathrm{Spd}\,\mathcal{O}_1\times\ldots\times\mathrm{Spd}\,\mathcal{O}_m}$$

factors over $\mathrm{Gr}_{\mathcal{G},\mathrm{Spd}\,\mathcal{O}_1\times\ldots\times\mathrm{Spd}\,\mathcal{O}_m,\leq\mu_\bullet}$, and induces a surjective map of v-sheaves

$$\widetilde{\mathrm{Gr}}_{\mathcal{G},\mathrm{Spd}\,\mathcal{O}_1\times\ldots\times\mathrm{Spd}\,\mathcal{O}_m,\leq\mu_\bullet} \to \mathrm{Gr}_{\mathcal{G},\mathrm{Spd}\,\mathcal{O}_1\times\ldots\times\mathrm{Spd}\,\mathcal{O}_m,\leq\mu_\bullet}\ .$$

3. The v-sheaf $\mathrm{Gr}_{\mathcal{G},\mathrm{Spd}\,\mathcal{O}_1\times\ldots\times\mathrm{Spd}\,\mathcal{O}_m,\leq\mu_\bullet}$ is proper over $\mathrm{Spd}\,\mathcal{O}_1 \times \ldots \times \mathrm{Spd}\,\mathcal{O}_m$, and

$$\mathrm{Gr}_{\mathcal{G},\mathrm{Spd}\,\mathcal{O}_1\times\ldots\times\mathrm{Spd}\,\mathcal{O}_m,\leq\mu_\bullet} \subset \mathrm{Gr}_{\mathcal{G},\mathrm{Spd}\,\mathcal{O}_1\times\ldots\times\mathrm{Spd}\,\mathcal{O}_m}$$

is a closed embedding.
4. Finally, $\mathrm{Gr}_{\mathcal{G},\mathrm{Spd}\,\mathcal{O}_1\times\ldots\times\mathrm{Spd}\,\mathcal{O}_m,\leq\mu_\bullet} \to \mathrm{Spd}\,\mathcal{O}_1 \times \ldots \times \mathrm{Spd}\,\mathcal{O}_m$ is representable in spatial diamonds.

Proof. The same arguments as before apply. $\square$

Lecture 21

Affine flag varieties

In this lecture, we generalize some of the previous results to the case where $\mathcal{G}$ over $\mathbf{Z}_p$ is a parahoric group scheme. In fact, slightly more generally, we allow the case that the special fiber is not connected, with connected component of the identity $\mathcal{G}^\circ$ being a parahoric group scheme. This case comes up naturally in the classical definition of Rapoport-Zink spaces.

21.1   OVER $\mathbf{F}_p$

First, we discuss the Witt vector affine flag variety $\mathrm{Gr}_{\mathcal{G}}^W$ over $\mathbf{F}_p$. Recall that this is an increasing union of perfections of quasiprojective varieties along closed immersions. In the case that $\mathcal{G}^\circ$ is parahoric, one gets ind-properness.

Theorem 21.1.1 ([Zhu17a]). If $\mathcal{G}^\circ$ is parahoric, the Witt vector affine flag variety $\mathrm{Gr}_{\mathcal{G}}^W$ is ind-proper. In particular, by Theorem 20.3.4, it is an increasing union of perfections of projective varieties along closed immersions.

Remark 21.1.2. Zhu proved this for the Iwahori subgroup. However, to check the valuative criterion of properness, one can first lift the generic point to the affine flag variety for an Iwahori subgroup of $\mathcal{G}$, at least after an extension of the fraction field. Then properness of the Iwahori affine flag variety gives the desired result.

As we will need this later, let us briefly discuss the case of tori. Recall that a torus has a unique parahoric model, which is the connected component of its Néron model.

Proposition 21.1.3. If $G = T$ is a torus and $\mathcal{G} = \mathcal{T} = \mathcal{T}^\circ$ is the connected component of its Néron model, the affine flag variety $\mathrm{Gr}_{\mathcal{T}}^W$ is an étale $\mathbf{F}_p$-scheme which corresponds to the $\mathrm{Gal}(\overline{\mathbf{F}}_p/\mathbf{F}_p)$-module $\pi_1(T_{\overline{\mathbf{Q}}_p})_I$, where $I \subset \Gamma = \mathrm{Gal}(\overline{\mathbf{Q}}_p/\mathbf{Q}_p)$ is the inertia group.

Proof. For any algebraically closed field k of characteristic p, the Kottwitz map gives a bijection

$$\kappa : T(W(k)[\tfrac{1}{p}])/\mathcal{T}(W(k)) \to \pi_1(T_{\overline{\mathbf{Q}}_p})_I$$

by [Rap05, Note 1 at the end]. Moreover, the bijection is Frobenius equivariant. This easily implies the result. □

We will also need the general description of the connected components. Note that if $\mathcal{G}$ is not connected, the group of connected components of its special fiber $\pi_0\mathcal{G}$ is an étale group scheme over $\mathbf{F}_p$ such that the Kottwitz map defines an injection

$$\kappa : \pi_0\mathcal{G}_{\overline{\mathbf{F}}_p} \hookrightarrow \pi_1(G_{\overline{\mathbf{Q}}_p})_I \; ;$$

cf. the appendix by Haines-Rapoport to [PR08].

Proposition 21.1.4 ([Zhu17a, Proposition 1.21]). *If $\mathcal{G}$ is parahoric, the geometric connected components $\pi_0 \operatorname{Gr}^W_{\mathcal{G},\overline{\mathbf{F}}_p}$ are given by $\pi_1(G_{\overline{\mathbf{Q}}_p})_I$, equivariantly for the Frobenius map, via the Kottwitz map. In general, the map*

$$\operatorname{Gr}^W_{\mathcal{G}^\circ} \to \operatorname{Gr}^W_{\mathcal{G}}$$

is a torsor under $\pi_0\mathcal{G}$. This induces a bijection between $\pi_0 \operatorname{Gr}^W_{\mathcal{G},\overline{\mathbf{F}}_p}$ and the quotient $\pi_1(G_{\overline{\mathbf{Q}}_p})_I/\pi_0\mathcal{G}_{\overline{\mathbf{F}}_p}$, and the map

$$\operatorname{Gr}^W_{\mathcal{G}^\circ,\overline{\mathbf{F}}_p} \to \operatorname{Gr}^W_{\mathcal{G},\overline{\mathbf{F}}_p}$$

induces an isomorphism on connected components.

21.2   OVER $\mathbf{Z}_p$

As in the last lecture, we have the Beilinson-Drinfeld Grassmannian $\operatorname{Gr}_{\mathcal{G},\operatorname{Spd}\mathbf{Z}_p} \to \operatorname{Spd}\mathbf{Z}_p$ whose special fiber is the v-sheaf associated with the perfect ind-scheme $\operatorname{Gr}^W_{\mathcal{G}}$ and whose generic fiber is $\operatorname{Gr}_{G,\operatorname{Spd}\mathbf{Q}_p}$. Our goal in this section is to prove the following theorem.

Theorem 21.2.1. *The map of v-sheaves $\operatorname{Gr}_{\mathcal{G},\operatorname{Spd}\mathbf{Z}_p} \to \operatorname{Spd}\mathbf{Z}_p$ is ind-proper. More precisely, one can write $\operatorname{Gr}_{\mathcal{G},\operatorname{Spd}\mathbf{Z}_p}$ as an increasing union of closed subfunctors that are proper over $\operatorname{Spd}\mathbf{Z}_p$.*

If $\rho : \mathcal{G} \hookrightarrow \operatorname{GL}_n$ is a closed immersion, the induced map $\operatorname{Gr}_{\mathcal{G},\operatorname{Spd}\mathbf{Z}_p} \to \operatorname{Gr}_{\operatorname{GL}_n,\operatorname{Spd}\mathbf{Z}_p}$ is a closed immersion.

Note that it is enough to prove the last part, as then the first part follows by pullback from the similar results for GL_n.

We will use the following result of Anschütz to establish something close to a valuative criterion of properness.

Theorem 21.2.2 ([Ans18]). Let C be an algebraically closed nonarchimedean field over $\mathbf{Z}_p$ with tilt $C^\flat$, and let $\theta\colon A_{\inf} = W(\mathcal{O}_{C^\flat}) \to \mathcal{O}_C$ with kernel generated by ξ as usual. Let $\mathcal{P}$ be a $\mathcal{G}$-torsor on $\operatorname{Spec} A_{\inf} \setminus \{\mathfrak{m}\}$ that is trivial on $\operatorname{Spec} A_{\inf}[\frac{1}{\xi}]$. Then $\mathcal{P}$ is trivial; in particular, it extends, necessarily uniquely by the Tannakian formalism and Theorem 14.2.3, to a $\mathcal{G}$-torsor on $\operatorname{Spec} A_{\inf}$.

Corollary 21.2.3. For any algebraically closed nonarchimedean field C over $\mathbf{Z}_p$, and every map $\operatorname{Spd} C \to \operatorname{Gr}_{\mathcal{G},\operatorname{Spd}\mathbf{Z}_p}$ over $\operatorname{Spd}\mathbf{Z}_p$ given by a $\mathcal{G}$-torsor P over $\operatorname{Spec} B_{\mathrm{dR}}^+(C)$ trivialized on $\operatorname{Spec} B_{\mathrm{dR}}(C)$, there is a unique (up to unique isomorphism) $\mathcal{G}$-torsor $\mathcal{P}$ over $\operatorname{Spec} A_{\inf}$ trivialized on $\operatorname{Spec} A_{\inf}[\frac{1}{\xi}]$ whose restriction to $\operatorname{Spec} B_{\mathrm{dR}}^+(C)$ is P (with given trivialization over $\operatorname{Spec} B_{\mathrm{dR}}(C)$).

Proof. By the Tannakian formalism and the Beauville-Laszlo lemma, the $\mathcal{G}$-torsor P with its trivialization is equivalent to a $\mathcal{G}$-torsor on $\operatorname{Spec} A_{\inf} \setminus \{\mathfrak{m}\}$ with a trivialization over $\operatorname{Spec} A_{\inf}[\frac{1}{\xi}]$. By Theorem 21.2.2, this extends uniquely to a $\mathcal{G}$-torsor over all of $\operatorname{Spec} A_{\inf}$, still trivialized over $\operatorname{Spec} A_{\inf}[\frac{1}{\xi}]$. $\qquad\square$

Unfortunately, the valuative criterion of properness in [Sch17, Proposition 18.3] is of a different sort, and in particular it requires quasicompactness as input, which is the hard statement in our case. Our idea now is to go back to classical topology, and establish quasicompactness by showing that sequences of points converge.

Proof of Theorem 21.2.1. Fix the closed immersion $\rho\colon \mathcal{G} \hookrightarrow \mathrm{GL}_n$. Fix a cocharacter μ of GL_n. It is enough to prove that the map

$$\operatorname{Gr}_{\mathcal{G},\operatorname{Spd}\mathbf{Z}_p} \times_{\operatorname{Gr}_{\mathrm{GL}_n,\operatorname{Spd}\mathbf{Z}_p}} \operatorname{Gr}_{\mathrm{GL}_n,\operatorname{Spd}\mathbf{Z}_p,\leq\mu} \to \operatorname{Gr}_{\mathrm{GL}_n,\operatorname{Spd}\mathbf{Z}_p,\leq\mu}$$

is a closed immersion. As the map is injective on geometric points and partially proper, it is in fact enough to show that it is quasicompact by Corollary 17.4.8. For this, pick a surjection $X \to \operatorname{Gr}_{\mathrm{GL}_n,\operatorname{Spd}\mathbf{Z}_p,\leq\mu}$ from an affinoid perfectoid space X, and let $Y \to X$ be the pullback of the displayed map. Now choose maps from $\operatorname{Spa}(C_i, C_i^+)$, $i \in I$, into Y, covering all points of $|Y|$, and choose pseudouniformizers $\varpi_i \in C_i$ for all $i \in I$. Let R^+ be the product of all C_i^+, let $\varpi = (\varpi_i)_i \in R^+$, and let $R = R^+[\frac{1}{\varpi}]$. We wish to show that there is a map from $\operatorname{Spa}(R, R^+)$ to Y that restricts to the given maps on all $\operatorname{Spa}(C_i, C_i^+)$. Then $\operatorname{Spa}(R, R^+)$ surjects onto Y, which is therefore quasicompact. Note that, for well-chosen ϖ_i, we get a unique map to X that restricts to the given maps on all $\operatorname{Spa}(C_i, C_i^+)$, by taking the product on the level of algebras. Thus, it remains to factor the resulting map from $\operatorname{Spa}(R, R^+)$ to $\operatorname{Gr}_{\mathrm{GL}_n,\operatorname{Spd}\mathbf{Z}_p}$ over $\operatorname{Gr}_{\mathcal{G},\operatorname{Spd}\mathbf{Z}_p}$.

Moreover, by partial properness we can replace all C_i^+ by $\mathcal{O}_{C_i}$, and R^+ by $R^\circ = \prod_i \mathcal{O}_{C_i}$. For each $i \in I$, Corollary 21.2.3 implies that we get $\mathcal{G}$-torsors $\mathcal{P}_i$ on $A_{\inf,i} = W(\mathcal{O}_{C^\flat,i})$, trivialized on $A_{\inf,i}[\frac{1}{\xi_i}]$. In the Tannakian interpretation, their product gives rise to a $\mathcal{G}$-torsor $\mathcal{P}$ on $\prod_i A_{\inf,i} = W(R^{\circ\flat})$, as for any family of finite free $A_{\mathrm{inf},i}$-modules M_i of constant rank r, their product $\prod_i M_i$ is a finite free $W(R^{\mathrm{ob}})$-module of rank r. Moreover, as by assumption the type of the modification is bounded by μ, the trivializations over $A_{\mathrm{inf},i}[\frac{1}{\xi_i}]$ extend to the product. Now base changing along $W(R^{\mathrm{ob}}) \to B_{\mathrm{dR}}^+(R)$, we get a $\mathrm{Spa}(R, R^\circ)$-point of $\mathrm{Gr}_{\mathcal{G},\mathrm{Spd}\,\mathbf{Z}_p}$ whose image in $\mathrm{Gr}_{\mathrm{GL}_n,\mathrm{Spd}\,\mathbf{Z}_p}$ is correct. $\qquad\square$

21.3 AFFINE FLAG VARIETIES FOR TORI

Let us analyze the affine flag varieties associated with tori. So assume that $G = T$ is a torus, and that $\mathcal{G} = \mathcal{T}$ is the connected component of the Néron model of T.

Consider the $\Gamma = \mathrm{Gal}(\overline{\mathbf{Q}}_p/\mathbf{Q}_p)$-module $X = X_*(T_{\overline{\mathbf{Q}}_p}) = \pi_1(T_{\overline{\mathbf{Q}}_p})$. This is equivalent to an étale $\mathbf{Q}_p$-scheme $\underline{X}$. The normalization of $\mathbf{Z}_p$ inside $\underline{X}$ defines an integral model $\underline{X}^{\mathrm{int}}$ of $\underline{X}$; locally, if $U = \mathrm{Spec}\,K \subset \underline{X}$ is an open subscheme, then $U^{\mathrm{int}} = \mathrm{Spec}\,\mathcal{O}_K \subset \underline{X}^{\mathrm{int}}$ is an open subspace. If $I \subset \Gamma$ is the inertia subgroup, then the reduced special fiber of $\underline{X}^{\mathrm{int}}$ is the étale $\mathbf{F}_p$-scheme corresponding to the $\Gamma/I = \mathrm{Gal}(\overline{\mathbf{F}}_p/\mathbf{F}_p)$-module $X_I = \pi_1(T_{\overline{\mathbf{Q}}_p})_I$.

Proposition 21.3.1. Assume that $\mathcal{T}$ is the connected component of the Néron model of T. Then there is a natural isomorphism

$$\mathrm{Gr}_{\mathcal{T},\mathrm{Spd}\,\mathbf{Z}_p} \cong \underline{X}^{\mathrm{int},\diamond}$$

over $\mathrm{Spd}\,\mathbf{Z}_p$.

Proof. The geometric generic fiber $\mathrm{Gr}_{T,\mathrm{Spd}\,\mathbf{C}_p}$ is given by $\underline{X}_{\mathbf{C}_p}$, by reduction to $T = \mathbf{G}_{\mathrm{m}}$; here $\mu \in X$ maps to $\mathrm{Gr}_{T,\mathrm{Spd}\,\mathbf{C}_p,\mu} \cong \mathrm{Spd}\,\mathbf{C}_p$. By Galois descent, this implies that $\mathrm{Gr}_{T,\mathrm{Spd}\,\mathbf{Q}_p} \cong \underline{X}$. By Theorem 21.2.2 and Lemma 18.1.2, this extends to a map $\underline{X}^{\mathrm{int},\diamond} \to \mathrm{Gr}_{\mathcal{T},\mathrm{Spd}\,\mathbf{Z}_p}$. This map is necessarily proper, as the source is locally proper over $\mathrm{Spd}\,\mathbf{Z}_p$ (in turn because each $\mathrm{Spd}\,\mathcal{O}_K$ is proper). Thus, to check that it is an isomorphism, it is enough to check that it is bijective on geometric points. For characteristic 0 points, this is clear, so it remains to understand the special fiber. By Proposition 21.1.3, $\mathrm{Gr}_{\mathcal{T},\mathrm{Spd}\,\mathbf{F}_p}$ is the étale $\mathbf{F}_p$-scheme corresponding to $\pi_1(T_{\overline{\mathbf{Q}}_p})_I$. It remains to see that the identifications agree. Choosing a surjection $\widetilde{T} \to T$ from an induced torus $\widetilde{T}$, this reduces to an induced torus, where the result is immediate. $\qquad\square$

21.4 LOCAL MODELS

The local models of Shimura varieties studied by Rapoport and collaborators, [Rap90], can be considered as closed subspaces of the affine flag variety

$\mathrm{Gr}_{\mathcal{G},\mathrm{Spd}\,\mathbf{Z}_p}$. A similar description using a more classical affine flag variety was used by Görtz, [Gör01], and others to obtain geometric results on local models. Our approach has the advantage that the intervening objects are canonical, contrary to the local models introduced by Pappas–Zhu, [PZ13], that depend on some auxiliary choices.

Using our machinery, we can state the following general conjecture about the existence of local models.

Conjecture 21.4.1. Let $\mathcal{G}$ be a group scheme over $\mathbf{Z}_p$ with reductive generic fiber and $\mathcal{G}^\circ$ parahoric, and let μ be a minuscule conjugacy class of cocharacters $\mathbf{G}_\mathrm{m} \to G_{\overline{\mathbf{Q}}_p}$, defined over the reflex field E. There is a flat projective $\mathcal{O}_E$-scheme $\mathbb{M}^{\mathrm{loc}}_{(\mathcal{G},\mu)}$ with reduced special fiber and with a closed immersion

$$\mathbb{M}^{\mathrm{loc},\Diamond}_{(\mathcal{G},\mu)} \hookrightarrow \mathrm{Gr}_{\mathcal{G},\mathrm{Spd}\,\mathcal{O}_E}$$

whose generic fiber is $\mathscr{Fl}^\Diamond_{G,\mu} \cong \mathrm{Gr}_{\mathcal{G},\mathrm{Spd}\,E,\mu}$.

Remark 21.4.2. Note that a flat scheme over $\mathcal{O}_E$ with reduced special fiber is necessarily normal by Serre's criterion.

We note that by Proposition 18.4.1, the scheme $\mathbb{M}^{\mathrm{loc}}_{(\mathcal{G},\mu)}$ is unique if it exists. Indeed, the closed subfunctor

$$\mathbb{M}^{\mathrm{loc},\Diamond}_{(\mathcal{G},\mu)} \hookrightarrow \mathrm{Gr}_{\mathcal{G},\mathrm{Spd}\,\mathcal{O}_E}$$

must be given by the closure of $\mathrm{Gr}_{\mathcal{G},\mathrm{Spd}\,E,\mu}$.[1]

In the appendix to this lecture, we will make contact with the local models as introduced in the book of Rapoport-Zink, [RZ96]. For this, it is necessary to include the case where $\mathcal{G}$ is not necessarily parahoric because the special fiber is disconnected, as we did. This does however not actually impact local models.

Proposition 21.4.3. Let $\mathcal{G}$ be a group scheme over $\mathbf{Z}_p$ with connected generic fiber G whose connected component $\mathcal{G}^\circ$ is parahoric. Let μ be a conjugacy class of minuscule cocharacters of G, defined over E. Let $\mathrm{Gr}_{\mathcal{G},\mathrm{Spd}\,\mathcal{O}_E,\mu} \subset \mathrm{Gr}_{\mathcal{G},\mathrm{Spd}\,\mathcal{O}_E}$ and $\mathrm{Gr}_{\mathcal{G}^\circ,\mathrm{Spd}\,\mathcal{O}_E,\mu} \subset \mathrm{Gr}_{\mathcal{G}^\circ,\mathrm{Spd}\,\mathcal{O}_E}$ denote the closure of $\mathrm{Gr}_{G,\mathrm{Spd}\,E,\mu} \cong \mathscr{Fl}^\Diamond_{G,\mu}$. The induced map

$$\mathrm{Gr}_{\mathcal{G}^\circ,\mathrm{Spd}\,\mathcal{O}_E,\mu} \to \mathrm{Gr}_{\mathcal{G},\mathrm{Spd}\,\mathcal{O}_E,\mu}$$

is an isomorphism.

Proof. Both v-sheaves are proper over $\mathrm{Spd}\,\mathcal{O}_E$, and thus so is the map. As

---

[1]We warn the reader that in general, taking the closure in the sense of v-sheaves does not correspond to the closure on the level of topological spaces; rather, one must look for the minimal closed superset whose pullback to any perfectoid space is stable under generalizations. We expect that this nuisance does not matter in this situation.

the image of $|\mathrm{Gr}_{\mathcal{G}^\circ, \mathrm{Spd}\, \mathcal{O}_{E}, \mu}| \to |\mathrm{Gr}_{\mathcal{G}, \mathrm{Spd}\, \mathcal{O}_{E}, \mu}|$ is closed (and the pullback to any perfectoid space is stable under generalizations) and contains the subset $|\mathrm{Gr}_{G, \mathrm{Spd}\, E, \mu}|$, it is all of $|\mathrm{Gr}_{\mathcal{G}, \mathrm{Spd}\, \mathcal{O}_{E}, \mu}|$. By Lemma 17.4.9, it follows that

$$\mathrm{Gr}_{\mathcal{G}^\circ, \mathrm{Spd}\, \mathcal{O}_{E}, \mu} \to \mathrm{Gr}_{\mathcal{G}, \mathrm{Spd}\, \mathcal{O}_{E}, \mu}$$

is surjective as a map of v-sheaves. By Lemma 17.4.1 and properness, it remains to see that it is injective on $(C, \mathcal{O}_C)$-valued points. This is clear on the generic fiber. But on the geometric special fiber,

$$\mathrm{Gr}^{W}_{\mathcal{G}^\circ, \overline{\mathbf{F}}_p} \to \mathrm{Gr}^{W}_{\mathcal{G}, \overline{\mathbf{F}}_p}$$

is a torsor under the finite group $\mathcal{G}/\mathcal{G}^\circ$. By Proposition 21.1.4, it suffices to see that the special fiber of $\mathrm{Gr}_{\mathcal{G}^\circ, \mathrm{Spd}\, \mathcal{O}_{E}, \mu}$ is completely contained in one connected component (with Kottwitz invariant determined by μ); this follows as usual by passage to z-extensions and reduction to tori from the description in the case of a torus.   □

21.5   DÉVISSAGE

Conjecture 21.4.1 can be approached by some dévissage techniques, for example by understanding the behavior of $\mathrm{Gr}_{\mathcal{G}, \mathrm{Spd}\, \mathbf{Z}_p}$ when one changes G by central isogenies. The general form of the following result is due to João Lourenço.

   More precisely, let $\widetilde{G} \to G$ be a surjective map of reductive groups over $\mathbf{Q}_p$, with kernel a central subgroup $Z \subset \widetilde{G}$. Let $\widetilde{\mathcal{G}}$ over $\mathbf{Z}_p$ be a quasiparahoric model of $\widetilde{G}$, stabilizing some facet F of the Bruhat-Tits building of $\widetilde{G}$. As the Bruhat-Tits building depends only on the adjoint group, F also defines a quasiparahoric model $\mathcal{G}$ of G such that $\mathcal{G}(\breve{\mathbf{Z}}_p)$ is the stabilizer of F. There is a natural map $\widetilde{\mathcal{G}} \to \mathcal{G}$ of group schemes over $\mathbf{Z}_p$.

Proposition 21.5.1. Let $\widetilde{\mathcal{G}} \to \mathcal{G}$ be a map of quasiparahoric groups over $\mathbf{Z}_p$ as described above. Let $\widetilde{\mu}$ be a minuscule conjugacy class of cocharacters $\mathbf{G}_{\mathrm{m}} \to \widetilde{G}_{\overline{\mathbf{Q}}_p}$ defined over $\widetilde{E}$, and let μ be the corresponding conjugacy class of G, defined over $E \subset \widetilde{E}$.

   Let

$$\mathrm{Gr}_{\widetilde{\mathcal{G}}, \mathrm{Spd}\, \mathcal{O}_{\widetilde{E}}, \widetilde{\mu}} \subset \mathrm{Gr}_{\widetilde{\mathcal{G}}, \mathrm{Spd}\, \mathcal{O}_{\widetilde{E}}}$$

be the closure of $\mathrm{Gr}_{\widetilde{G}, \mathrm{Spd}\, \widetilde{E}, \widetilde{\mu}}$. Similarly, let

$$\mathrm{Gr}_{\mathcal{G}, \mathrm{Spd}\, \mathcal{O}_{E}, \mu} \subset \mathrm{Gr}_{\mathcal{G}, \mathrm{Spd}\, \mathcal{O}_{E}}$$

be the closure of $\mathrm{Gr}_{G,\mathrm{Spd}\, E,\mu}$. Then the natural map

$$\mathrm{Gr}_{\widetilde{\mathcal{G}},\mathrm{Spd}\,\mathcal{O}_{\widetilde{E}},\widetilde{\mu}} \to \mathrm{Gr}_{\mathcal{G},\mathrm{Spd}\,\mathcal{O}_E,\mu} \times_{\mathrm{Spd}\,\mathcal{O}_E} \mathrm{Spd}\,\mathcal{O}_{\widetilde{E}}$$

is an isomorphism.

Proof. First, one checks that $\mathrm{Gr}_{G,\mathrm{Spd}\,\widetilde{E},\mu} \subset \mathrm{Gr}_{\mathcal{G},\mathrm{Spd}\,\mathcal{O}_E,\mu} \times_{\mathrm{Spd}\,\mathcal{O}_E} \mathrm{Spd}\,\mathcal{O}_{\widetilde{E}}$ is still dense; this is easy if $\widetilde{E}/E$ is unramified, and in the totally ramified case one uses the fact that the special fiber does not change, and that $\mathrm{Spd}\,\mathcal{O}_{\widetilde{E}} \to \mathrm{Spd}\,\mathcal{O}_E$ is proper. This then implies that the map

$$\mathrm{Gr}_{\widetilde{\mathcal{G}},\mathrm{Spd}\,\mathcal{O}_{\widetilde{E}},\widetilde{\mu}} \to \mathrm{Gr}_{\mathcal{G},\mathrm{Spd}\,\mathcal{O}_E,\mu} \times_{\mathrm{Spd}\,\mathcal{O}_E} \mathrm{Spd}\,\mathcal{O}_{\widetilde{E}}$$

is surjective as a map of v-sheaves by Lemma 17.4.9.

For injectivity, it suffices to see that the map

$$\mathrm{Gr}_{\widetilde{\mathcal{G}},\mathbf{F}_p} \to \mathrm{Gr}_{\mathcal{G},\mathbf{F}_p}$$

is an injection on connected components. It suffices to prove injectivity on k-valued points, where k is any algebraically closed field of characteristic p. As both sides are homogenous spaces, it suffices to see that if $\widetilde{g} \in \widetilde{G}(W(k)[1/p])$ is any element with $\kappa_{\widetilde{G}}(\widetilde{g})$ in the image of $\pi_0\widetilde{G} \hookrightarrow \pi_1(G_{\overline{\mathbf{Q}}_p})_\Gamma$ (corresponding to lying in the connected component of the identity) and whose image $g \in G(W(K)[1/p])$ lies in $\mathcal{G}(W(K))$, then actually $\widetilde{g} \in \widetilde{\mathcal{G}}(W(k))$. By the appendix of Haines–Rapoport to [PR08], we know that $\widetilde{\mathcal{G}}(W(k)) \subset \widetilde{G}(W(k)[1/p])$ is the intersection of the stabilizer of F with $\kappa_{\widetilde{G}}^{-1}(\pi_0\widetilde{\mathcal{G}})$. But $\widetilde{g}$ stabilizes F as g does, so the result follows. $\qquad\square$

In particular, if Conjecture 21.4.1 holds true for $(\mathcal{G},\mu)$, then it also holds true for $(\widetilde{\mathcal{G}},\widetilde{\mu})$, and the converse holds if $\widetilde{E}/E$ is unramified.

Appendix to Lecture 21:
Examples of local models

In this appendix, we make explicit the relation to the local models discussed in [RZ96]. They start with the following rational data; cf. [RZ96, Section 1.38]. Here EL stands for "endomorphism + level," and PEL stands for "polarization + endomorphism + level"; the level structure will be encoded in the integral data.

- (Case EL) Let B be a finite-dimensional semisimple $\mathbf{Q}_p$-algebra, and let V be a finite-dimensional B-module.
- (Case PEL) Assume in this case that $p \neq 2$. In addition to (B, V) as in the EL case, let $(\, , \,)$ be a nondegenerate alternating $\mathbf{Q}_p$-bilinear form on V and let $b \mapsto b^*$ be an involution on B that satisfies

$$(bv, w) = (v, b^*w)$$

for all $b \in B$, $v, w \in V$.

Note that the center of B is an étale $\mathbf{Q}_p$-algebra F, i.e., a finite product of finite field extensions. In the PEL case, the involution $*$ acts on F, and we have a similar decomposition of $F_0 = F^{*=1}$. As in the following, everything will accordingly decompose into a product, we assume from the start that F_0 is a field. In fact, there is only one case in which F is it not itself a field (cf. case (I) in [RZ96, Section A.6]), and this PEL case is essentially equivalent to an EL case by Morita equivalence. Thus, for simplicity we assume from the start that F is a field, and leave the (trivial) modifications in the general case to the reader.

In the EL case, we let $G = \mathrm{GL}_B(V)$ be the reductive group over $\mathbf{Q}_p$. In the PEL case, we let G be the group whose R-valued points for a $\mathbf{Q}_p$-algebra R are given by

$$G(R) = \{g \in \mathrm{GL}_B(V \otimes_{\mathbf{Q}_p} R) \mid (gv, gw) = c(g)(v, w) \, , \; c(g) \in R^\times\} \, .$$

Thus, $c : G \to \mathbf{G}_m$ defines the similitude character of G. In general G need not be connected. In order to be properly in the setup studied above, we assume that G is connected.

Moreover, we fix integral structures as follows. First, we fix a maximal order

$\mathcal{O}_B \subset B$. Moreover, we fix a chain of lattices $\mathcal{L}$ according to [RZ96, Definition 3.1], i.e., a set $\mathcal{L}$ of $\mathcal{O}_B$-lattices $\Lambda \subset V$ that form a chain, i.e., for any two lattices $\Lambda, \Lambda' \in \mathcal{L}$, either $\Lambda \subset \Lambda'$ or $\Lambda' \subset \Lambda$, and for any $x \in B^\times$ that normalizes $\mathcal{O}_B$, the relation $x\Lambda \in \mathcal{L}$ holds for all $\Lambda \in \mathcal{L}$.[2]

In the PEL case, we moreover assume that $\mathcal{L}$ is selfdual, i.e., for all $\Lambda \in \mathcal{L}$, the dual lattice Λ^* is also in $\mathcal{L}$.

We want to define the corresponding group scheme $\mathcal{G}$ as the group of compatible isomorphisms of all Λ for $\Lambda \in \mathcal{L}$. In order to say what "compatible" means, it is convenient to introduce the notion of a chain of lattices of type $(\mathcal{L})$; cf. [RZ96, Definition 3.6].

Definition 21.6.1. Let R be a $\mathbf{Z}_p$-algebra. A chain of $\mathcal{O}_B \otimes_{\mathbf{Z}_p} R$-lattices is a functor $\Lambda \mapsto M_\Lambda$ from $\mathcal{L}$ (considered as a category with inclusions as morphisms) to $\mathcal{O}_B \otimes_{\mathbf{Z}_p} R$-modules, together with an isomorphism $\theta_b : M_\Lambda^b \cong M_{b\Lambda}$ for all $b \in B^\times$ normalizing $\mathcal{O}_B$, where $M_\Lambda^b = M_\Lambda$ with $\mathcal{O}_B$-action conjugated by b, satisfying the following conditions.

1. Locally on $\operatorname{Spec} R$, there are isomorphisms $M_\Lambda \cong \Lambda \otimes_{\mathbf{Z}_p} R$ of $\mathcal{O}_B \otimes_{\mathbf{Z}_p} R$-modules.
2. For any two adjacent lattices $\Lambda' \subset \Lambda$, there is an isomorphism

$$M_\Lambda / M_{\Lambda'} \cong \Lambda/\Lambda' \otimes_{\mathbf{Z}_p} R$$

   of $\mathcal{O}_B \otimes_{\mathbf{Z}_p} R$-modules, locally on $\operatorname{Spec} R$.
3. The periodicity isomorphisms θ_b commute with the transition maps $M_\Lambda \to M_{\Lambda'}$.
4. For all $b \in B^\times \cap \mathcal{O}_B$ normalizing $\mathcal{O}_B$, the composition $M_\Lambda^b \cong M_{b\Lambda} \to M_\Lambda$ is multiplication by b.

In the PEL case, we consider the following structure.

Definition 21.6.2. Consider data of PEL type, including a selfdual chain $\mathcal{L}$ of $\mathcal{O}_B$-lattices. A polarized chain of lattices of type $(\mathcal{L})$ over a $\mathbf{Z}_p$-algebra R is a chain M_Λ of type $(\mathcal{L})$ together with an antisymmetric isomorphism of multichains $M_\Lambda \cong M_{\Lambda^*}^* \otimes_R L$ for some invertible R-module L.

Remark 21.6.3. The invertible R-module L does not appear in [RZ96]. It is related to the similitude factor, and if one wants to describe torsors under a parahoric model of G, it is necessary to include it.

Theorem 21.6.4 ([RZ96, Theorem 3.11, Theorem 3.16]). *The automorphisms of the chain $\mathcal{L}$ of $\mathcal{O}_B$-lattices, resp. the automorphisms of the polarized chain $\mathcal{L}$ of $\mathcal{O}_B$-lattices in the PEL case, is a smooth group scheme $\mathcal{G}$ over $\mathbf{Z}_p$. If M_Λ is any chain of $\mathcal{O}_B \otimes_{\mathbf{Z}_p} R$-lattices of type $(\mathcal{L})$, resp. any polarized chain of $\mathcal{O}_B \otimes_{\mathbf{Z}_p} R$-lattices of type $(\mathcal{L})$, then étale locally on $\operatorname{Spec} R$ there is an isomorphism of*

---

[2]If F is not a field, one has to consider multichains instead.

$M_\Lambda \cong \Lambda \otimes_{\mathbf{Z}_p} R$ of chains, resp. of polarized chains.

Remark 21.6.5. The result is stated in [RZ96] only when p is nilpotent in R. This easily implies the result when R is p-adically complete. If p is invertible in R, the result is easy to verify (as then all chains become equalities); combining these two results, one easily deduces the result in general.

Importantly, we get a linear-algebraic description of $\mathcal{G}$-torsors.

Corollary 21.6.6. For any $\mathbf{Z}_p$-algebra R, there is a natural equivalence between $\mathcal{G}$-torsors over R and the groupoid of chains of $\mathcal{O}_B \otimes_{\mathbf{Z}_p} R$-lattices of type $(\mathcal{L})$, resp. of polarized chains of $\mathcal{O}_B \otimes_{\mathbf{Z}_p} R$-lattices of type $(\mathcal{L})$.

Proof. Any $\mathcal{G}$-torsor defines a chain, resp. a polarized chain, of modules, by taking the corresponding twisted form of the standard chain. As $\mathcal{G}$ is the group of automorphisms of the standard chain, it is clear that this functor is fully faithful. As any (polarized) chain is étale-locally isomorphic to the standard chain, the functor is essentially surjective. □

Remark 21.6.7. The group $\mathcal{G}$ may not have connected fibers. We refer to [PR08, Section 4] for a discussion of when $\mathcal{G}$ has connected fibers. This includes the EL case and the symplectic and unitary cases in the PEL situation, except possibly some even unitary cases when the quadratic extension is ramified. In general $\mathcal{G}^\circ$ is a parahoric group, so we are in the situation considered above.

Finally, we fix conjugacy class μ of minuscule cocharacters $\mathbf{G}_m \to G_{\overline{\mathbf{Q}}_p}$ defined over the reflex field E. We warn the reader that the normalization in [RZ96] is incompatible with our normalization. The group G has a natural central $\chi : \mathbf{G}_m \to G$ acting via diagonal multiplication on V. Then in [RZ96], one considers instead $\mu_{RZ} = \chi \mu^{-1}$. The conditions imposed in [RZ96, Definition 3.18] translate into the following condition on μ.

1. The weights of μ on $V \otimes_{\mathbf{Q}_p} \overline{\mathbf{Q}}_p$ are only 0 and 1, so we get a weight decomposition $V \otimes_{\mathbf{Q}_p} \overline{\mathbf{Q}}_p = V_0 \oplus V_1$.[3]
2. In the PEL case, the composite of μ and $c : G \to \mathbf{G}_m$ is the identity $\mathbf{G}_m \to \mathbf{G}_m$.

We recall the definition of the local model. As usual, E is the field of definition of μ.

Definition 21.6.8 ([RZ96, Definition 3.27]). The $\mathcal{O}_E$-scheme $\mathrm{M}^{\mathrm{loc,naive}}_{(\mathcal{G},\mu)}$ represents the functor taking an $\mathcal{O}_E$-algebra R to the following data.

1. A functor $\Lambda \mapsto t_\Lambda$ from the category $\mathcal{L}$ to the category of $\mathcal{O}_B \otimes_{\mathbf{Z}_p} \mathcal{O}_S$-modules that are finite projective over $\mathcal{O}_S$, and

---

[3] Beware that V_0 and V_1 interchanged with respect to [RZ96]!

2. A morphism of functors

$$\varphi_\Lambda : \Lambda \otimes_{\mathbf{Z}_p} \mathcal{O}_S \to t_\Lambda \,,$$

where each φ_Λ is surjective,

satisfying the following conditions.

1. For any $b \in B^\times$ normalizing $\mathcal{O}_B$, the periodicity isomorphism $\theta_b : \Lambda^b \otimes_{\mathbf{Z}_p} \mathcal{O}_S \cong b\Lambda \otimes_{\mathbf{Z}_p} \mathcal{O}_S$ is compatible with a necessarily unique periodicity isomorphism $t_\Lambda^b \cong t_{b\Lambda}$ along the surjections φ_Λ, $\varphi_{b\Lambda}$.[4]
2. We have an identity of polynomial functions

$$\det{}_R(a; t_\Lambda) = \det{}_{\overline{\mathbf{Q}}_p}(a; V_1)$$

for $a \in \mathcal{O}_B$. Here, both sides define polynomial maps from $\mathcal{O}_B$ to R (the second a priori to $\overline{\mathbf{Q}}_p$, but in fact maps to $\mathcal{O}_E$, which maps to R).[5]
3. In case PEL, the composite

$$t_\Lambda^* \xrightarrow{\varphi_\Lambda^*} \Lambda^* \otimes_{\mathbf{Z}_p} \mathcal{O}_S \xrightarrow{\varphi_{\Lambda^*}} t_{\Lambda^*}$$

is zero for all $\Lambda \in \mathcal{L}$.

Clearly, $\mathbb{M}^{\mathrm{loc,naive}}_{(\mathcal{G},\mu)}$ is a projective scheme over $\mathcal{O}_E$. To get the relation with affine flag varieties, we need the following proposition.

Proposition 21.6.9. Let R be a perfectoid Tate-$\mathcal{O}_E$-algebra with a map $\mathrm{Spec}\, R \to \mathbb{M}^{\mathrm{loc,naive}}_{(\mathcal{G},\mu)}$ given by $(t_\Lambda, \varphi_\Lambda)$ as above. The functor

$$\Lambda \mapsto M_\Lambda = \ker(\Lambda \otimes_{\mathbf{Z}_p} B^+_{\mathrm{dR}}(R) \to \Lambda \otimes_{\mathbf{Z}_p} R \xrightarrow{\varphi_\Lambda} t_\Lambda)$$

defines a (polarized) chain of $\mathcal{O}_B \otimes_{\mathbf{Z}_p} B^+_{\mathrm{dR}}(R)$-lattices of type $(\mathcal{L})$.

Proof. As R is of projective dimension 1 over $B^+_{\mathrm{dR}}(R)$, it follows that M_Λ is a finite projective $B^+_{\mathrm{dR}}(R)$-module. Each t_Λ is of fixed dimension over R. Thus, in the PEL case, condition (3) in the definition of $\mathbb{M}^{\mathrm{loc,naive}}$ implies easily that $M_\Lambda \cong M^*_{\Lambda^*} \otimes_{B^+_{\mathrm{dR}}(R)} \xi B^+_{\mathrm{dR}}(R)$ compatibly with the pairing on $\Lambda \otimes_{\mathbf{Z}_p} B^+_{\mathrm{dR}}(R)$; here ξ is a generator of $B^+_{\mathrm{dR}}(R) \to R$ as usual. This reduces us to the EL case. We need to see that locally on $\mathrm{Spec}\, B^+_{\mathrm{dR}}(R)$, there is an isomorphism $M_\Lambda \cong \Lambda \otimes_{\mathbf{Z}_p} B^+_{\mathrm{dR}}(R)$ and an isomorphism $M_\Lambda/M_{\Lambda'} \cong \Lambda/\Lambda' \otimes_{\mathbf{Z}_p} B^+_{\mathrm{dR}}(R)$ of $\mathcal{O}_B \otimes_{\mathbf{Z}_p} B^+_{\mathrm{dR}}(R)$-modules, for any two adjacent lattices Λ, Λ'. In fact, by adic descent, it is enough to check this locally on $\mathrm{Spa}(R, R^+)$. Moreover, if it is true

---

[4]This condition seems to have been overlooked in [RZ96].
[5]Recall that a polynomial map $M \to N$ between two abelian groups M and N is a natural transformation of functors $M \otimes A \to N \otimes A$ on the category of $\mathbf{Z}$-algebras A.

at one point of $\mathrm{Spa}(K, K^+)$ of $\mathrm{Spa}(R, R^+)$, it is true in an open neighborhood, by spreading a local isomorphism. This reduces us to the case that $R = K$ is a perfectoid field. If K is of characteristic 0, the result is clear, so it remains to consider the case that K is a perfect field, in which case $B_{\mathrm{dR}}^+(K) = W(K)$ is the ring of Witt vectors.

Note that $\mathcal{O}_B = M_n(\mathcal{O}_D)$ for the ring of integers $\mathcal{O}_D$ in a division algebra D/F. By Morita equivalence, we can assume that $\mathcal{O}_B = \mathcal{O}_D$. Let $\Pi \in \mathcal{O}_D$ be a uniformizer. Then M_Λ is Π-torsion free and Π-adically complete, so its isomorphism class as $\mathcal{O}_D \otimes_{\mathbf{Z}_p} W(K)$-module is determined by the isomorphism class of M_Λ/Π as $\mathcal{O}_D/\Pi \otimes_{\mathbf{F}_p} K = \prod_{\mathcal{O}_D/\Pi \hookrightarrow K} K$-module. This reduces to the dimensions of the various components, and we have to see that they all agree. Similarly, $M_\Lambda/M_{\Lambda'}$ is a module over $\mathcal{O}_D/\Pi \otimes_{\mathbf{F}_p} K$. To see that this is isomorphic to $\Lambda/\Lambda' \otimes_{\mathbf{F}_p} K$, we have to check equality of some dimensions again. In fact, this second condition implies the first, by inductively deducing a similar result for $M_\Lambda/\Pi M_\Lambda$.

Now we look at the exact sequence

$$0 \to \ker(t_{\Lambda'} \to t_\Lambda) \to M_\Lambda/M_{\Lambda'} \to \Lambda/\Lambda' \otimes_{\mathbf{F}_p} K \to \mathrm{coker}(t_{\Lambda'} \to t_\Lambda) \to 0$$

from the snake lemma and the definitions. Using Jordan-Hölder multiplicities, we see that it suffices to see that the Jordan-Hölder multiplicities of t_Λ and $t_{\Lambda'}$ as $\mathcal{O}_D \otimes_{\mathbf{Z}_p} K$-modules are the same. This follows from the determinant condition. $\square$

Corollary 21.6.10. There is a natural closed immersion

$$\mathbb{M}_{(\mathcal{G},\mu)}^{\mathrm{loc,naive},\diamond} \hookrightarrow \mathrm{Gr}_{\mathcal{G},\mathrm{Spd}\,\mathcal{O}_E}$$

over $\mathrm{Spd}\,\mathcal{O}_E$. If $S = \mathrm{Spa}(R, R^+) \in \mathrm{Perf}$ with untilt $(R^\sharp, R^{\sharp+})$ over $\mathcal{O}_E$, this sends an $R^\sharp$-valued point $(t_\Lambda, \varphi_\Lambda)$ of $\mathbb{M}_{(\mathcal{G},\mu)}^{\mathrm{loc,naive},\diamond}$ to the $\mathcal{G}$-torsor over $B_{\mathrm{dR}}^+(R^\sharp)$ given by the (polarized) chain of $\mathcal{O}_B \otimes_{\mathbf{Z}_p} B_{\mathrm{dR}}^+(R^\sharp)$-lattices

$$\Lambda \mapsto M_\Lambda = \ker(\Lambda \otimes_{\mathbf{Z}_p} B_{\mathrm{dR}}^+(R^\sharp) \to \Lambda \otimes_{\mathbf{Z}_p} R^\sharp \xrightarrow{\varphi_\Lambda} t_\Lambda)$$

together with the isomorphism of $\mathcal{G}$-torsors over $B_{\mathrm{dR}}(R^\sharp)$ given by the isomorphism of (polarized) chains of $\mathcal{O}_B \otimes_{\mathbf{Z}_p} B_{\mathrm{dR}}(R^\sharp)$-lattices

$$M_\Lambda \otimes_{B_{\mathrm{dR}}^+(R^\sharp)} B_{\mathrm{dR}}(R^\sharp) \cong \Lambda \otimes_{\mathbf{Z}_p} B_{\mathrm{dR}}(R^\sharp) \ .$$

After base change to the generic fiber $\mathrm{Spd}\,E$, we get an isomorphism

$$\mathbb{M}_{(\mathcal{G},\mu)}^{\mathrm{loc,naive},\diamond} \cong \mathrm{Gr}_{G,\mathrm{Spd}\,E,\mu} \ .$$

Proof. The map is an injection as t_Λ can be recovered as $\Lambda \otimes_{\mathbf{Z}_p} B_{\mathrm{dR}}^+(R^\sharp)/M_\Lambda$. As $\mathbb{M}_{(\mathcal{G},\mu)}^{\mathrm{loc,naive}}$ is proper over $\mathcal{O}_E$, also $\mathbb{M}_{(\mathcal{G},\mu)}^{\mathrm{loc,naive},\diamond}$ is proper over $\mathrm{Spd}\,\mathcal{O}_E$, and hence the map is a closed immersion.

For the final statement, as $\mathrm{Gr}_{G,\mathrm{Spd}\,E,\mu}$ is given by $(G/P_\mu)^\diamond$, it is enough to check that $\mathbb{M}^{\mathrm{loc,naive}}_{(\mathcal{G},\mu),E} \cong G/P_\mu$ (compatibly with the map), which is standard. □

The problem of local models is that in general $\mathbb{M}^{\mathrm{loc,naive}}_{(\mathcal{G},\mu)}$ is not flat over $\mathcal{O}_E$. Pappas-Rapoport, [PR03], [PR05], and [PR09], and collaborators have since studied the problem of imposing additional linear-algebraic conditions to define a closed subscheme

$$\mathbb{M}^{\mathrm{loc}}_{(\mathcal{G},\mu)} \subset \mathbb{M}^{\mathrm{loc,naive}}_{(\mathcal{G},\mu)}$$

with the same generic fiber, that is flat over $\mathcal{O}_E$. Of course, abstractly this can be done by taking the flat closure of the generic fiber. One expects that $\mathbb{M}^{\mathrm{loc}}_{(\mathcal{G},\mu)}$ is normal; in fact, with reduced special fiber. This brings the connection to Conjecture 21.4.1.

Finally, we discuss some explicit examples that appear in the work of Rapoport-Zink, [RZ17], and Kudla-Rapoport-Zink, [KRZ], on a moduli-theoretic proof of Čerednik's p-adic uniformization theorem.

In both cases, an alternative description of the Drinfeld moduli problem is proposed. Our aim in this appendix and Lecture 25 is to prove their conjectures in full generality. In this appendix, we prove the conjectures of [RZ17] and [KRZ] about local models.

21.7 AN EL CASE

First, we consider the situation in [RZ17]. Fix a finite extension F of $\mathbf{Q}_p$, and a central division algebra D over F of invariant $1/n$, $n \geq 2$. We take $B = D$ and $\mathcal{O}_B = \mathcal{O}_D$. Moreover, we fix $V = D$ and the lattice chain $\mathcal{L} = \{\Pi^n \mathcal{O}_D; n \in \mathbf{Z}\}$ where $\Pi \in \mathcal{O}_D$ is a uniformizer. The associated group G is the algebraic group with $\mathbf{Q}_p$-valued points $D^\times$, and its parahoric model $\mathcal{G}$ has $\mathcal{G}(\mathbf{Z}_p) = \mathcal{O}_D^\times$. In fact, $\mathcal{G} = \mathrm{Res}_{\mathcal{O}_F/\mathbf{Z}_p}\,\mathcal{G}_0$ for a parahoric group $\mathcal{G}_0$ over $\mathcal{O}_F$.

Note that

$$G_{\overline{\mathbf{Q}}_p} \cong \prod_{\varphi:F\hookrightarrow\overline{\mathbf{Q}}_p} \mathrm{GL}_n \ .$$

To define the cocharacter μ, we fix one embedding $\varphi_0 : F \hookrightarrow \overline{\mathbf{Q}}_p$, and let μ be the conjugacy class of cocharacters whose component μ_φ for an embedding $\varphi : F \hookrightarrow \overline{\mathbf{Q}}_p$ is given by $(1,0,\dots,0)$ if $\varphi = \varphi_0$, and by one of $(0,\dots,0)$ or $(1,\dots,1)$ at all other places. We let E be its field of definition as usual, which contains F through the embedding φ_0.

In this situation, the usual local model $\mathbb{M}^{\mathrm{loc,naive}}_{(\mathcal{G},\mu)}$ is usually not flat. However, Rapoport-Zink [RZ17, Section 5] adds further linear-algebraic conditions, the so-called Eisenstein conditions, to define a closed subscheme

$$\mathbb{M}^{\mathrm{loc}}_{(\mathcal{G},\mu)} \subset \mathbb{M}^{\mathrm{loc,naive}}_{(\mathcal{G},\mu)}$$

that is flat over $\mathcal{O}_E$, with the same generic fiber but reduced special fiber; in particular, it is normal; cf. [RZ17, Corollary 5.6]. There is one case where one can understand this explicitly; this is the case where $\mu_\varphi = (0,\ldots,0)$ for all $\varphi \neq \varphi_0$. In this case, $E = F$ and for an $\mathcal{O}_E = \mathcal{O}_F$-algebra R, the R-valued points of $\mathbb{M}^{\mathrm{loc}}_{(\mathcal{G},\mu)}$ parametrize $\mathcal{O}_D \otimes_{\mathcal{O}_F} R$-linear quotients $\mathcal{O}_D \otimes_{\mathcal{O}_F} R \to M$ where M is a finite projective R-module of rank n. This can be analyzed directly— cf. [RZ96, Section 3.76]—and one sees in particular that it is a regular scheme, which we will denote by $\mathbb{M}^{\mathrm{loc}}_{\mathrm{Dr}}$, called the Drinfeld local model.

The following result proves a conjecture of Rapoport-Zink, [RZ17, Remark 5.7].

Theorem 21.7.1. For general μ as in [RZ17], there is a natural $\mathcal{G}$-equivariant isomorphism

$$\mathbb{M}^{\mathrm{loc}}_{(\mathcal{G},\mu)} \cong \mathbb{M}^{\mathrm{loc}}_{\mathrm{Dr}} \times_{\operatorname{Spec}\mathcal{O}_F} \operatorname{Spec}\mathcal{O}_E$$

of $\mathcal{O}_E$-schemes.

Remark 21.7.2. To prove the theorem, it would be enough to produce a $\mathcal{G}$-equivariant map

$$\mathbb{M}^{\mathrm{loc}}_{\mathrm{Dr}} \times_{\operatorname{Spec}\mathcal{O}_F} \operatorname{Spec}\mathcal{O}_E \to \mathbb{M}^{\mathrm{loc,naive}}_{(\mathcal{G},\mu)} \ .$$

Indeed, this would necessarily factor over $\mathbb{M}^{\mathrm{loc}}_{(\mathcal{G},\mu)}$, and then one could easily check the map to be an isomorphism. Note that both of these schemes are very explicit linear-algebraic moduli problems. However, it appears to be impossible to write down the map. In our proof, the most subtle step is in Proposition 21.5.1, where a local model is lifted to an extension; this lift is very inexplicit.

Proof. We consider the exact sequence

$$1 \to \operatorname{Res}_{\mathcal{O}_F/\mathbf{Z}_p} \mathbf{G}_{\mathrm{m}} \to \mathcal{G} \to \overline{\mathcal{G}} \to 1 \ ;$$

then $\overline{\mathcal{G}}$ is a parahoric model of $\overline{G}$. Any two μ as considered above project to the same conjugacy class of cocharacters $\overline{\mu}$ of $\overline{G}$. Applying Proposition 21.5.1 twice, we see that $(\mathbb{M}^{\mathrm{loc}}_{(\mathcal{G},\mu)})^\diamond$ is independent of the choice of μ. As all of them are flat and normal, Proposition 18.4.1 implies that the formal schemes are isomorphic, and then by GAGA the schemes are also isomorphic. □

21.8 A PEL CASE

Now we consider the PEL case considered in [KRZ]. For the moment, we allow the case $p = 2$. We fix a finite field extension F_0 of $\mathbf{Q}_p$ of degree d with uniformizer $\pi_0 \in F_0$ and residue field $\mathbf{F}_{p^f}$, and a quadratic extension F of F_0, and set $B = F$ and $\mathcal{O}_B = \mathcal{O}_F$. If $p = 2$, we demand that F/F_0 be unramified; cf. [Kir18] for some work in the ramified 2-adic situation. Let $a \mapsto a^*$ be the nontrivial automorphism of F over F_0. We let V be a 2-dimensional F-vector space equipped with a specific alternating perfect form

$$(\, , \,) : V \times V \to \mathbf{Q}_p$$

satisfying $(av, w) = (v, a^* w)$ for all $v, w \in V$, $a \in F$, as specified now. First, any such pairing $(\, , \,)$ is of the form $(v, w) = \mathrm{tr}_{F_0/\mathbf{Q}_p} \vartheta_{F_0}^{-1}(v, w)_{F_0}$ for a unique alternating perfect form

$$(\, , \,)_{F_0} : V \times V \to F_0 \, ,$$

satisfying $(av, w)_{F_0} = (v, a^* w)_{F_0}$ for all $v, w \in V$, $a \in F_0$; here $\vartheta_{F_0} \in \mathcal{O}_{F_0}$ is a generator for the different ideal of F_0 over $\mathbf{Q}_p$, so that an element $x \in F_0$ lies in $\vartheta_{F_0}^{-1}\mathcal{O}_{F_0}$ if and only if $\mathrm{tr}_{F_0/\mathbf{Q}_p}(x\mathcal{O}_{F_0}) \subset \mathbf{Z}_p$. In particular, the dual of an $\mathcal{O}_F$-lattice $\Lambda \subset V$ with respect to $(\, , \,)$ agrees with the dual with respect to $(\, , \,)_{F_0}$. Next, there is a unique perfect hermitian form

$$\langle \, , \, \rangle : V \times V \to F$$

such that $(v, w)_{F_0} = \mathrm{tr}_{F/F_0} \delta^{-1}\langle v, w \rangle$ for all $v, w \in V$. Here $\delta \in F^\times$ satisfies $\delta^* = -\delta$. If F/F_0 is unramified, we assume that $\delta \in \mathcal{O}_F^\times$; otherwise, we assume that $\delta = \pi \in F$ is a uniformizer of F and $\pi_0 = \pi^2$. Then again, the dual of Λ with respect to $(\, , \,)_{F_0}$ agrees with the dual with respect to $\langle \, , \, \rangle$, so we can unambiguously talk about the dual of a lattice.

Now there are two isomorphism classes of alternating forms $\langle \, , \, \rangle$, the split form and the nonsplit form. We assume that the form is nonsplit. In other words, the associated group G is not split. Note that G sits in an exact sequence

$$1 \to \mathrm{Res}_{F_0/\mathbf{Q}_p} U \to G \to \mathbf{G}_{\mathrm{m}} \to 1 \, ,$$

where U is the nonsplit unitary group over F_0 associated with the extension F and the hermitian F-vector space $(V, \langle \, , \, \rangle)$. The adjoint group G_{ad} is given by $\mathrm{Res}_{F_0/\mathbf{Q}_p} U_{\mathrm{ad}}$, where U_{ad} is the nonsplit form of PGL_2 over F_0. The group $U(F_0)$ is compact.

Next, we fix the lattice chain $\mathcal{L}$. If F is a ramified extension of F_0, then there is a unique selfdual lattice $\Lambda \subset V$ with respect to $(\, , \,)$. In this case we fix $\mathcal{L} = \{\pi^n \Lambda; n \in \mathbf{Z}\}$ for some selfdual lattice Λ. On the other hand, if F is the unramified extension of F_0, then there are no selfdual lattices $\Lambda \subset V$. Instead, there are *almost selfdual* lattices, meaning lattices $\Lambda \subset V$ such that $\pi_0 \Lambda \subset \Lambda^* \subset \Lambda$ where

$$\dim_{\mathbf{F}_{p^f}} \Lambda/\Lambda^* = \dim_{\mathbf{F}_{p^f}} \Lambda^*/\pi_0\Lambda = 2 \, .$$

Note that the dimension is at least 2, as the modules are in fact modules over $\mathcal{O}_F/\pi_0 \cong \mathbf{F}_{p^{2f}}$. In fact, again there is a unique such lattice Λ. In this case, we let $\mathcal{L}$ be the chain of all $\{\pi_0^n \Lambda, \pi_0^n \Lambda^*; n \in \mathbf{Z}\}$.

In the unramified case, the corresponding group $\mathcal{G}$ over $\mathbf{Z}_p$ is parahoric. In the ramified case, it is not parahoric; in fact, the special fiber has two connected components.

As in the EL case, we fix an embedding $\varphi_0 : F_0 \hookrightarrow \overline{\mathbf{Q}}_p$. We consider a conjugacy class μ of cocharacters $\mathbf{G}_m \to G_{\overline{\mathbf{Q}}_p}$ whose image under

$$G_{\overline{\mathbf{Q}}_p} \hookrightarrow \prod_{\varphi:F_0 \hookrightarrow \overline{\mathbf{Q}}_p} GU(1,1) \hookrightarrow \prod_{\widetilde{\varphi}:F \hookrightarrow \overline{\mathbf{Q}}_p} GL_2$$

has components $\mu_{\widetilde{\varphi}} = (1,1)$ if $\widetilde{\varphi}$ lies over φ_0, and $\mu_{\widetilde{\varphi}} = (0,0)$ or $(2,2)$ otherwise, where above each φ, both possibilities happen once. Note that the field of definition E of μ contains F_0 via the embedding φ_0.

Again, the usual local model $\mathrm{M}_{(\mathcal{G},\mu)}^{\mathrm{loc,naive}}$ is not flat, but Kudla-Rapoport-Zink pose additional linear-algebraic conditions, again called the Eisenstein conditions, to define a closed subscheme

$$\mathrm{M}_{(\mathcal{G},\mu)}^{\mathrm{loc}} \subset \mathrm{M}_{(\mathcal{G},\mu)}^{\mathrm{loc,naive}}$$

with the same generic fiber and reduced special fiber; cf. [KRZ, Sections 5.2, 6.2].

Theorem 21.8.1. There is a natural $\mathcal{G}$-equivariant isomorphism

$$\mathrm{M}_{(\mathcal{G},\mu)}^{\mathrm{loc}} \cong \mathrm{M}_{\mathrm{Dr}}^{\mathrm{loc}} \times_{\mathrm{Spec}\,\mathcal{O}_{F_0}} \mathrm{Spec}\,\mathcal{O}_E$$

of $\mathcal{O}_E$-schemes, where $\mathrm{M}_{\mathrm{Dr}}^{\mathrm{loc}}$ is the Drinfeld local model from the EL case for the field F_0 and $n = 2$.

Proof. Using Proposition 21.5.1 and Proposition 21.4.3, this follows from the observation that G_{ad}° is the same group as the adjoint group appearing in the EL case, with the same cocharacter μ_{ad}; also, we use Proposition 18.4.1 to get the isomorphism on the level of formal schemes, and then by GAGA on schemes. $\qquad\square$

Lecture 22

Vector bundles and G-torsors on the relative Fargues-Fontaine curve

In preparation for the discussion of moduli spaces of shtukas, we need to discuss the relative Fargues-Fontaine curve, and vector bundles and G-torsors on it.

22.1 VECTOR BUNDLES

We wish to study vector bundles on the relative Fargues-Fontaine curve $\mathcal{X}_{\mathrm{FF},S}$. Analogously to the discussion in Lecture 13, these can also be studied in terms of φ-modules on (subspaces of) $\mathcal{Y}_{(0,\infty)}(S)$, or through φ-modules on the relative Robba ring.

For simplicity, we assume that $S = \mathrm{Spa}(R, R^+)$ is affinoid, and fix a pseudouniformizer $\varpi \in R$, so that we get the function

$$\kappa : \mathcal{Y}_{[0,\infty)}(S) = S \,\dot\times\, \mathrm{Spa}\,\mathbf{Z}_p \to [0, \infty) .$$

For a rational number $r \in (0, \infty)$, we consider the subspaces

$$\mathcal{Y}_{[0,r]}(S) = \{|p|^r \leq |[\varpi]|\} \subset \mathcal{Y}_{(0,\infty)}(S) \ , \ \mathcal{Y}_{(0,r]}(S) = \mathcal{Y}_{[0,r]}(S) \cap \mathcal{Y}_{(0,\infty)}(S) \ ,$$

and

$$\mathcal{Y}_{[r,\infty)}(S) = \{|[\varpi]| \leq |p|^r \neq 0\} \subset \mathcal{Y}_{(0,\infty)}(S) .$$

On these subspaces, κ maps to the corresponding interval, and on rank 1-points these are precisely the preimages under κ of the given intervals; however, something more subtle may happen on higher rank points. Note that φ induces isomorphisms $\mathcal{Y}_{(0,r]}(S) \cong \mathcal{Y}_{(0,pr]}(S)$ and $\mathcal{Y}_{[r,\infty)}(S) \cong \mathcal{Y}_{[pr,\infty)}(S)$. In particular, φ acts on $\mathcal{Y}_{[r,\infty)}(S)$, and φ^{-1} acts on $\mathcal{Y}_{(0,r]}(S)$.

Proposition 22.1.1. The following categories are naturally equivalent.

1. The category of vector bundles on $\mathcal{X}_{\mathrm{FF},S}$.
2. The category of φ-modules on $\mathcal{Y}_{(0,\infty)}(S)$.
3. The category of φ-modules on $\mathcal{Y}_{[r,\infty)}(S)$.
4. The category of φ^{-1}-modules on $\mathcal{Y}_{(0,r]}(S)$.

Proof. The first two are equivalent by descent, and the other two are equivalent to (2) by the standard spreading out argument. $\qquad\square$

Definition 22.1.2. The relative Robba ring is the ring

$$\widetilde{\mathcal{R}}_R = \varinjlim_{r>0} H^0(\mathcal{Y}_{(0,r]}(S), \mathcal{O}_{\mathcal{Y}_{(0,r]}(S)})$$

on which φ is an automorphism.

Note that any φ-module over $\widetilde{\mathcal{R}}_R$, or equivalently φ^{-1}-module, descends to a φ^{-1}-module over

$$\widetilde{\mathcal{R}}_R^r = H^0(\mathcal{Y}_{(0,r]}(S), \mathcal{O}_{\mathcal{Y}_{(0,r]}(S)})$$

for r small enough. In particular, we get a φ^{-1}-module on $\mathcal{Y}_{(0,r]}(S)$. Kedlaya-Liu prove that this gives an equivalence.

Theorem 22.1.3 ([KL15]). The category of vector bundles on $\mathcal{X}_{\mathrm{FF},S}$ is equivalent to the category of φ-equivariant finite projective $\widetilde{\mathcal{R}}_R$-modules.

22.2   SEMICONTINUITY OF THE NEWTON POLYGON

Kedlaya-Liu prove two important foundational theorems about vector bundles on the Fargues-Fontaine curve. The first is the semicontinuity of the Newton polygon.

Let S be a perfectoid space of characteristic p, and let $\mathcal{E}$ be a vector bundle on the relative Fargues-Fontaine curve $\mathcal{X}_{\mathrm{FF},S}$. By passing to an open and closed cover of S, we can assume that the rank of $\mathcal{E}$ is constant. Now for any $s \in S$, we can choose a geometric point $\bar{s} = \mathrm{Spa}(C, C^+) \to S$ whose closed point maps to s, and pullback $\mathcal{E}$ to a vector bundle $\mathcal{E}_{\bar{s}}$ on $\mathcal{X}_{\mathrm{FF},C}$. By the classification theorem, Theorem 13.5.3, $\mathcal{E}_{\bar{s}}$ is classified by a Newton polygon, which gives a map

$$\nu_{\mathcal{E}} : |S| \to (\mathbf{Q}^n)^+ = \{(\lambda_1, \ldots, \lambda_n) \in \mathbf{Q}^n \mid \lambda_1 \geq \ldots \geq \lambda_n\}.$$

We endow the target with the dominance order for which

$$(\lambda_1, \ldots, \lambda_n) \geq (\mu_1, \ldots, \mu_n)$$

if and only if for all $i = 1, \ldots, n$, one has $\lambda_1 + \ldots + \lambda_i \geq \mu_1 + \ldots + \mu_i$, with equality for $i = n$.

Theorem 22.2.1 ([KL15]). The map $\nu_{\mathcal{E}}$ is upper semicontinuous.

22.3   THE ÉTALE LOCUS

The second theorem of Kedlaya-Liu concerns the open locus where the Newton polygon is constant 0. Note that one can build examples of such vector bundles by taking a pro-étale $\mathbf{Q}_p$-local system $\mathbb{L}$ on S and looking at $\mathcal{E} = \mathbb{L} \otimes_{\mathbf{Q}_p} \mathcal{O}_{\mathcal{X}_{\mathrm{FF},S}}$. By the pro-étale (or even v-)descent of vector bundles on $\mathcal{Y}_{(0,\infty)}(S)$, and thus on $\mathcal{X}_{\mathrm{FF},S}$, given by Proposition 19.5.3, one sees that this defines a vector bundle on $\mathcal{X}_{\mathrm{FF},S}$, by descending to the case where $\mathbb{L}$ is trivial.

Theorem 22.3.1 ([KL15]). This construction defines an equivalence between the category of pro-étale $\mathbf{Q}_p$-local systems on S and the category of vector bundles $\mathcal{E}$ on $\mathcal{X}_{\mathrm{FF},S}$ which are trivial at every geometric point of S (which is to say, the function $\nu_{\mathcal{E}}$ is identically 0).

We will also need to understand the relation to φ^{-1}-modules on $\mathcal{Y}_{[0,r]}(S)$ (note that we are now including characteristic p points). By a φ^{-1}-module on $\mathcal{Y}_{[0,r]}(S)$, we mean a locally free $\mathcal{Y}_{[0,r]}(S)$-module $\mathcal{E}$ equipped with an isomorphism $(\varphi^{-1})^* \mathcal{E}|_{\mathcal{Y}_{[0,r]}(S)} \overset{\sim}{\to} \mathcal{E}$. If $S = \mathrm{Spa}(R, R^+)$ is affinoid, the category of φ^{-1}-modules on $\mathcal{Y}_{[0,r]}(S)$ is equivalent to the category of φ^{-1}-modules (equivalently, φ-modules) over the *integral Robba ring*

$$\widetilde{\mathcal{R}}_R^{\mathrm{int}} = \varinjlim_{r>0} H^0(\mathcal{Y}_{[0,r]}(S), \mathcal{O}_{\mathcal{Y}_{[0,r]}(S)}).$$

If $\mathbb{L}$ is a pro-étale $\mathbf{Z}_p$-local system on S, then $\mathbb{L} \otimes_{\mathbf{Z}_p} \mathcal{O}_{\mathcal{Y}_{[0,r]}(S)}$ defines such an object, again using Proposition 19.5.3.

Proposition 22.3.2 ([KL15]). The functor $\mathbb{L} \mapsto \mathbb{L} \otimes_{\mathbf{Z}_p} \mathcal{O}_{\mathcal{Y}_{[0,r]}(S)}$ defines an equivalence between the category of pro-étale $\mathbf{Z}_p$-local systems on S and the category of φ^{-1}-modules on $\mathcal{Y}_{[0,r]}(S)$. In particular, any φ^{-1}-module on $\mathcal{Y}_{[0,r]}(S)$ has constant Newton polygon equal to 0 when restricted to $\mathcal{Y}_{(0,r]}(S)$. The functor $\mathcal{E} \mapsto \mathcal{E}|_{\mathcal{Y}_{(0,r]}(S)}$ on φ^{-1}-modules over $\mathcal{Y}_{[0,r]}(S)$ corresponds to the functor $\mathbb{L} \mapsto \mathbb{L}[1/p]$ on pro-étale $\mathbf{Z}_p$-local systems on S.

The following corollary sums up the situation.

Corollary 22.3.3. Let $\mathcal{E}_\eta$ be a φ^{-1}-module on $\mathcal{Y}_{(0,r]}(S)$. Consider the functor $\mathrm{Latt}(\mathcal{E}_\eta)$ sending any $S' \to S$ to the set of φ^{-1}-modules $\mathcal{E}'$ on $\mathcal{Y}_{[0,r]}(S')$ together with an φ^{-1}-equivariant identification of $\mathcal{E}'|_{\mathcal{Y}_{(0,r]}(S')}$ with the pullback of $\mathcal{E}_\eta$ to $\mathcal{Y}_{(0,r]}(S')$. Then $\mathrm{Latt}(\mathcal{E}_\eta)$ is representable by a perfectoid space étale over S. Its image in S is the open subset $S^a \subset S$ where the Newton polygon of $\mathcal{E}_\eta$ is identically 0, called the admissible locus. Over S^a, one has a corresponding $\mathbf{Q}_p$-local system $\mathbb{L}_\eta$, and $\mathrm{Latt}(\mathcal{E}_\eta)$ parametrizes $\mathbf{Z}_p$-lattices $\mathbb{L} \subset \mathbb{L}_\eta$.

Proof. As in Proposition 22.1.1, $\mathcal{E}_\eta$ is the restriction of a unique φ-module on all of $\mathcal{Y}_{(0,\infty)}(S)$, which is the same thing as a vector bundle $\mathcal{F}$ on $\mathcal{X}_{\mathrm{FF},S}$. By

Theorem 22.2.1, the locus $S^a \subset S$ where $\nu_{\mathcal{F}}$ is identically 0 is open. By Theorem 22.3.1, the restriction of $\mathcal{F}$ to $\mathcal{X}_{\mathrm{FF},S^a}$ corresponds to a $\mathbf{Q}_p$-local system $\mathbb{L}_\eta$.

For a perfectoid space $S' \to S$, the S'-points of $\mathrm{Latt}(\overline{\mathcal{E}_\eta})$ are the extensions of $\mathcal{E}_\eta$ to $\mathcal{Y}_{[0,r]}(S')$. By Proposition 22.3.2, these correspond to $\mathbf{Z}_p$-lattices $\mathbb{L} \subset \mathbb{L}_\eta$ over S'. This shows that $\mathrm{Latt}(\mathcal{E}_\eta) \to S^a$ is étale, as it becomes isomorphic to $\mathrm{GL}_n(\mathbf{Q}_p)/\mathrm{GL}_n(\mathbf{Z}_p) \times S' \to S'$ after passing to a pro-étale cover $S' \to S^a$ which trivializes $\mathbb{L}$. $\qquad\qquad\qquad\qquad\qquad\qquad\qquad\qquad\qquad\qquad\qquad\qquad$ $\square$

For the applications to the moduli spaces of shtukas, we need to generalize these results to the case of G-torsors for a general reductive group G.

22.4   CLASSIFICATION OF G-TORSORS

First, we'll discuss the classification of G-torsors on $\mathcal{X}_{\mathrm{FF},S}$ in case $S = \mathrm{Spa}(C, C^+)$ is a geometric point, so that $\mathcal{X}_{\mathrm{FF},S} = \mathcal{X}_{\mathrm{FF}}$ is the absolute Fargues-Fontaine curve discussed in Lecture 13.

Theorem 13.5.4 relates isocrystals to vector bundles on $\mathcal{X}_{\mathrm{FF}}$. We intend to upgrade this to a relation between isocrystals with G-structure and G-torsors, using a Tannakian formalism.

Let us fix an algebraically closed discrete field k (e.g. $k = \overline{\mathbf{F}}_p$) and an inclusion $k \hookrightarrow C$. Let $L = W(k)[\frac{1}{p}]$, and let φ be the Frobenius automorphism of L. Recall that an isocrystal over k is a pair (V, φ_V), where V is a finite-dimensional L-vector space and $\varphi_V \colon V \xrightarrow{\sim} V$ is a φ-linear isomorphism. Let Isoc_k be the category of isocrystals over k.

Recall the following definition of Kottwitz, [Kot85].

Definition 22.4.1. An isocrystal with G-structure is an exact $\otimes$-functor $\mathrm{Rep}_{\mathbf{Q}_p} G \to \mathrm{Isoc}_k$.

Let $\mathrm{Isoc}_k \to \mathrm{Vec}_L$ be the forgetful functor towards the category of L-vector spaces. Given an isocrystal with G-structure $\mathrm{Rep}_{\mathbf{Q}_p} G \to \mathrm{Isoc}_k$, the composite functor $\mathrm{Rep}_{\mathbf{Q}_p} G \to \mathrm{Vec}_L$ is a fiber functor. Such fiber functors are a torsor under $H^1(L, G)$, which is trivial by Steinberg's theorem [Ste65, Theorem 1.9]. Thus $\mathrm{Rep}_{\mathbf{Q}_p} G \to \mathrm{Vec}_L$ is isomorphic to the standard fiber functor on Rep_G base changed to L.

After fixing such a trivialization, our isocrystal is a φ-linear automorphism of the standard fiber functor $\mathrm{Rep}_{\mathbf{Q}_p} G \to \mathrm{Vec}_L$, which determines an element $b \in G(L)$. In other words, the isocrystal sends a $\mathbf{Q}_p$-linear representation V to the isocrystal $(V \otimes_{\mathbf{Q}_p} L, b \otimes \varphi)$. Changing the trivialization of the fiber functor replaces b with $\varphi(y)by^{-1}$ for some $y \in G(L)$. The relation $b \sim \varphi(y)by^{-1}$ is called σ-conjugacy by Kottwitz. We let $B(G)$ be the set of equivalence classes of $G(L)$ under σ-conjugacy. Thus isomorphism classes of isocrystals with G-structure are in bijection with $B(G)$.

Kottwitz gives a combinatorial description of $B(G)$, which is analogous to the Dieudonné-Manin classification. This description is based on two invariants. Let $T \subset B \subset G_{\overline{\mathbf{Q}}_p}$ be a maximal torus and a Borel, and let $\Gamma = \mathrm{Gal}(\overline{\mathbf{Q}}_p/\mathbf{Q}_p)$. Kottwitz constructs a *Newton map*

$$\nu \colon B(G) \to ((X_*(T) \otimes \mathbf{Q})^+)^\Gamma,$$

where the right-hand side is the set of Γ-invariants in the set of dominant rational cocharacters. In the case of $G = \mathrm{GL}_n$ we may identify $X_*(T)$ with $\mathbf{Z}^n$, and then $\nu(b) \in (\mathbf{Q}^n)^+$ is the Newton point of b.

Kottwitz also constructs a map

$$\kappa \colon B(G) \to \pi_1(G_{\overline{\mathbf{Q}}_p})_\Gamma,$$

where π_1 denotes Borovoi's fundamental group of $G_{\overline{\mathbf{Q}}_p}$ [Bor98]. If $G = \mathrm{GL}_n$, this is already determined by the Newton point, intuitively as the endpoint of the Newton polygon, i.e., $\kappa(b) = v_p(\det b) \in \mathbf{Z}$.

Remark 22.4.2. In the equal characteristic analogue, κ is a map

$$G(k((t))) \to \pi_1(G_{\overline{\mathbf{Q}}_p})_\Gamma$$

that maps an algebraic loop to a topological loop.

The map

$$(\lambda, \kappa) \colon B(G) \hookrightarrow ((X_*(T) \otimes \mathbf{Q})^+)^\Gamma \times \pi_1(G_{\overline{\mathbf{Q}}_p})_\Gamma$$

is an injection, and one can describe the image. In particular, $B(G)$ is independent of the choice of k.

Returning now to the Fargues-Fontaine curve, we let $\mathrm{Bun}(\mathcal{X}_{\mathrm{FF}})$ be the category of vector bundles on $\mathcal{X}_{\mathrm{FF}}$. We have an exact $\otimes$-functor

$$\mathrm{Isoc}_k \to \mathrm{Bun}(\mathcal{X}_{\mathrm{FF}})$$

sending (V, φ_V) to the descent of $V \otimes_L \mathcal{Y}_{(0,\infty)}$ to $\mathcal{X}_{\mathrm{FF}}$ formed using $\varphi_V \otimes \varphi$. Composing with $\mathrm{Isoc}_k \to \mathrm{Bun}(\mathcal{X}_{\mathrm{FF}})$ defines a functor from the category of isocrystals with G-structure to the category of G-torsors on $\mathcal{X}_{\mathrm{FF}}$.

Theorem 22.4.3 (Fargues, [Far17])**.** The map from $B(G)$ to the set of isomorphism classes of G-torsors on $\mathcal{X}_{\mathrm{FF}}$ is a bijection.

22.5   SEMICONTINUITY OF THE NEWTON POINT

Now we consider the relative situation, so let $\mathcal{P}$ be a G-torsor on $\mathcal{X}_{FF,S}$ for some perfectoid space S of characteristic p. The invariants ν, κ defined in the last section define maps

$$\nu_{\mathcal{P}} : |S| \to ((X_*(T) \otimes \mathbf{Q})^+)^{\Gamma} ,$$

and

$$\kappa_{\mathcal{P}} : |S| \to \pi_1(G_{\overline{\mathbf{Q}}_p})_{\Gamma} .$$

On the set $((X_*(T) \otimes \mathbf{Q})^+)^{\Gamma}$, one has a natural dominance order, as discussed in [RR96, Section 2]. In particular, Theorem 22.2.1 and [RR96, Lemma 2.2 (iv)] imply the following result.

Corollary 22.5.1. The map

$$\nu_{\mathcal{P}} : |S| \to ((X_*(T) \otimes \mathbf{Q})^+)^{\Gamma}$$

is upper semicontinuous.

Moreover, the map $\kappa_{\mathcal{P}}$ should be locally constant. This result will appear in [FS]. We will be able to get our desired results without knowing this (but we will prove a slightly weaker version of this result below).

We record here the analogue of Theorem 22.3.1 for G-torsors on the Fargues-Fontaine curve.

Theorem 22.5.2. The following categories are equivalent.

- Pro-étale $G(\mathbf{Q}_p)$-torsors on S, and
- G-torsors on $\mathcal{X}_{FF,S}$ which are trivial at every geometric point of S.

Proof. This will follow from Theorem 22.3.1 by the Tannakian formalism. Suppose $\mathbb{P}$ is a pro-étale $G(\mathbf{Q}_p)$-torsor on S. For every representation $\rho\colon G \to \mathrm{GL}_n$, the pushfoward $\rho_*(\mathbf{P})$ is a $\mathbf{Q}_p$-local system on S, and so corresponds to a vector bundle $\mathcal{E}_\rho$ on $\mathcal{X}_{FF,S}$ by Theorem 22.3.1. The functor $\rho \mapsto \mathcal{E}_\rho$ corresponds to the required G-torsor on $\mathcal{X}_{FF,S}$.

For the converse, it suffices to see that any G-torsor on $\mathcal{X}_{FF,S}$ is pro-étale locally constant, as then the torsor of trivializations is the desired $G(\mathbf{Q}_p)$-torsor; so assume that S is strictly totally disconnected. Given a G-torsor $\mathcal{E}$ on $\mathcal{X}_{FF,S}$ that is trivial at every geometric point, we get in particular for any representation $\rho : G \to \mathrm{GL}_n$ a trivial rank n-vector bundle over $\mathcal{X}_{FF,S}$, or equivalently a pro-étale $\mathbf{Q}_p$-local system on S. As S is strictly totally disconnected, these are equivalent to finite projective $C(S, \mathbf{Q}_p)$-modules. In other words, we get a fibre functor on $\mathrm{Rep}_{\mathbf{Q}_p} G$ with values in $C(S, \mathbf{Q}_p)$, i.e., a G-torsor over $C(S, \mathbf{Q}_p)$. We know that for all $s \in S$, the corresponding fibre is the trivial G-torsor over $\mathbf{Q}_p$. The local rings of $C(S, \mathbf{Q}_p)$ at all points of S are henselian: Indeed, the local rings of $C(S, \mathbf{Z}_p)$ at all points of S are henselian along (p) as a filtered colimit of p-complete rings, and thus henselian along the kernel of the evaluation map to $\mathbf{Z}_p$. Inverting p then shows that the local rings of $C(S, \mathbf{Q}_p)$ at all points of S are henselian. As the G-torsor on $C(S, \mathbf{Q}_p)$ is trivial at all points of S, this implies that it is locally trivial, as desired. $\square$

22.6 EXTENDING G-TORSORS

We want to understand all possible extensions of a φ^{-1}-equivariant G-torsor $\mathcal{P}_\eta$ on $\mathcal{Y}_{(0,r]}(S)$ to a φ^{-1}-equivariant $\mathcal{G}$-torsor $\mathcal{P}$ on $\mathcal{Y}_{[0,r]}(S)$, where we fix some smooth integral model $\mathcal{G}/\mathbf{Z}_p$ of G. We assume that $\mathcal{G}$ has connected fibers. As usual, we assume that $S = \mathrm{Spa}(R, R^+)$ is affinoid and we fix a pseudo-uniformizer $\varpi \in R$ to define $\mathcal{Y}_{[0,r]}(S)$.

First, we classify φ^{-1}-equivariant $\mathcal{G}$-torsors $\mathcal{P}$ on $\mathcal{Y}_{[0,r]}(S)$. Note that if $\mathbb{P}$ is a pro-étale $\mathcal{G}(\mathbf{Z}_p)$-torsor on S, then

$$\mathcal{P} = \mathbb{P} \times^{\mathcal{G}(\mathbf{Z}_p)} (\mathcal{G} \times_{\mathrm{Spa}\,\mathbf{Z}_p} \mathcal{Y}_{[0,r]}(S))$$

defines such a φ^{-1}-equivariant $\mathcal{G}$-torsor on $\mathcal{Y}_{[0,r]}(S)$. In the Tannakian language, for any representation $\rho : G \to \mathrm{GL}_n$, the pushforward $\rho_*(\mathbb{P})$ defines a $\underline{\mathbf{Z}_p}$-local system of rank n, which gives rise to a φ^{-1}-equivariant vector bundle of rank n by the construction of the previous sections.

Proposition 22.6.1. This construction defines an equivalence of categories between pro-étale $\underline{\mathcal{G}(\mathbf{Z}_p)}$-torsors on S and φ^{-1}-equivariant $\mathcal{G}$-torsors on $\mathcal{Y}_{[0,r]}(S)$.

Proof. From Proposition 22.3.2 and the Tannakian formalism of Theorem 19.5.2, we see that the target category is equivalent to the category of exact $\otimes$-functors from $\mathrm{Rep}_{\mathbf{Z}_p} \mathcal{G}$ towards pro-étale $\underline{\mathbf{Z}_p}$-local systems on S. To see that this defines a pro-étale $\underline{\mathcal{G}(\mathbf{Z}_p)}$-torsor, we need to see that pro-étale locally on S, this fiber functor is equivalent to the forgetful functor. At each geometric point of S, the corresponding $\mathbf{Z}_p$-valued fiber functor on $\mathrm{Rep}_{\mathbf{Z}_p} \mathcal{G}$ is equivalent to the forgetful functor, as $H^1_{\text{ét}}(\mathrm{Spec}\,\mathbf{Z}_p, \mathcal{G}) = 0$ by Lang's lemma. In general, we can assume that S is strictly totally disconnected, in which case $\underline{\mathbf{Z}_p}$-local systems on S are equivalent to finite projective $C^0(\pi_0 S, \mathbf{Z}_p)$-modules, and exact $\otimes$-functors from $\mathrm{Rep}_{\mathbf{Z}_p} \mathcal{G}$ to $\underline{\mathbf{Z}_p}$-local systems on S are equivalent to $\mathcal{G}$-torsors on $C^0(\pi_0 S, \mathbf{Z}_p)$. At each point $s \in S$, we know that the torsor is trivial. But the local rings of $C^0(\pi_0 S, \mathbf{Z}_p)$ are henselian along their maximal ideal (p) (as filtered colimits of p-adically complete rings), so any $\mathcal{G}$-torsor is locally trivial. $\square$

In particular, over a geometric point, all φ^{-1}-equivariant $\mathcal{G}$-torsors on $\mathcal{Y}_{[0,r]}(S)$ are trivial, and thus their restrictions to $\mathcal{Y}_{(0,r]}(S)$ are also trivial.

Theorem 22.6.2. Let $\mathcal{P}_\eta$ be a G-torsor on $\mathcal{Y}_{(0,r]}(S)$ together with an isomorphism with its φ^{-1}-pullback. Consider the functor $\mathrm{Latt}(\mathcal{P}_\eta)$ sending any $S' \to S$

to the set of $\mathcal{G}$-torsors $\mathcal{P}'$ on $\mathcal{Y}_{[0,r]}(S')$ with an isomorphism with their φ^{-1}-pullback and a φ^{-1}-equivariant identification of their restriction to $\mathcal{Y}_{(0,r]}(S')$ with the pullback of $\mathcal{P}_\eta$. Then $\mathrm{Latt}(\mathcal{P}_\eta)$ is representable by a perfectoid space étale over S. Its open image in S is the open subset $S^a \subset S$ where $\nu_{\mathcal{P}_\eta}$ and $\kappa_{\mathcal{P}_\eta}$ are identically 0, called the admissible locus. Over S^a, one has a corresponding pro-étale $G(\mathbf{Q}_p)$-torsor $\mathbb{P}_\eta$, and $\mathrm{Latt}(\mathcal{P}_\eta)$ parametrizes pro-étale $\mathcal{G}(\mathbf{Z}_p)$-torsors $\mathbb{P}$ with an identification

$$\mathbb{P} \times^{\underline{\mathcal{G}(\mathbf{Z}_p)}} G(\mathbf{Q}_p) = \mathbb{P}_\eta \ .$$

Proof. From the previous proposition, we see that $\mathrm{Latt}(\mathcal{P}_\eta)$ maps into $S^a \subset S$. First, we claim that $S^a \subset S$ is open. By Corollary 22.5.1, we can assume that $\nu_{\mathcal{P}_\eta}$ is identically 0. Moreover, we can work pro-étale locally, and so assume that S is strictly totally disconnected. As in the proof of Theorem 22.5.2, we get a G-torsor on $C^0(\pi_0 S, \mathbf{Q}_p)$. If this G-torsor is trivial at some $s \in \pi_0 S$, then it is trivial in an open neighborhood, as the local rings of $C^0(\pi_0 S, \mathbf{Q}_p)$ are henselian. By the classification of G-torsors on the Fargues-Fontaine curve, S^a is the locus where the G-torsor is trivial. Thus, this implies that $S^a \subset S$ is open.

For the rest, we can assume that $S = S^a$. The previous discussion shows that we then get an exact $\otimes$-functor from $\mathrm{Rep}_{\mathbf{Q}_p} G$ to pro-étale $\mathbf{Q}_p$-local systems on S, which is pro-étale locally isomorphic to the forgetful functor. The pro-étale sheaf of isomorphisms with the forgetful $\otimes$-functor defines a pro-étale $G(\mathbf{Q}_p)$-torsor $\mathbb{P}_\eta$, and unraveling the definitions we see that

$$\mathcal{P}_\eta = \mathbb{P}_\eta \times^{\underline{G(\mathbf{Q}_p)}} (G \times \mathcal{Y}_{(0,r]}(S)) \ .$$

Comparing with Proposition 22.6.1, we see that $\mathrm{Latt}(\mathcal{P}_\eta)$ parametrizes pro-étale $\mathcal{G}(\mathbf{Z}_p)$-torsors $\mathbb{P}$ with an identification

$$\mathbb{P} \times^{\underline{\mathcal{G}(\mathbf{Z}_p)}} G(\mathbf{Q}_p) = \mathbb{P}_\eta \ .$$

Using Theorem 9.1.3, one sees that this is indeed étale over S, as fixing a trivialization of $\mathbb{P}_\eta$, one sees that pro-étale locally on S it is the product $S \times G(\mathbf{Q}_p)/\mathcal{G}(\mathbf{Z}_p)$. $\qquad\square$

Lecture 23

Moduli spaces of shtukas

Today we define the moduli spaces of mixed-characteristic local G-shtukas and show that they are representable by locally spatial diamonds.

These will be the mixed-characteristic local analogues of the moduli spaces of global equal-characteristic shtukas introduced by Varshavsky [Var04]. It may be helpful to briefly review the construction in the latter setting. The ingredients are a smooth projective geometrically connected curve X defined over a finite field $\mathbf{F}_q$ and a reductive group $G/\mathbf{F}_q$. The moduli space of G-shtukas with m legs (called F-bundles in [Var04]) is a stack $\mathrm{Sht}_{G,m}$ equipped with a morphism to X^m. For an $\mathbf{F}_q$-scheme S, the S-points of $\mathrm{Sht}_{G,m}$ classify triples consisting of a G-torsor $\mathcal{P}$ on $S \times_{\mathbf{F}_q} X$, an m-tuple $x_1, \ldots, x_m \in X(S)$, and an isomorphism

$$\varphi_{\mathcal{P}} \colon (\mathrm{Frob}_S^* \mathcal{P})|_{(S\times_{\mathbf{F}_q} X)\setminus \bigcup_{i=1}^m \Gamma_{x_i}} \xrightarrow{\sim} \mathcal{P}|_{(S\times_{\mathbf{F}_q} X)\setminus \bigcup_{i=1}^m \Gamma_{x_i}}.$$

The stack $\mathrm{Sht}_{G,m}$ is not locally of finite type. But if in addition we are given an m-tuple of conjugacy classes of cocharacters $\mu_1, \ldots, \mu_m$ of G, we may define a closed substack $\mathrm{Sht}_{G,\{\mu_i\}}$ of $\mathrm{Sht}_{G,m}$ classifying shtukas where $\varphi_{\mathcal{P}}$ is bounded by μ_i at Γ_{x_i} for $i = 1, \ldots, m$. (Caveat: the boundedness condition is more subtle than what we describe when the x_i intersect, see below.) Then $\mathrm{Sht}_{G,\{\mu_i\}}$ is a Deligne-Mumford stack, locally of finite type over X^m.

Furthermore, if we choose a rational dominant cocharacter $\nu \in (X_*(T) \times \mathbf{Q})^+$ of a maximal torus of G, then we may restrict the stack of shtukas further, to obtain an open substack $\mathrm{Sht}_{G,\{\mu_i\},\leq\nu} \subset \mathrm{Sht}_{G,\{\mu_i\}}$ classifying shtukas where $\mathcal{P}$ is bounded by ν in the appropriate sense. Then $\mathrm{Sht}_{G,\{\mu_i\},\leq\nu}$ is particularly nice: each connected component is a quotient of a quasi-projective scheme by a finite group.

From there, it is possible to add level structures to the spaces of shtukas, to obtain a tower of moduli spaces admitting an action of the adelic group $G(\mathbf{A}_F)$, where F is the function field of X. The cohomology of these towers of moduli spaces is the primary means by which V. Lafforgue [Laf18] constructs the "automorphic to Galois" direction of the Langlands correspondence for G over F.

23.1   DEFINITION OF MIXED-CHARACTERISTIC LOCAL SHTUKAS

Now let G be a reductive group over $\mathbf{Q}_p$. We will mimic Varshavsky's definition of G-shtukas, replacing X with $\mathbf{Z}_p$ and products $X \times S$ with $X \dot\times S$. In the equal-characteristic setting, G naturally becomes a group scheme over X by base change, so that we can talk about G-torsors on $X \times S$. But in the mixed-characteristic setting, G does not naturally live over $\mathbf{Z}_p$. Therefore we choose a smooth group scheme $\mathcal{G}$ over $\mathbf{Z}_p$ with generic fiber G and connected special fiber.

Let $S = \mathrm{Spa}(R, R^+)$ be an affinoid perfectoid space of characteristic p, with pseudouniformizer ϖ. Suppose we are given a triple $(\mathcal{P}, \{S_i^\sharp\}, \varphi_{\mathcal{P}})$, where $\mathcal{P}$ is a $\mathcal{G}$-torsor on $S \dot\times \mathrm{Spa}\,\mathbf{Z}_p$, where $S_1^\sharp, \ldots, S_m^\sharp$ are untilts of S to $\mathbf{Q}_p$, and where

$$\varphi_{\mathcal{P}} \colon (\mathrm{Frob}_S^* \mathcal{P})|_{(S \dot\times X) \setminus \bigcup_{i=1}^m \Gamma_{x_i}} \xrightarrow{\sim} \mathcal{P}|_{(S \dot\times X) \setminus \bigcup_{i=1}^m \Gamma_{x_i}}$$

is an isomorphism which is meromorphic along the closed Cartier divisor $\bigcup_{i=1}^m S_i^\sharp$ of $S \dot\times \mathbf{Z}_p$. For sufficiently large r, the open subset

$$\mathcal{Y}_{[r,\infty)}(S) = \{|\varpi| \le |p|^r \ne 0\} \subset S \dot\times \mathrm{Spa}\,\mathbf{Z}_p$$

will be disjoint from the $S_i^\sharp$, so that the restriction of $\varphi_{\mathcal{P}}$ to $\mathcal{Y}_{[r,\infty)}(S)$ is an isomorphism. We can use this restriction to descend $\mathcal{P}$ to a G-torsor $\mathcal{E}$ over the relative Fargues-Fontaine curve $\mathcal{X}_{\mathrm{FF},S}$. Then $\mathcal{E}$ is independent of the choice of ϖ and r, and is recording what happens far away from $p = 0$.

We have already seen in Theorem 22.4.3 that there is a bijection $b \mapsto \mathcal{E}^b$ between Kottwitz's set $B(G)$ and the set of isomorphism classes of G-torsors on the absolute Fargues-Fontaine curve. In the setup of local shtukas, we choose in advance an element $b \in B(G)$, and enforce an isomorphism $\mathcal{E} \xrightarrow{\sim} \mathcal{E}^b$.

We are now ready to define the space of local mixed-characteristic G-shtukas. As in the previous section, we let k be a discrete algebraically closed field, and $L = W(k)[1/p]$.

Definition 23.1.1. Let $(\mathcal{G}, b, \{\mu_i\})$ be a triple consisting of a smooth group scheme $\mathcal{G}$ with reductive generic fiber G and connected special fiber, an element $b \in G(L)$, and a collection $\mu_1, \ldots, \mu_m$ of conjugacy classes of cocharacters $\mathbf{G}_m \to G_{\overline{\mathbf{Q}}_p}$. For $i = 1, \ldots, m$, let $E_i/\mathbf{Q}_p$ be the field of definition of μ_i, and let $\breve{E}_i = E_i \cdot L$.

The moduli space

$$\mathrm{Sht}_{\mathcal{G},b,\{\mu_i\}} \to \mathrm{Spd}\,\breve{E}_1 \times_{\mathrm{Spd}\,k} \cdots \times_{\mathrm{Spd}\,k} \mathrm{Spd}\,\breve{E}_m$$

of shtukas associated with $(\mathcal{G}, b, \{\mu_i\})$ is the presheaf on Perf_k sending $S = \mathrm{Spa}(R, R^+)$ to the set of quadruples $(\mathcal{P}, \{S_i^\sharp\}, \varphi_{\mathcal{P}}, \iota_r)$, where:

- $\mathcal{P}$ is a $\mathcal{G}$-torsor on $S \dot{\times} \operatorname{Spa} \mathbf{Z}_p$,
- $S_i^\sharp$ is an untilt of S to $\breve{E}_i$, for $i = 1, \ldots, m$,
- $\varphi_\mathcal{P}$ is an isomorphism

$$\varphi_\mathcal{P} \colon (\operatorname{Frob}_S^* \mathcal{P})|_{(S \dot{\times} X) \setminus \bigcup_{i=1}^m \Gamma_{x_i}} \xrightarrow{\sim} \mathcal{P}|_{(S \dot{\times} X) \setminus \bigcup_{i=1}^m \Gamma_{x_i}},$$

and finally
- ι_r is an isomorphism

$$\iota_r \colon \mathcal{P}|_{\mathcal{Y}_{[r,\infty)}(S)} \xrightarrow{\sim} G \times \mathcal{Y}_{[r,\infty)}(S)$$

for large enough r, under which $\varphi_\mathcal{P}$ gets identified with $b \times \operatorname{Frob}_S$.

The isomorphism $\varphi_\mathcal{P}$ is required to be meromorphic along the closed Cartier divisor $\bigcup_{i=1}^m S_i^\sharp \subset S \dot{\times} \operatorname{Spa} \mathbf{Z}_p$, and it is subject to the following boundedness condition: At all geometric rank 1 points of S, the relative position of $\operatorname{Frob}_S^* \mathcal{P}$ and $\mathcal{P}$ at $S_i^\sharp$ is bounded by $\sum_{j \mid S_j^\sharp = S_i^\sharp} \mu_j$ in the Bruhat order.

Remark 23.1.2. The boundedness condition appearing in this definition is essentially the same as the one that appears in the definition of the Schubert variety $\operatorname{Gr}_{G, \operatorname{Spd} E_1 \times \ldots \times \operatorname{Spd} E_m, \leq \mu_\bullet}$.

Remark 23.1.3. The isomorphism class of $\operatorname{Sht}_{\mathcal{G}, b, \{\mu_i\}}$ only depends on the class of b in $B(G)$; replacing b by $\varphi(y) b y^{-1}$ corresponds to composing ι_r with $y \times \operatorname{id}$, for $y \in G(L)$.

The main theorem of the course is the following.

Theorem 23.1.4. The moduli space $\operatorname{Sht}_{\mathcal{G}, b, \{\mu_i\}}$ is a locally spatial diamond.

Before embarking on the proof, let us note that the descent result of Proposition 19.5.3 already implies that $\operatorname{Sht}_{(\mathcal{G}, b, \mu_\bullet)}$ is a v-sheaf.

23.2 THE CASE OF NO LEGS

Let us dispense with the case of no legs (that is, $m = 0$).

Proposition 23.2.1. The moduli space $\operatorname{Sht}_{\mathcal{G}, b, \emptyset}$ is empty if b does not lie in the trivial class in $B(G)$, and $\operatorname{Sht}_{\mathcal{G}, 1, \emptyset}$ is isomorphic to the constant perfectoid space $\underline{G(\mathbf{Q}_p)/\mathcal{G}(\mathbf{Z}_p)}$.

Proof. Let S be an affinoid perfectoid space over k. An S-point of $\operatorname{Sht}_{\mathcal{G}, b, \emptyset}$ is a φ_S-equivariant $\mathcal{G}$-torsor $\mathcal{P}$ on $\mathcal{Y}_{[0,\infty)}(S)$, together with a trivialization ι_r of $\mathcal{P}$ over $\mathcal{Y}_{[r,\infty)}(S)$ which identifies $\varphi_\mathcal{P}$ with $b \times \operatorname{Frob}_S$. Since there are no legs to interfere, repeated applications of $\varphi_\mathcal{P}^{-1}$ can be used to extend ι_r to a trivialization of $\mathcal{P}$ on all of $\mathcal{Y}_{(0,\infty)}(S)$. Thus there is an isomorphism $\mathcal{P}|_{\mathcal{Y}_{(0,\infty)}(S)} \cong G \times \mathcal{Y}_{(0,\infty)}(S)$, which identifies $\varphi_{\mathcal{P}}$ with $b \times \mathrm{Frob}_S$ on this locus. In other words, the φ-equivariant G-torsor $\mathcal{P}|_{\mathcal{Y}_{(0,\infty)}(S)}$ is the pullback of the G-torsor $\mathcal{E}^b$ on $\mathcal{X}_{\mathrm{FF},S}$.

By Theorem 22.6.2, a φ-equivariant G-torsor on $\mathcal{Y}_{(0,\infty)}(S)$ can be extended to a $\mathcal{G}$-torsor on $\mathcal{Y}_{[0,\infty)}(S)$ if and only if its descent to $\mathcal{X}_{\mathrm{FF},S}$ is trivial at all geometric points of S. In the present situation, we have a $\mathcal{G}$-torsor on $\mathcal{Y}_{[0,\infty)}(S)$ whose restriction to $\mathcal{Y}_{(0,\infty)}(S)$ is the pullback of $\mathcal{E}^b$. Therefore the class of b in $B(G)$ is trivial. Again by Theorem 22.6.2, φ-equivariant extensions of the trivial φ-equivariant G-torsor on $\mathcal{Y}_{(0,\infty)}(S)$ to a φ-equivariant $\mathcal{G}$-torsor on $\mathcal{Y}_{[0,\infty)}(S)$ are in correspondence with $\mathcal{G}(\mathbf{Z}_p)$-lattices in the trivial pro-étale $G(\mathbf{Q}_p)$-torsor on S. Such lattices in turn correspond to S-points of $G(\mathbf{Q}_p)/\mathcal{G}(\mathbf{Z}_p)$. $\square$

23.3   THE CASE OF ONE LEG

We now prove Theorem 23.1.4 in the case of one leg. In this case, the space of shtukas can be compared with a space of modifications of G-torsors at one point on the Fargues-Fontaine curve. Let μ be a cocharacter of G, with field of definition E.

Proposition 23.3.1. Let $S \in \mathrm{Perf}_k$. The S-points of the moduli space $\mathrm{Sht}_{\mathcal{G},b,\mu}$ are in bijection with isomorphism classes of quadruples $(S^\sharp, \mathcal{E}, \alpha, \mathbb{P})$, where

- $S^\sharp$ is an untilt of S to E,
- $\mathcal{E}$ is a G-torsor on $\mathcal{X}_{\mathrm{FF},S}$, which is trivial at every geometric point of S,
- α is an isomorphism of G-torsors

$$\mathcal{E}|_{\mathcal{X}_{\mathrm{FF},S}\setminus S^\sharp} \xrightarrow{\sim} \mathcal{E}^b|_{\mathcal{X}_{\mathrm{FF},S}\setminus S^\sharp}$$

which is meromorphic along $S^\sharp$ and bounded by μ, and finally
- $\mathbb{P}$ is $\mathcal{G}(\mathbf{Z}_p)$-lattice in the pro-étale $G(\mathbf{Q}_p)$-torsor corresponding to $\mathcal{E}$ under Theorem 22.5.2.

Proof. Assume that S is affinoid, and suppose we are given an S-point of $\mathrm{Sht}_{\mathcal{G},b,\mu}$, corresponding to a quadruple $(\mathcal{P}, S^\sharp, \varphi_{\mathcal{P}}, \iota_r)$. Let $\epsilon > 0$ be such that $S^\sharp \subset \mathcal{Y}_{(0,\infty)}(S)$ is disjoint from $\mathcal{Y}_{(0,\epsilon)}(S)$. Then the restriction of $\mathcal{P}$ to $\mathcal{Y}_{(0,\epsilon)}(S)$ is φ^{-1}-equivariant, and so descends to a G-torsor $\mathcal{E}$ on $\mathcal{X}_{\mathrm{FF},S}$. The restriction of $\mathcal{P}$ is a φ^{-1}-equivariant $\mathcal{G}$-torsor on $\mathcal{Y}_{[0,\epsilon]}(S)$, and so corresponds by Proposition 22.6.1 to a pro-étale $\mathcal{G}(\mathbf{Z}_p)$-torsor $\mathbb{P}$ on S, whose generic fiber $\mathbb{P}_\eta$ corresponds to $\mathcal{E}$ under Theorem 22.5.2.

By our constructions, the pullbacks of $\mathcal{E}$ and $\mathcal{E}^b$ to $\mathcal{Y}_{(0,\infty)}(S)$ are both identified φ-equivariantly with $\mathcal{P}$ away from $\bigcup_{n\in\mathbf{Z}} \varphi^n(S^\sharp)$; thus we get an isomorphism α between $\mathcal{E}$ and $\mathcal{E}^b$ away from the image of $S^\sharp$ in $\mathcal{X}_{\mathrm{FF},S}$. $\square$

The description of $\mathrm{Sht}_{\mathcal{G},b,\mu}$ in Proposition 23.3.1 shows that it only depends on the triple (G, b, μ) and the subgroup $\mathcal{G}(\mathbf{Z}_p) \subset G(\mathbf{Q}_p)$, and not necessarily on the model $\mathcal{G}$. We can define a space of shtukas $\mathrm{Sht}_{G,b,\mu,K}$ with K-structure, where $K \subset G(\mathbf{Q}_p)$ is any compact open subgroup: this will parametrize quadruples $(S^\sharp, \mathcal{E}, \alpha, \mathbb{P})$, where $\mathbb{P}$ is instead a pro-étale $\underline{K}$-torsor, such that $\mathcal{P}_\eta = \mathcal{P} \times^{\underline{K}} G(\mathbf{Q}_p)$ is the pro-étale $\underline{G(\mathbf{Q}_p)}$-torsor corresponding to $\mathcal{E}$. If $K' \subset K$ are two such subgroups, then

$$\mathrm{Sht}_{G,b,\mu,K'} \to \mathrm{Sht}_{G,b,\mu,K}$$

is finite étale, and a K/K'-torsor in case $K' \subset K$ is normal. All spaces in this tower carry compatible continuous actions by $J_b(\mathbf{Q}_p)$, since this group operates on $\mathcal{E}^b$.

The limit

$$\mathrm{Sht}_{G,b,\mu,\infty} = \varprojlim_K \mathrm{Sht}_{G,b,\mu,K}$$

has a simple description. Its S-points parametrize pairs $(S^\sharp, \alpha)$, where $S^\sharp$ is an untilt of S to E, and α is an isomorphism of G-torsors

$$\mathcal{E}^1|_{\mathcal{X}_{\mathrm{FF},S} \setminus S^\sharp} \xrightarrow{\sim} \mathcal{E}^b|_{\mathcal{X}_{FF,S} \setminus S^\sharp}$$

which is meromorphic along $S^\sharp$ and bounded by μ. We remark that $\mathrm{Sht}_{G,b,\mu,\infty}$ admits an continuous action of the product $G(\mathbf{Q}_p) \times J_b(\mathbf{Q}_p)$, via the actions of those groups on $\mathcal{E}^1$ and $\mathcal{E}^b$.

This description of $\mathrm{Sht}_{G,b,\mu,\infty}$ implies a general duality result for moduli spaces of shtukas, proving a conjecture of Rapoport-Zink, [RZ96, Section 5.54], generalizing the duality between the Lubin-Tate and Drinfeld towers, [Fal02b], [FGL08].

Corollary 23.3.2. Assume that (G, b, μ) is a local shtuka datum such that b is basic. Define a dual local shtuka datum $(\check{G}, \check{b}, \check{\mu})$ via $\check{G} = J_b$ the σ-centralizer of b, which is an inner form of G, $\check{b} = b^{-1} \in \check{G} = J_b$ and $\check{\mu} = \mu^{-1}$ under the identification $\check{G}_{\overline{\mathbf{Q}}_p} \cong G_{\overline{\mathbf{Q}}_p}$. Then there is a natural $G(\mathbf{Q}_p) \times J_b(\mathbf{Q}_p)$-equivariant isomorphism

$$\mathrm{Sht}_{G,b,\mu,\infty} \cong \mathrm{Sht}_{(\check{G}, \check{b}, \check{\mu}),\infty}$$

over $\mathrm{Spd}\, \check{E}$.

Proof. As J_b is an inner form of G, the category of G-torsors is equivalent to the category of J_b-torsors, via sending a G-torsor $\mathcal{P}$ to the J_b-torsor of G-equivariant identifications between $\mathcal{P}$ and $\mathcal{E}_b$. Under this identification, $\mathcal{E}_b$ goes to the trivial J_b-torsor, and $\mathcal{E}_1$ goes to the J_b-torsor corresponding to $\check{b} = b^{-1}$. The condition that the original modification is bounded by μ gets translated into the boundedness by $\check{\mu} = \mu^{-1}$. $\qquad\square$

We have yet to show that $\mathrm{Sht}_{G,b,\mu,K}$ is a locally spatial diamond. We will do this by showing that $\mathrm{Sht}_{G,b,\mu,K}$ admits an étale morphism to a Beilinson-Drinfeld

Grassmanniann. Recall that $\mathrm{Gr}_{G,\mathrm{Spd}\,E,\mu}$ is the functor sending S to the set of untilts $S^\sharp$ over E together with a G-torsor $\mathcal{P}$ on $S \dot\times \mathbf{Q}_p$ and a trivialization of $\mathcal{P}$ away from $S^\sharp$, which is meromorphic and bounded by μ at $S^\sharp$. The alternate description given in Proposition 20.2.2 shows that the S-points of $\mathrm{Gr}_{G,\mathrm{Spd}\,E,\mu}$ are in correspondence with untilts $S^\sharp = (R^\sharp, R^{\sharp+})$ over E together with G-bundles over $B_{\mathrm{dR}}^+(R^\sharp)$ which come with a trivialization over $B_{\mathrm{dR}}(R^\sharp)$.

Suppose we are given an S-point of $\mathrm{Sht}_{G,b,\mu,K}$, corresponding to a quadruple $(S^\sharp, \mathcal{E}, \alpha, \mathbb{P})$ as in Proposition 23.3.1. The G-torsor $\mathcal{E}^b$ pulls back to the trivial G-torsor on $S \dot\times \mathrm{Spa}\,\mathbf{Q}_p$ by definition. Therefore the pullback of $\mathcal{E}$ to $B_{\mathrm{dR}}^+(S^\sharp)$ can be regarded as a G-torsor $\mathcal{E}_{S^\sharp}$ over $B_{\mathrm{dR}}^+(R^\sharp)$ with a trivialization over $B_{\mathrm{dR}}(R^\sharp)$ induced by α, and thus an S-point of $\mathrm{Gr}_{G,\mathrm{Spd}\,\breve{E},\mu}$. (We need the field $\breve{E} = E \cdot L$ because S lives over k.) We arrive at a *period morphism*

$$\pi_{GM} : \mathrm{Sht}_{G,b,\mu} \to \mathrm{Gr}_{G,\mathrm{Spd}\,\breve{E},\leq\mu}.$$

The subscript GM stands for Grothendieck-Messing. In the case of moduli spaces of p-divisible groups, this period map is closely related to the Grothendieck-Messing deformation theory of p-divisible groups.

Proposition 23.3.3. The period morphism

$$\pi_{GM} : \mathrm{Sht}_{G,b,\mu,K} \to \mathrm{Gr}_{G,\mathrm{Spd}\,\breve{E},\leq\mu}$$

is étale.

Remark 23.3.4. Since $\mathrm{Gr}_{G,\mathrm{Spd}\,\breve{E},\leq\mu}$ is a spatial diamond (Proposition 20.2.3), this implies that $\mathrm{Sht}_{G,b,\mu,K}$ is a locally spatial diamond.

Proof. Let $S \in \mathrm{Perf}_k$, and suppose we are given an S-point of $\mathrm{Gr}_{G,\mathrm{Spd}\,\breve{E},\leq\mu}$. This point corresponds to a modification of $\mathcal{E}^b$ on $\mathcal{X}_{\mathrm{FF},S}$, which results in a G-torsor $\mathcal{E}$ on $\mathcal{X}_{\mathrm{FF},S}$. By Theorem 22.6.2, the locus $S^a \subset S$ where $\mathcal{E}$ is trivial is an open subset. (Thus there exists an open subset $\mathrm{Gr}_{G,\mathrm{Spd}\,\breve{E},\leq\mu}^a \subset \mathrm{Gr}_{G,\mathrm{Spd}\,\breve{E},\leq\mu}$, the *admissible locus*, through which the morphism from S^a must factor.)

The G-torsor $\mathcal{E}|_{S^a \dot\times \mathrm{Spa}\,\mathbf{Q}_p}$ corresponds via Theorem 22.5.2 to a pro-étale $G(\mathbf{Q}_p)$-torsor $\mathbb{P}_\eta$ on S^a. The fiber of π_{GM} over S is the space of $\underline{K}$-lattices in $\overline{\mathbb{P}_\eta}$. This is indeed étale, as pro-étale locally on S, this fiber is the projection $S^a \times \underline{G(\mathbf{Q}_p)/K} \to S$. $\qquad\square$

23.4 THE CASE OF TWO LEGS

Much of our discussion of the moduli space of one-legged shtukas carries over to the general case. For every compact open subgroup $K \subset G(\mathbf{Q}_p)$, there is a functor $\mathrm{Sht}_{G,b,\{\mu_i\},K}$, of which $\mathrm{Sht}_{G,b,\{\mu_i\}} = \mathrm{Sht}_{G,b,\{\mu_i\},G(\mathbf{Z}_p)}$ is a special case.

The limit

$$\mathrm{Sht}_{G,b,\{\mu_i\}} = \varprojlim_K \mathrm{Sht}_{G,b,\{\mu_i\},K}$$

has essentially the following description: for $S \in \mathrm{Perf}_k$, an S-point of $\mathrm{Sht}_{G,b,\{\mu_i\}}$ corresponds to a tuple of untilts $S_i^\sharp/E_i$ for $i = 1, \ldots, m$, together with an iso-morphism

$$\alpha: \mathcal{E}^1|_{\mathcal{X}_{\mathrm{FF},S} \setminus \bigcup_{i=1}^m S_i^\sharp} \xrightarrow{\sim} \mathcal{E}^b|_{\mathcal{X}_{\mathrm{FF},S} \setminus \bigcup_{i=1}^m S_i^\sharp}$$

subject to a boundedness condition. Unfortunately, the precise boundedness condition turns out to be somewhat subtle to specify, depending on the specific position of the untilts and their Frobenius translates, and in fact in general is not merely a condition but includes extra data specifying intermediate modifications.

To prove that $\mathrm{Sht}_{G,b,\{\mu_i\},K}$ is a locally spatial diamond, we will again use a period morphism π_{GM}, which presents $\mathrm{Sht}_{G,b,\{\mu_i\},\infty}$ as a pro-étale $G(\mathbf{Q}_p)$-torsor over an open subset of a Grassmannian, which records what is happening at the legs. However, one has to be careful with the target of the period morphism. Over the locus where all Frobenius translates of all legs are disjoint from each other, the required Grassmannian is simply the product of the $\mathrm{Gr}_{G,\mathrm{Spd}\,E_i,\mu_i}$, but in general one needs a twisted form of the Beilinson-Drinfeld Grassmannian.

It may be helpful to spell out the case of two legs first. Inside $\mathrm{Spd}\,\mathbf{Q}_p \times \mathrm{Spd}\,\mathbf{Q}_p$, we have the diagonal $\Delta = \mathrm{Spd}\,\mathbf{Q}_p \subset \mathrm{Spd}\,\mathbf{Q}_p \times \mathrm{Spd}\,\mathbf{Q}_p$, and its translates $(\varphi \times 1)^n(\Delta) \subset \mathrm{Spd}\,\mathbf{Q}_p \times \mathrm{Spd}\,\mathbf{Q}_p$, where $\varphi \times 1$ acts only on the first copy of $\mathrm{Spd}\,\mathbf{Q}_p$.

Definition 23.4.1. Let $\mathrm{Gr}^{\mathrm{tw}}_{G,(\mathrm{Spd}\,\mathbf{Q}_p)^2} \to (\mathrm{Spd}\,\mathbf{Q}_p)^2$ be the following version of the Beilinson-Drinfeld Grassmannian.

1. Over the open subset $(\mathrm{Spd}\,\mathbf{Q}_p)^2 \setminus \bigcup_{n \neq 0}(\varphi \times 1)^n(\Delta)$, it is the Beilinson-Drinfeld Grassmannian $\mathrm{Gr}_{G,(\mathrm{Spd}\,\mathbf{Q}_p)^2}$.
2. Over the open subset $(\mathrm{Spd}\,\mathbf{Q}_p)^2 \setminus \bigcup_{n \neq m}(\varphi \times 1)^n(\Delta)$ with $m > 0$, it is the pullback of the convolution Beilinson-Drinfeld Grassmannian $\widetilde{\mathrm{Gr}}_{G,(\mathrm{Spd}\,\mathbf{Q}_p)^2}$ under $(\varphi \times 1)^{-m}$.
3. Over the open subset $(\mathrm{Spd}\,\mathbf{Q}_p)^2 \setminus \bigcup_{n \neq m}(\varphi \times 1)^n(\Delta)$ with $m < 0$, it is the pullback of the convolution Beilinson-Drinfeld Grassmannian $\widetilde{\mathrm{Gr}}_{G,(\mathrm{Spd}\,\mathbf{Q}_p)^2}$ under the switching of the two factors $(\mathrm{Spd}\,\mathbf{Q}_p)^2 \cong (\mathrm{Spd}\,\mathbf{Q}_p)^2$ composed with $(\varphi \times 1)^m$.

Moreover, the Schubert variety $\mathrm{Gr}^{\mathrm{tw}}_{G,\mathrm{Spd}\,E_1 \times \mathrm{Spd}\,E_2, \leq(\mu_1,\mu_2)}$ within

$$\mathrm{Gr}^{\mathrm{tw}}_{G,\mathrm{Spd}\,E_1 \times \mathrm{Spd}\,E_2} = \mathrm{Gr}^{\mathrm{tw}}_{G,(\mathrm{Spd}\,\mathbf{Q}_p)^2} \times_{(\mathrm{Spd}\,\mathbf{Q}_p)^2} (\mathrm{Spd}\,E_1 \times \mathrm{Spd}\,E_2)$$

is the closed subfunctor given by the corresponding Schubert varieties in all cases.

These spaces are glued via the obvious identification over $(\mathrm{Spd}\,\mathbf{Q}_p)^2 \setminus \bigcup_n(\varphi \times 1)^n(\Delta)$, where they are all identified with two copies of the usual affine Grassmannian $\mathrm{Gr}_{G,\mathrm{Spd}\,\mathbf{Q}_p}$. The following gives a more direct definition.

Proposition 23.4.2. The functor $\mathrm{Gr}^{\mathrm{tw}}_{G,\mathrm{Spd}\,E_1\times\mathrm{Spd}\,E_2,\leq(\mu_1,\mu_2)} \to \mathrm{Spd}\,E_1\times\mathrm{Spd}\,E_2$ parametrizes over a perfectoid space S of characteristic p with two untilts $S_i^\sharp = \mathrm{Spa}(R_i^\sharp, R_i^{\sharp+})$, $i = 1, 2$, over E_i, the set of G-torsors $\mathcal{P}_\eta$ on $S\dot\times\mathrm{Spa}\,\mathbf{Q}_p$ together with an isomorphism

$$\varphi_{\mathcal{P}_\eta} : (\mathrm{Frob}_S^*\,\mathcal{P}_\eta)|_{S\dot\times\,\mathrm{Spa}\,\mathbf{Q}_p\setminus\bigcup_{i=1}^2 S_i^\sharp} \cong \mathcal{P}_\eta|_{S\dot\times\,\mathrm{Spa}\,\mathbf{Q}_p\setminus\bigcup_{i=1}^2 S_i^\sharp}$$

that is meromorphic along the closed Cartier divisor $\bigcup_{i=1}^2 S_i^\sharp \subset S\dot\times\mathrm{Spa}\,\mathbf{Q}_p$, and an isomorphism

$$\iota_r : \mathcal{P}_\eta|_{\mathcal{Y}_{[r,\infty)}(S)} \cong G\times\mathcal{Y}_{[r,\infty)}(S)$$

for large enough r under which $\varphi_{\mathcal{P}_\eta}$ gets identified with $b\times\mathrm{Frob}_S$, satisfying the following boundedness condition:

At all geometric rank 1 points, the relative position of $\mathrm{Frob}_S^*\,\mathcal{P}$ and $\mathcal{P}$ at $S_i^\sharp$ is bounded by $\sum_{j|S_j^\sharp=S_i^\sharp}\mu_j$ in the Bruhat order.

Remark 23.4.3. This description seems to depend on b, but in fact it is canonically independent of it: One can change the isomorphism $\varphi_{\mathcal{P}_\eta}$ by any element of $G(L)$.

Proof. We note that ι_r extends to an isomorphism

$$\mathcal{P}|_{\mathcal{Y}_{(0,\infty)}(S)\setminus\bigcup_{i=1,2,n\geq 0}\varphi^{-n}(S_i^\sharp)} \cong G\times\Big(\mathcal{Y}_{(0,\infty)}(S)\setminus\bigcup_{i=1,2,n\geq 0}\varphi^{-n}(S_i^\sharp)\Big)$$

that is meromorphic along $\varphi^{-n}(S_i^\sharp)$. To describe $\mathcal{P}$ as a modification of the trivial G-torsor, we have to prescribe the modification. Over the open subset $(\mathrm{Spd}\,\mathbf{Q}_p)^2\setminus\bigcup_{n\neq 0}(\varphi\times 1)^n(\Delta)$, the modification at $S_1^\sharp$ and $S_2^\sharp$ gives rise to a section of $\mathrm{Gr}_{G,(\mathrm{Spd}\,\mathbf{Q}_p)^2}$, which can then be φ^{-1}-periodically continued. Over the open subset $(\mathrm{Spd}\,\mathbf{Q}_p)^2\setminus\bigcup_{n\neq m}(\varphi\times 1)^n(\Delta)$ with $m > 0$, we can first modify the trivial G-torsor at $S_1^\sharp$ (which is different from $\varphi^{-n}(S_2^\sharp)$ for $n \geq 0$ at all geometric points) and continue this modification φ^{-1}-periodically, and afterwards modify at $S_2^\sharp$ and continue φ^{-1}-periodically. This gives rise to an object of the Beilinson-Drinfeld Grassmannian $\widetilde{\mathrm{Gr}}_{G,(\mathrm{Spd}\,\mathbf{Q}_p)^2}$. Symmetrically, the same happens in the third case.

Finally, the boundedness conditions correspond to Schubert varieties in all cases. $\square$

Therefore, the arguments in the case of one leg give the following result.

Corollary 23.4.4. There is an étale period map

$$\pi_{GM} : \mathrm{Sht}_{\mathcal{G},b,\{\mu_1,\mu_2\}} \to \mathrm{Gr}^{\mathrm{tw}}_{G,\mathrm{Spd}\,\breve{E}_1\times_k\mathrm{Spd}\,\breve{E}_2,\leq(\mu_1,\mu_2)}\,.$$

Its open image is the admissible locus

$$\mathrm{Gr}^{\mathrm{tw},a}_{G,\mathrm{Spd}\,\check{E}_1\times_k\mathrm{Spd}\,\check{E}_2,\leq(\mu_1,\mu_2)} \subset \mathrm{Gr}^{\mathrm{tw}}_{G,\mathrm{Spd}\,\check{E}_1\times_k\mathrm{Spd}\,\check{E}_2,\leq(\mu_1,\mu_2)}$$

over which there is a $G(\mathbf{Q}_p)$-torsor $\mathbb{P}_\eta$. The space $\mathrm{Sht}_{\mathcal{G},b,\{\mu_1,\mu_2\}}$ parametrizes $\mathcal{G}(\mathbf{Z}_p)$-lattices $\mathbb{P}$ inside $\overline{\mathbb{P}}_\eta$.

23.5   THE GENERAL CASE

Now we discuss the general case. In order to avoid an extensive discussion of various cases, we make the following definition.

Definition 23.5.1. The functor $\mathrm{Gr}^{\mathrm{tw}}_{G,\mathrm{Spd}\,E_1\times\ldots\times\mathrm{Spd}\,E_m,\leq\mu_\bullet} \to \mathrm{Spd}\,E_1 \times \ldots \times$ $\mathrm{Spd}\,E_m$ parametrizes over a perfectoid space S of characteristic p with untilts $S^\sharp_i = \mathrm{Spa}(R^\sharp_i, R^{\sharp+}_i)$, $i = 1, \ldots, m$, over E_i, the set of G-torsors $\mathcal{P}_\eta$ on $S\dot\times\mathrm{Spa}\,\mathbf{Q}_p$ together with an isomorphism

$$\varphi_{\mathcal{P}_\eta} : (\mathrm{Frob}^*_S\,\mathcal{P}_\eta)|_{S\dot\times\,\mathrm{Spa}\,\mathbf{Q}_p\backslash\bigcup_{i=1}^m S^\sharp_i} \cong \mathcal{P}_\eta|_{S\dot\times\,\mathrm{Spa}\,\mathbf{Q}_p\backslash\bigcup_{i=1}^m S^\sharp_i}$$

that is meromorphic along the closed Cartier divisor $\bigcup_{i=1}^m S^\sharp_i \subset S\dot\times\mathrm{Spa}\,\mathbf{Q}_p$, and an isomorphism

$$\iota_r : \mathcal{P}_\eta|_{\mathcal{Y}_{[r,\infty)}(S)} \cong G \times \mathcal{Y}_{[r,\infty)}(S)$$

for large enough r under which $\varphi_{\mathcal{P}_\eta}$ gets identified with $b \times \mathrm{Frob}_S$, satisfying the following boundedness condition:

At all geometric rank 1 points, the relative position of $\mathrm{Frob}^*_S\,\mathcal{P}$ and $\mathcal{P}$ at $S^\sharp_i$ is bounded by $\sum_{j|S^\sharp_j=S^\sharp_i}\mu_j$ in the Bruhat order.

We get the following qualitative result.

Proposition 23.5.2. The map

$$\mathrm{Gr}^{\mathrm{tw}}_{G,\mathrm{Spd}\,E_1\times\ldots\times\mathrm{Spd}\,E_m,\leq\mu_\bullet} \to \mathrm{Spd}\,E_1 \times \ldots \times \mathrm{Spd}\,E_m$$

is proper and representable in spatial diamonds.

Proof. We base change to a quasicompact open subspace U of the locally spatial diamond $\mathrm{Spd}\,E_1 \times \ldots \times \mathrm{Spd}\,E_m$. As before, the isomorphism ι_r extends to the open subspace

$$S\dot\times\mathrm{Spa}\,\mathbf{Q}_p \setminus \bigcup_{i=1,\ldots,m;n\geq0} \varphi^{-n}(S^\sharp_i)\,,$$

and is meromorphic along the closed Cartier divisor $\bigcup_{i=1,\ldots,m;n\geq0}\varphi^{-n}(S^\sharp_i)$. Over U, there is some $\epsilon > 0$ such that all $S^\sharp_i$ are contained in $\mathcal{Y}_{[\epsilon,\infty)}(S)$, and then some large enough n_0 so that for all $n \geq n_0$, all $\varphi^{-n}(S_i^\sharp)$ are contained in $\mathcal{Y}_{(0,\epsilon/p]}(S)$. In that case, the G-torsor $\mathcal{P}_\eta$ is determined by the modification of the trivial G-torsor at all $\varphi^{-n}(S_i^\sharp)$ for $i = 1, \ldots, m$ and $n = 0, \ldots, n_0 - 1$ (as afterwards it is continued φ^{-1}-periodically). This gives an embedding

$$\mathrm{Gr}^{\mathrm{tw}}_{G,U,\leq\mu_\bullet} \hookrightarrow \mathrm{Gr}_{G,U\times\varphi^{-1}(U)\times\ldots\times\varphi^{-n_0+1}(U)} \ .$$

Moreover, the image is contained in some Schubert variety, and is closed in there (as the boundedness condition is a closed condition). The result follows. $\qquad\square$

Corollary 23.5.3. There is an étale period map

$$\pi_{GM} : \mathrm{Sht}_{\mathcal{G},b,\{\mu_i\}} \to \mathrm{Gr}^{\mathrm{tw}}_{G,\mathrm{Spd}\,\breve{E}_1\times_k\ldots\times_k\mathrm{Spd}\,\breve{E}_m,\leq\mu_\bullet} \ ;$$

in particular, $\mathrm{Sht}_{\mathcal{G},b,\{\mu_i\}}$ is a locally spatial diamond. Its open image is the admissible locus

$$\mathrm{Gr}^{\mathrm{tw},a}_{G,\mathrm{Spd}\,\breve{E}_1\times_k\ldots\times_k\mathrm{Spd}\,\breve{E}_m,\leq\mu_\bullet} \subset \mathrm{Gr}^{\mathrm{tw}}_{G,\mathrm{Spd}\,\breve{E}_1\times_k\ldots\times_k\mathrm{Spd}\,\breve{E}_m,\leq\mu_\bullet}$$

over which there is a $G(\mathbf{Q}_p)$-torsor $\mathbb{P}_\eta$. The space $\mathrm{Sht}_{\mathcal{G},b,\{\mu_i\}}$ parametrizes $\mathcal{G}(\mathbf{Z}_p)$-lattices $\mathbb{P}$ inside $\overline{\mathbb{P}_\eta}$.

In particular, $\mathrm{Sht}_{\mathcal{G},b,\{\mu_i\}}$ depends on $\mathcal{G}$ only through G and $\mathcal{G}(\mathbf{Z}_p)$, and for any compact open subgroup $K \subset G(\mathbf{Q}_p)$, we can define

$$\mathrm{Sht}_{G,b,\{\mu_i\},K}$$

as parametrizing $\underline{K}$-lattices $\mathbb{P}$ inside $\mathbb{P}_\eta$. This gives rise to a tower

$$(\mathrm{Sht}_{G,b,\{\mu_i\},K})_{K\subset G(\mathbf{Q}_p)}$$

of finite étale covers whose inverse limit is the $G(\mathbf{Q}_p)$-torsor

$$\mathrm{Sht}_{G,b,\{\mu_i\},\infty} = \varprojlim_K \mathrm{Sht}_{G,b,\{\mu_i\},K} \to \mathrm{Gr}^{\mathrm{tw},a}_{G,\mathrm{Spd}\,\breve{E}_1\times_k\ldots\times_k\mathrm{Spd}\,\breve{E}_m,\leq\mu_\bullet} \ .$$

Lecture 24

Local Shimura varieties

In this lecture, we specialize the theory back to the case of local Shimura varieties, and explain the relation with Rapoport-Zink spaces.

24.1 DEFINITION OF LOCAL SHIMURA VARIETIES

We start with a local Shimura datum.

Definition 24.1.1. A local Shimura datum is a triple (G, b, μ) consisting of a reductive group G over $\mathbf{Q}_p$, a conjugacy class μ of minuscule cocharacters $\mathbf{G}_m \to G_{\overline{\mathbf{Q}}_p}$, and $b \in B(G, \mu^{-1})$, i.e., $\nu_b \le (\mu^{-1})^\diamond$ and $\kappa(b) = -\mu^\natural$.

The final condition is explained by the following result.

Proposition 24.1.2 ([Rap18]). Let C be a complete algebraically closed extension of $\mathbf{Q}_p$ with tilt $C^\flat$, and let $\mathcal{E}_b$ be the G-torsor on the Fargues-Fontaine curve $X_{\mathrm{FF},C^\flat}$ corresponding to $b \in B(G)$. Then there is a trivialization of $\mathcal{E}_b|_{X_{\mathrm{FF},C^\flat} \setminus \operatorname{Spec} C}$ whose relative position at $\operatorname{Spec} C$ is given by μ if and only if $b \in B(G, \mu^{-1})$.

Under Proposition 23.3.3, this means that $\mathrm{Sht}_{G,b,\mu,K}$ is nonempty if and only if $b \in B(G, \mu^{-1})$. Thus, we make this assumption.

Recall that E is the field of definition of μ and $\check{E} = E \cdot L$. From Proposition 23.3.3 and Proposition 19.4.2, we see that

$$\pi_{GM} : \mathrm{Sht}_{G,b,\mu,K} \to \mathrm{Gr}_{G,\operatorname{Spd} \check{E}, \le \mu} \cong \mathcal{F}\ell^\diamond_{G,\mu,\check{E}}$$

is étale, where the right-hand side is the diamond corresponding to a smooth rigid space over $\check{E}$. By Theorem 10.4.2, we see that

$$\mathrm{Sht}_{G,b,\mu,K} = \mathcal{M}^\diamond_{G,b,\mu,K}$$

for a unique smooth rigid space $\mathcal{M}_{G,b,\mu,K}$ over $\check{E}$ with an étale map towards $\mathcal{F}\ell_{G,\mu,\check{E}}$. Moreover, the transition maps are finite étale.

Definition 24.1.3. The local Shimura variety associated with (G, b, μ) is the tower

$$(\mathcal{M}_{G,b,\mu,K})_{K \subset G(\mathbf{Q}_p)}$$

of smooth rigid spaces over $\breve{E}$, together with its étale period map to $\mathcal{F}\ell_{G,\mu,\breve{E}}$.

24.2   RELATION TO RAPOPORT-ZINK SPACES

Rapoport-Zink spaces are moduli of deformations of a fixed p-divisible group. After reviewing these, we will show that the diamond associated with the generic fiber of a Rapoport-Zink space is isomorphic to a moduli space of shtukas of the form $\mathrm{Sht}_{G,b,\mu}$ with μ minuscule.

We start with the simplest Rapoport-Zink spaces. In this case, $G = \mathrm{GL}_n$, the cocharacter μ is given by $(1, 1, \ldots, 1, 0, \ldots, 0)$ with d occurences of 1, $0 \le d \le n$, and b corresponds to a p-divisible group $\mathbb{X} = \mathbb{X}_b$ over $\overline{\mathbf{F}}_p$ of characteristic p of dimension d and height n.

We recall from [RZ96] the definition and main properties of Rapoport-Zink spaces.

Definition 24.2.1. Let $\mathrm{Def}_{\mathbb{X}}$ be the functor which assigns to a formal scheme $S/\mathrm{Spf}\,\breve{\mathbf{Z}}_p$ the set of isomorphism classes of pairs (X, ρ), where X/S is a p-divisible group, and $\rho\colon X \times_S \overline{S} \xrightarrow{\sim} \mathbb{X} \times_{\overline{\mathbf{F}}_p} \overline{S}$ is a quasi-isogeny, where $\overline{S} = S \times_{\mathrm{Spf}\,\breve{\mathbf{Z}}_p} \mathrm{Spec}\,\overline{\mathbf{F}}_p$.

Theorem 24.2.2 ([RZ96]). The functor $\mathrm{Def}_{\mathbb{X}}$ is representable by a formal scheme $\mathcal{M}_{\mathbb{X}}$ over $\mathrm{Spf}\,\breve{\mathbf{Z}}_p$, which is formally smooth and locally formally of finite type. Furthermore, all irreducible components of the special fiber of $\mathcal{M}_{\mathbb{X}}$ are proper over $\mathrm{Spec}\,\overline{\mathbf{F}}_p$.

Remark 24.2.3. "Locally formally of finite type" means that locally $\mathcal{M}_{\mathbb{X}}$ is isomorphic to a formal scheme of the form $\mathrm{Spf}\,\breve{\mathbf{Z}}_p[\![T_1, \ldots, T_m]\!]\langle U_1, \ldots, U_r \rangle / I$.

Let $\mathcal{M}_{\mathbb{X},\breve{\mathbf{Q}}_p}$ be the generic fiber of $\mathcal{M}_{\mathbb{X}}$ as an adic space. It is useful to have a moduli interpretation of $\mathcal{M}_{\mathbb{X},\breve{\mathbf{Q}}_p}$.

Proposition 24.2.4 ([SW13, Proposition 2.2.2]). Let $\mathrm{CAff}_{\breve{\mathbf{Q}}_p}^{op}$ be the category opposite to the category of complete Huber pairs over $(\breve{\mathbf{Q}}_p, \breve{\mathbf{Z}}_p)$, equipped with the analytic topology. Then $\mathcal{M}_{\mathbb{X},\breve{\mathbf{Q}}_p}$ is the sheafification of the presheaf on $\mathrm{CAff}_{\breve{\mathbf{Q}}_p}^{op}$ defined by

$$(R, R^+) \mapsto \varinjlim_{R_0 \subset R^+} \mathrm{Def}_{\mathbb{X}}(R_0),$$

where the colimit runs over open and bounded $\breve{\mathbf{Z}}_p$-subalgebras $R_0 \subset R^+$.

Thus, to give a section of $\mathcal{M}_{\mathbb{X}}$ over (R, R^+) is to give a covering of $\mathrm{Spa}(R, R^+)$ by rational subsets $\mathrm{Spa}(R_i, R_i^+)$, and for each i a deformation $(X_i, \rho_i) \in \mathrm{Def}_{\mathbb{X}}(R_{i0})$ over an open and bounded $\check{\mathbf{Z}}_p$-subalgebra $R_{i0} \subset R_i^+$, such that the (X_i, ρ_i) are compatible on overlaps.

Theorem 24.2.5 ([SW13]). There is a natural isomorphism

$$\mathcal{M}_{\mathbb{X}, \check{\mathbf{Q}}_p}^{\diamond} \cong \mathrm{Sht}_{(\mathrm{GL}_n, b, \mu)}$$

as diamonds over $\mathrm{Spd}\, \check{\mathbf{Q}}_p$.

Here, GL_n denotes the group scheme over $\mathbf{Z}_p$, so the right-hand side corresponds to level $\mathrm{GL}_n(\mathbf{Z}_p)$.

Proof. We give a proof using the étale period maps. A different proof is given in the next lecture. First we construct a map

$$\mathcal{M}_{\mathbb{X}, \check{\mathbf{Q}}_p}^{\diamond} \to \mathrm{Sht}_{(\mathrm{GL}_r, b, \mu)} \ .$$

For this, let $S = \mathrm{Spa}(R, R^+)$ be any affinoid perfectoid space over $\check{\mathbf{Q}}_p$, and assume that (X, ρ) is a pair of a p-divisible group X/R^+ and a quasi-isogeny $\rho \colon X \times_{\mathrm{Spf}\, R^+} \mathrm{Spec}\, R^+/p \to \mathbb{X} \times_{\mathrm{Spec}\, \overline{\mathbf{F}}_p} \mathrm{Spec}\, R^+/p$. Note that $R^+ \subset R$ is automatically open and bounded, and it is enough to define the map before sheafification in the sense of the previous proposition.

By Grothendieck-Messing theory, if $EX \to X$ is the universal vector extension, ρ induces an isomorphism

$$\mathrm{Lie}\, EX[\tfrac{1}{p}] \cong M(\mathbb{X}) \otimes_{\check{\mathbf{Q}}_p} R \ ,$$

where $M(\mathbb{X}) \cong \check{\mathbf{Q}}_p^n$ is the covariant rational Dieudonné module of $\mathbb{X}$, on which Frobenius acts as $b\varphi$. On the other hand, we have the natural surjection $\mathrm{Lie}\, EX \to \mathrm{Lie}\, X$, which defines a point of the usual Grassmannian $\mathrm{Gr}(d, n)_{\check{\mathbf{Q}}_p}$ of d-dimensional quotients of an n-dimensional vector space.

As in the previous lecture, we can use this to build a vector bundle $\mathcal{E}_\eta$ on $S^\flat \dot\times \mathrm{Spa}\, \mathbf{Q}_p$ with an isomorphism

$$(\mathrm{Frob}_{S^\flat}^* \mathcal{E}_\eta)|_{(S^\flat \dot\times \mathrm{Spa}\, \mathbf{Q}_p) \setminus S} \cong \mathcal{E}_\eta|_{(S^\flat \dot\times \mathrm{Spa}\, \mathbf{Q}_p) \setminus S}$$

and an isomorphism with the equivariant vector bundle corresponding to b over $\mathcal{Y}_{[r, \infty)}(S)$ for r large. For r small, the φ^{-1}-equivariant vector bundle $\mathcal{E}_\eta|_{\mathcal{Y}_{(0, r]}(S)}$ gives rise to another vector bundle $\mathcal{F}$ on the Fargues-Fontaine curve $\mathcal{X}_{\mathrm{FF}, S^\flat}$, which is a modification

$$0 \to \mathcal{F} \to \mathcal{E}^b \to i_{\infty *} \mathrm{Lie}\, X \to 0 \ ,$$

where $i_\infty \colon S \hookrightarrow \mathcal{X}_{\mathrm{FF}, S^\flat}$ is the closed Cartier divisor corresponding to the untilt

S.

On the other hand, considering the pro-étale $\mathbf{Z}_p$-local system $\mathbb{T} = \mathbb{T}(X)$ given by the p-adic Tate module of X, we have a natural map

$$\mathbb{T} \times^{\underline{\mathbf{Z}_p}} \mathcal{O}_{X_{\mathrm{FF},S^\flat}} \to \mathcal{E}^b \ .$$

Indeed, it suffices to construct this pro-étale locally on S, so we can assume $\mathbb{T}$ is trivial. For any $t \in \mathbb{T}$, we get a map $\mathbf{Q}_p/\mathbf{Z}_p \to X$ of p-divisible groups over R^+, which gives a corresponding map of covariant Dieudonné modules over A_{crys}, giving the desired map $\mathcal{O}_{X_{\mathrm{FF},S^\flat}} \to \mathcal{E}^b$.

By [SW13, Proposition 5.1.6], the map

$$\mathbb{T} \times^{\underline{\mathbf{Z}_p}} \mathcal{O}_{X_{\mathrm{FF},S^\flat}} \to \mathcal{E}^b$$

factors over $\mathcal{F} \subset \mathcal{E}^b$, and induces an isomorphism

$$\mathcal{F} \cong \mathbb{T} \times^{\underline{\mathbf{Z}_p}} \mathcal{O}_{X_{\mathrm{FF},S^\flat}} \ .$$

More precisely, the reference gives this result over geometric points, which implies the factorization. Moreover, the reference implies that $\mathcal{F}$ is trivial at all geometric fibers, and thus corresponds under Theorem 22.3.1 to a pro-étale $\mathbf{Q}_p$-local system $\mathbb{F}$. But then we get a map $\mathbb{T}[\frac{1}{p}] \to \mathbb{F}$ which is an isomorphism at all geometric points, and thus an isomorphism.

In particular, $\mathbb{T}$ gives a pro-étale $\mathbf{Z}_p$-lattice in the pro-étale local system corresponding to $\mathcal{E}_\eta|_{\mathcal{Y}_{(0,r]}(S)}$, so Proposition 23.3.3 gives a natural map

$$\mathcal{M}^{\diamondsuit}_{\mathbb{X},\breve{\mathbf{Q}}_p} \to \mathrm{Sht}_{(\mathrm{GL}_r,b,\mu)} \ ,$$

which in fact by definition commutes with the period maps to

$$\mathrm{Gr}(d,n)^{\diamondsuit}_{\breve{\mathbf{Q}}_p} \cong \mathrm{Gr}_{\mathrm{GL}_n,\mathrm{Spd}\,\breve{\mathbf{Q}}_p,\leq\mu} \ .$$

By Grothendieck-Messing theory, the period map

$$\pi_{GM} : \mathcal{M}_{\mathbb{X},\breve{\mathbf{Q}}_p} \to \mathrm{Gr}(d,n)_{\breve{\mathbf{Q}}_p}$$

is étale; cf. [RZ96, Proposition 5.17]. By [SW13, Theorem 6.2.1], the image of the map

$$\mathcal{M}_{\mathbb{X},\breve{\mathbf{Q}}_p} \to \mathrm{Gr}(d,n)_{\breve{\mathbf{Q}}_p}$$

is precisely the admissible open subspace $\mathrm{Gr}(d,n)^a_{\breve{\mathbf{Q}}_p} \subset \mathrm{Gr}(d,n)_{\breve{\mathbf{Q}}_p}$; this was first proved by Faltings, [Fal10]. By [RZ96, Proposition 5.37], the nonempty geometric fibers are given by $\mathrm{GL}_n(\mathbf{Q}_p)/\mathrm{GL}_n(\mathbf{Z}_p)$ by parametrizing lattices in the universal $\mathbf{Q}_p$-local system. These facts together imply the theorem. $\square$

Moreover, the proof shows that the tower of coverings

$$(\mathcal{M}^{\diamond}_{\mathbb{X},\breve{\mathbf{Q}}_p,K})_{K \subset \mathrm{GL}_n(\mathbf{Q}_p)} \cong (\mathrm{Sht}_{\mathrm{GL}_n,b,\mu,K})_{K \subset \mathrm{GL}_n(\mathbf{Q}_p)}$$

are also naturally identified, where the tower on the left is defined in [RZ96, Section 5.34]; indeed, we have identified the two universal pro-étale $\mathbf{Q}_p$-local systems on the admissible locus in the Grassmannian. Thus, we have an isomorphism of towers

$$(\mathcal{M}_{\mathbb{X},\breve{\mathbf{Q}}_p,K})_{K \subset \mathrm{GL}_n(\mathbf{Q}_p)} \cong (\mathcal{M}_{\mathrm{GL}_n,b,\mu,K})_{K \subset \mathrm{GL}_n(\mathbf{Q}_p)}$$

between the generic fiber of the Rapoport-Zink space and the local Shimura variety.

24.3   GENERAL EL AND PEL DATA

Now we extend the results to general EL and PEL data. We fix EL data $(B, V, \mathcal{O}_B, \mathcal{L})$ as well as $(\, , \,)$ and an involution $b \mapsto b^*$ on B in the PEL case, as in the appendix to Lecture 21.

It is convenient to introduce the following definition, which is an analogue of Definition 21.6.1.

Definition 24.3.1. Let R be a $\mathbf{Z}_p$-algebra on which p is nilpotent. A chain of $\mathcal{O}_B$-p-divisible groups of type $(\mathcal{L})$ over R is a functor $\Lambda \mapsto X_\Lambda$ from $\mathcal{L}$ to p-divisible groups over R with $\mathcal{O}_B$-action, together with an isomorphism $\theta_b : X_\Lambda^b \cong X_{b\Lambda}$ for all $b \in B^\times$ normalizing $\mathcal{O}_B$, where $X_\Lambda^b = X_\Lambda$ with $\mathcal{O}_B$-action conjugated by b, satisfying the following conditions.

1. The periodicity isomorphisms θ_b commutes with the transition maps $X_\Lambda \to X_{\Lambda'}$.
2. For all $b \in B^\times \cap \mathcal{O}_B$ normalizing $\mathcal{O}_B$, the composition $M_\Lambda^b \cong M_{b\Lambda} \to M_\Lambda$ is multiplication by b.
3. If M_Λ is the Lie algebra of the universal vector extension of X_Λ, then the M_Λ define a chain of $\mathcal{O}_B \otimes_{\mathbf{Z}_p} R$-modules of type $(\mathcal{L})$.

Moreover, in the case of PEL data, a polarized chain of $\mathcal{O}_B$-p-divisible groups of type $(\mathcal{L})$ over R is a chain $\Lambda \mapsto X_\Lambda$ of type $(\mathcal{L})$ as before, together with a $\mathbf{Z}_p$-local system $\mathbb{L}$ on $\mathrm{Spec}\, R$ and an antisymmetric isomorphism of chains

$$X_\Lambda \cong X_{\Lambda^*}^* \otimes_{\mathbf{Z}_p} \mathbb{L} \, .$$

There is an obvious notion of quasi-isogeny between (polarized) chains of $\mathcal{O}_B$-p-divisible groups of type $(\mathcal{L})$. If X_Λ is a (polarized) chain of $\mathcal{O}_B$-p-divisible groups of type $(\mathcal{L})$, then M_Λ defines a (polarized) chain of $\mathcal{O}_B \otimes_{\mathbf{Z}_p} R$-modules, where the auxiliary line bundle L is given by $R \otimes_{\mathbf{Z}_p} \mathbb{L}$.

Moreover, we fix a conjugacy class μ of minuscule cocharacters $\mu : \mathbf{G}_{\mathrm{m}} \to G_{\overline{\mathbf{Q}}_p}$, defined over the reflex field E, satisfying the conditions of Section 21.4. In this case, we have the local model $\mathrm{M}^{\mathrm{loc,naive}}_{(\mathcal{G},\mu)}$.

Definition 24.3.2. Let R be a $\mathbf{Z}_p$-algebra on which p is nilpotent, and let X_Λ be a chain of $\mathcal{O}_B$-p-divisible groups of type $(\mathcal{L})$ over R. Then X_Λ is $\mathrm{M}^{\mathrm{loc,naive}}_{(\mathcal{G},\mu)}$-admissible if the system of quotients $t_\Lambda = \mathrm{Lie}\, X_\Lambda$ of M_Λ defines a point of $\mathrm{M}^{\mathrm{loc,naive}}_{(\mathcal{G},\mu)}$, after locally fixing an isomorphism of chains $M_\Lambda \cong \Lambda \otimes_{\mathbf{Z}_p} R$. Equivalently, for all $\Lambda \in \mathcal{L}$, the determinant condition holds, i.e., there is an identity

$$\det{}_R(a; \mathrm{Lie}\, X_\Lambda) = \det{}_{\overline{\mathbf{Q}}_p}(a; V_1)$$

of polynomial maps $\mathcal{O}_B \to R$.

Finally, we assume that G is connected and $\mathcal{G}$ is parahoric, and we fix an element $b \in B(G, \mu^{-1})$.[1] This is slightly stronger than the assumption made by Rapoport-Zink: We demand $\nu_b \leq (\mu^{-1})^\diamond$ and $\kappa(b) = -\mu^\natural$, where Rapoport-Zink only assume $\nu_b \leq (\mu^{-1})^\diamond$. However, it follows from [RZ96, Proposition 1.20, Proposition 5.33] that the generic fibers of their moduli problems are actually empty unless in addition $\kappa(b) = -\mu^\natural$, so we assume this from the start.

Now b defines a chain $\mathbb{X}_{b,\Lambda}$ of $\mathcal{O}_B$-p-divisible groups of type $(\mathcal{L})$ over $\overline{\mathbf{F}}_p$ up to quasi-isogeny, in the weak sense that each $\mathbb{X}_{b,\Lambda}$ (and $\mathbb{L}_b$) is only defined up to quasi-isogeny, rather than up to simultaneous quasi-isogeny. In particular, no condition on M_Λ is imposed.

Let $\mathcal{D} = (B, V, \mathcal{O}_B, \mathcal{L}, (\ ,\ ), *, b, \mu)$ denote the set of all data fixed.

Definition 24.3.3 ([RZ96, Definition 3.21]). The moduli problem $\mathcal{M}_\mathcal{D}$ over $\mathrm{Spf}\, \breve{\mathcal{O}}$ parametrizes over a p-torsion $\breve{\mathcal{O}}$-algebra R the following data.

1. A chain of $\mathcal{O}_B$-p-divisible groups X_Λ of type $(\mathcal{L})$ over R that is $\mathrm{M}^{\mathrm{loc,naive}}_{(\mathcal{G},\mu)}$-admissible, and
2. a quasi-isogeny

$$\rho_\Lambda : \mathbb{X}_{b,\Lambda} \times_{\mathrm{Spec}\,\overline{\mathbf{F}}_p} \mathrm{Spec}\, R/p \to X_\Lambda \times_{\mathrm{Spec}\, R} \mathrm{Spec}\, R/p$$

of chains of $\mathcal{O}_B$-p-divisible groups of type $(\mathcal{L})$.

Proposition 24.3.4. The forgetful map $\mathcal{M}_\mathcal{D} \to \prod_{\Lambda \in \mathcal{L}} \mathcal{M}_{\mathbb{X}_{b,\Lambda}}$ is a closed immersion. In fact, the same holds true when the product on the right-hand side is replaced by a large enough finite product. In particular, $\mathcal{M}_\mathcal{D}$ is representable by a formal scheme locally formally of finite type over $\mathrm{Spf}\, \breve{\mathcal{O}}$.

---

[1] In the different normalization $b_{RZ} = \chi(p)b$, $\mu_{RZ} = \chi \mu^{-1}$ of [RZ96], this corresponds to $b_{RZ} \in B(G, \mu_{RZ}^{-1})$.

Proof. This follows from the proof of [RZ96, Theorem 3.25].   □

Corollary 24.3.5. There is a natural isomorphism

$$\mathcal{M}_{\mathcal{D},\breve{E}} \cong \mathcal{M}_{G,b,\mu,\mathcal{G}(\mathbf{Z}_p)}$$

of smooth rigid spaces over $\breve{E}$, identifying the natural $\mathcal{G}(\mathbf{Z}_p)$-torsors.

Proof. It is enough to give an isomorphism of their diamonds, over Spd E. As both sides admit natural closed immersions into

$$\prod_\Lambda \mathcal{M}_{\mathbb{X}_{b,\Lambda},\breve{E}} \cong \prod_\Lambda \mathcal{M}_{\mathrm{GL}(\Lambda),\rho_\Lambda(b),\rho_\Lambda(\mu)} ,$$

where $\rho_\Lambda : \mathcal{G} \to \mathrm{GL}(\Lambda)$ is the natural representation, it is enough to see that their geometric rank 1 points agree under this embedding.

Thus, let $\mathcal{O}_C$ be the ring of integers in an algebraically closed nonarchimedean field $C/\breve{E}$, and let $A_{\mathrm{inf}} = W(\mathcal{O}_{C^\flat})$ as usual. A chain of $\mathcal{O}_B$-p-divisible groups X_Λ of type $(\mathcal{L})$ over $\mathcal{O}_C$ defines, via the Breuil-Kisin-Fargues module, a chain of $\mathcal{O}_B \otimes_{\mathbf{Z}_p} A_{\mathrm{inf}}$-modules N_Λ. We claim that N_Λ is automatically of type $(\mathcal{L})$. By assumption, $M_\Lambda = N_\Lambda/\xi$ is of type $(\mathcal{L})$. In particular, each M_Λ is isomorphic to $\Lambda \otimes_{\mathbf{Z}_p} \mathcal{O}_C$. This implies that $N_\Lambda \cong \Lambda \otimes_{\mathbf{Z}_p} A_{\mathrm{inf}}$ by lifting generators. Similarly, for consecutive lattices $\Lambda' \subset \Lambda$, we have by assumption $M_\Lambda/M_{\Lambda'} \cong \Lambda/\Lambda' \otimes_{\mathbf{Z}_p} \mathcal{O}_C$. On the other hand, ξ is a nonzerodivisor on $N_\Lambda/N_{\Lambda'}$, as the latter is a p-torsion module of projective dimension 1 and (p,ξ) is a regular sequence. Thus, similarly, we can lift generators to get the isomorphism $N_\Lambda/N_{\Lambda'} \cong \Lambda/\Lambda' \otimes_{\mathbf{Z}_p} A_{\mathrm{inf}}$.

Thus, we see that a $\mathbb{M}^{\mathrm{loc,naive}}_{(\mathcal{G},\mu)}$-admissible chain of $\mathcal{O}_B$-p-divisible groups X_Λ of type $(\mathcal{L})$ over $\mathcal{O}_C$ is equivalent to a $\mathcal{G}$-torsor $\mathcal{P}$ over A_{inf} together with an isomorphism

$$\mathrm{Frob}^*_{\mathcal{O}_{C^\flat}} \mathcal{P}|_{\mathrm{Spec}\, A_{\mathrm{inf}}[\frac{1}{\varphi(\xi)}]} \cong \mathcal{P}|_{\mathrm{Spec}\, A_{\mathrm{inf}}[\frac{1}{\varphi(\xi)}]}$$

such that the relative position at the untilt C is given by μ. The quasi-isogeny over $\mathcal{O}_C/p$ is equivalent to a φ-equivariant isomorphism of $\mathcal{G}$-torsors over B^+_{crys}.

On the other hand, the moduli spaces of shtukas parametrize the similar structures, noting that everything extends uniquely from $\mathrm{Spa}\, C^\flat \dot{\times} \mathrm{Spa}\, \mathbf{Z}_p$ to $\mathrm{Spec}\, A_{\mathrm{inf}}$ by Theorem 14.1.1 and Theorem 21.2.2.   □

Lecture 25

Integral models of local Shimura varieties

In this final lecture, we explain an application of the theory developed in these lectures towards the problem of understanding integral models of local Shimura varieties. As a specific example, we will resolve conjectures of Kudla-Rapoport-Zink, [KRZ], and Rapoport-Zink, [RZ17], that two Rapoport-Zink spaces associated with very different PEL data are isomorphic. The basic reason is that the corresponding group-theoretic data are related by an exceptional isomorphism of groups, so such results follow once one has a group-theoretic characterization of Rapoport-Zink spaces. The interest in these conjectures comes from the observation of Kudla-Rapoport-Zink [KRZ] that one can obtain a moduli-theoretic proof of Čerednik's p-adic uniformization for Shimura curves, [Cer76], using these exceptional isomorphisms.

25.1 DEFINITION OF THE INTEGRAL MODELS

We define integral models of local Shimura varieties as v-sheaves. Fix local Shimura data (G, b, μ) along with a parahoric model $\mathcal{G}$ of G; later, we will discuss the case when only $\mathcal{G}^\circ$ is parahoric. There is the problem of local models; more precisely, we need to extend the closed subspace $\mathrm{Gr}_{G,\mathrm{Spd}\,E,\mu} \subset \mathrm{Gr}_{G,\mathrm{Spd}\,E}$ to a closed subspace of $\mathrm{Gr}_{\mathcal{G},\mathrm{Spd}\,\mathcal{O}_E}$. Conjecture 21.4.1 would give us a canonical choice. For the moment, we allow an arbitrary choice, so we fix some closed subfunctor

$$\mathcal{M}^{\mathrm{loc}}_{(\mathcal{G},\mu)} \subset \mathrm{Gr}_{\mathcal{G},\mathrm{Spd}\,\mathcal{O}_E}$$

proper over $\mathrm{Spd}\,\mathcal{O}_E$, with generic fiber $\mathrm{Gr}_{G,\mathrm{Spd}\,E,\mu}$. We assume that $\mathcal{M}^{\mathrm{loc}}_{(\mathcal{G},\mu)}$ is stable under the $L^+\mathcal{G}$-action. In other words, whenever we have two $\mathcal{G}$-torsors $\mathcal{P}_1$, $\mathcal{P}_2$ over $B^+_{\mathrm{dR}}(R^\sharp)$ together with an isomorphism over $B_{\mathrm{dR}}(R^\sharp)$ for some perfectoid Tate-$\mathcal{O}_E$-algebra $R^\sharp$, we can ask whether the relative position of $\mathcal{P}_1$ and $\mathcal{P}_2$ is bounded by $\mathcal{M}^{\mathrm{loc}}_{(\mathcal{G},\mu)}$.

Definition 25.1.1. Let (G, b, μ) be a local Shimura datum and let $\mathcal{G}$ be a parahoric model of G over $\mathbf{Z}_p$. The integral model $\mathcal{M}^{\mathrm{int}}_{(\mathcal{G},b,\mu)}$ over $\mathrm{Spd}\,\breve{\mathcal{O}}$ of the local Shimura variety $\mathcal{M}_{(\mathcal{G},b,\mu)}$ is the functor sending $S = \mathrm{Spa}(R, R^+) \in \mathrm{Perf}_k$

to the set of untilts $S^\sharp = \mathrm{Spa}(R^\sharp, R^{\sharp+})$ over $\breve{\mathcal{O}}$ together with a $\mathcal{G}$-torsor $\mathcal{P}$ on $S \dot\times \mathrm{Spa}\,\mathbf{Z}_p$ and an isomorphism

$$\varphi_{\mathcal{P}} : (\mathrm{Frob}_S^* \, \mathcal{P})|_{S \dot\times \mathrm{Spa}\,\mathbf{Z}_p \setminus S^\sharp} \cong \mathcal{P}|_{S \dot\times \mathrm{Spa}\,\mathbf{Z}_p \setminus S^\sharp}$$

that is meromorphic along the closed Cartier divisor $S^\sharp \subset S \dot\times \mathrm{Spa}\,\mathbf{Z}_p$, and an isomorphism

$$\iota_r : \mathcal{P}|_{\mathcal{Y}_{[r,\infty)}(S)} \cong G \times \mathcal{Y}_{[r,\infty)}(S)$$

for large enough r (for an implicit choice of pseudouniformizer $\varpi \in R$) under which $\varphi_{\mathcal{P}}$ gets identified with $b \times \mathrm{Frob}_S$, such that the relative position of $\mathrm{Frob}_S^* \, \mathcal{P}$ and $\mathcal{P}$ at $S^\sharp$ is bounded by $\mathcal{M}_{(\mathcal{G},\mu)}^{\mathrm{loc}}$.

The final boundedness condition can be checked on geometric rank 1 points. The first order of business is to show that in the simplest case of GL_n, this agrees with the moduli space defined by Rapoport-Zink.

Theorem 25.1.2. Let $G = \mathrm{GL}_n$, $\mu = (1, \ldots, 1, 0, \ldots, 0) \in (\mathbf{Z}^n)^+$ and b correspond to a p-divisible group $\mathbb{X} = \mathbb{X}_b$ of dimension d and height n over $\overline{\mathbf{F}}_p$. Choose the local model

$$\mathcal{M}_{(\mathrm{GL}_n,\mu)}^{\mathrm{loc}} = \mathrm{Gr}_{\mathrm{GL}_n, \mathrm{Spd}\,\mathbf{Z}_p, \mu} \ .$$

Then there is a natural isomorphism

$$\mathcal{M}_{\mathbb{X}}^{\Diamond} \cong \mathcal{M}_{(\mathrm{GL}_n,b,\mu)}^{\mathrm{int}}$$

of v-sheaves over $\mathrm{Spd}\,\breve{\mathbf{Z}}_p$.

Proof. First, we construct a map. Take a perfectoid Tate-Huber pair (R, R^+) over $\breve{\mathbf{Z}}_p$ with a pseudo-uniformizer $\varpi \in R^+$ dividing p, and let X/R^+ be a p-divisible group with a quasi-isogeny $\rho : X \times_{\mathrm{Spec}\,R^+} \mathrm{Spec}\,R^+/\varpi \to \mathbb{X} \times_{\mathrm{Spec}\,\overline{\mathbf{F}}_p} \mathrm{Spec}\,R^+/\varpi$.[1] By Theorem 17.5.2, the p-divisible group X is equivalent to a Breuil-Kisin-Fargues module M over $A_{\inf}(R^+)$, with $\varphi_M : M[\frac{1}{\xi}] \cong M[\frac{1}{\varphi(\xi)}]$ satisfying $M \subset \varphi_M(M) \subset \frac{1}{\varphi(\xi)} M$. The quasi-isogeny is equivalent to a φ-equivariant isomorphism

$$M \otimes_{A_{\inf}(R^+)} B_{\mathrm{crys}}^+(R^+/\varpi) \cong M(\mathbb{X}) \otimes_{W(\overline{\mathbf{F}}_p)} B_{\mathrm{crys}}^+(R^+/\varpi)$$

by Theorem 15.2.3. Restricting from $A_{\inf}(R^+)$ to $\mathrm{Spa}\,R \dot\times \mathrm{Spa}\,\mathbf{Z}_p$, we get the desired map to $\mathcal{M}_{(\mathrm{GL}_n,b,\mu)}^{\mathrm{int}}$.

Now we check that this induces a bijection on geometric points, so assume

---

[1]The quasi-isogeny is defined only over R^+/ϖ, not over R^+/p, as a $\mathrm{Spf}\,R^+$-valued point is the same as a compatible system of $\mathrm{Spec}\,R^+/\varpi^n$-valued points, and the quasi-isogeny deforms uniquely from R^+/ϖ to $R^+/(\varpi^n, p)$.

that $R^+ = \mathcal{O}_C$ is the ring of integers in some algebraically closed nonar-chimedean field $C/\mathbf{Z}_p$. Then the result follows from Theorem 14.1.1 if C is of characteristic 0. In general, the same argument shows that the shtuka over $\operatorname{Spa} C^\flat \dot\times \operatorname{Spa} \mathbf{Z}_p$ extends uniquely to a finite free A_{inf}-module with Frobenius, and the result follows.

Thus, the map is an injection. To prove surjectivity, it is enough to prove that if C_i, $i \in I$, is any set of algebraically closed nonarchimedean fields over $\mathbf{Z}_p$, and one takes $R^+ = \prod_i \mathcal{O}_{C_i}$ and $R = R^+[\frac{1}{\varpi}]$ for some choice of pseudo-uniformizers $\varpi = (\varpi_i)_i$ dividing p, then for any map $\operatorname{Spa}(R, R^+) \to \mathcal{M}^{\mathrm{int}}_{(\mathrm{GL}_n, b, \mu)}$ factors over $\mathcal{M}^{\Diamond}_{\mathbb{X}}$. By the case of geometric points, we get pairs (X_i, ρ_i) over $\mathcal{O}_{C_i}$. This defines a p-divisible group X over $R^+ = \prod \mathcal{O}_{C_i}$, which has some Breuil-Kisin-Fargues module $M(X)$ over $A_{\mathrm{inf}}(R^+)$. On the other hand, over $\operatorname{Spa} A_{\mathrm{inf}}(R^+) \setminus \{p = [\varpi] = 0\}$, we have a natural vector bundle $\mathcal{E}$ given by gluing the given shtuka over $\mathcal{Y}_{[0,\infty)} = \operatorname{Spa} A_{\mathrm{inf}}(R^+) \setminus \{[\varpi] = 0\}$ with the vector bundle $\mathcal{E}^\flat$ over $\mathcal{Y}_{[r,\infty]}$ along ι_r. Note that $\mathcal{E}$ produces naturally a finite projective $A_{\mathrm{inf}}(R^+)[\frac{1}{p}]$-module N as in Section 14.3, together with an open and bounded submodule $M' \subset N$ for the p-adic topology on $A_{\mathrm{inf}}(R^+)$, where M' is the set of global sections of $\mathcal{E}$. Now for each i, M' maps into $M(X_i)$, giving a map $M' \to M(X)$ by taking the product. On the other hand, $M(X)$ maps back to N, as for each i, $M(X_i)$ maps into the corresponding component M'_i (and is in fact equal to it), and these components are bounded in N, so their product defines an element of N. Thus, $M' \subset M(X) \subset N$, and in particular $M(X)[\frac{1}{p}] = N$ as finite projective $A_{\mathrm{inf}}(R^+)[\frac{1}{p}]$-modules.[2]

Now the trivialization ι_r defines an element of

$$(N \otimes B^+_{\mathrm{crys}})^{\varphi=1} = (M(X) \otimes B^+_{\mathrm{crys}})^{\varphi=1},$$

which by Theorem 15.2.3 gives a quasi-isogeny $\rho : \mathbb{X} \times_{\operatorname{Spec} \mathbf{F}_p} \operatorname{Spec} R^+/\varpi \to X \times_{\operatorname{Spf} R^+} \operatorname{Spec} R^+/\varpi$, as desired. $\qquad\square$

Now we can prove the same result in the case of EL and PEL data. Fix data $\mathcal{D}$ as in the last lecture, including a choice of local model, and assume that $\mathcal{G}$ is parahoric.

Corollary 25.1.3. There is a natural isomorphism

$$\mathcal{M}^{\Diamond}_{\mathcal{D}} \cong \mathcal{M}^{\mathrm{int}}_{(\mathcal{G}, b, \mu)}$$

of v-sheaves over $\operatorname{Spd} \breve{\mathcal{O}}$.

---

[2]It follows a posteriori from the theorem that in fact $M' = M(X)$, but we do not prove this a priori.

Proof. The construction of the proof of the previous theorem gives a map

$$\mathcal{M}_{\mathcal{D}}^{\diamond} \to \mathcal{M}_{(\mathcal{G},b,\mu)}^{\text{int}} .$$

The left-hand side admits a closed embedding $\prod_{\Lambda \in \mathcal{L}} \mathcal{M}_{\mathbb{X}_\Lambda}^{\diamond}$, and the right-hand side maps compatibly to this space via the isomorphism of the previous theorem. In particular, the displayed map is a proper injection. Therefore it remains to see that it is surjective on geometric rank 1 points $S = \text{Spa}(C, \mathcal{O}_C)$.

Given a $\mathcal{G}$-torsor $\mathcal{P}$ over $\mathcal{Y}_{[0,\infty)} = \text{Spa}\, C \mathbin{\dot{\times}} \text{Spa}\, \mathbf{Z}_p$ with a trivialization to $\mathcal{P}_b$ over $\mathcal{Y}_{[r,\infty)}$, we can extend uniquely to a $\mathcal{G}$-torsor $\mathcal{P}_Y$ over $\mathcal{Y}_{[0,\infty]}$ together with a trivialization to $\mathcal{P}_b$ over $\mathcal{Y}_{[r,\infty]}$. Now any $\mathcal{G}$-torsor over $\mathcal{Y}_{[0,\infty]}$ extends uniquely to a $\mathcal{G}$-torsor over A_{inf}, by Corollary 21.6.6 and Theorem 14.2.1. But now this defines precisely a point of $\mathcal{M}_{\mathcal{D}}^{\diamond}$, as desired. $\qquad\square$

25.2 THE CASE OF TORI

Assume that $G = T$ is a torus and $\mathcal{T}$ is the connected component of the Néron model of T. In this case $b \in B(T) = \pi_1(T_{\overline{\mathbf{Q}}_p})_\Gamma = X_\Gamma$ is determined by $\mu \in X = \pi_1(T_{\overline{\mathbf{Q}}_p})$. For the local model, we take the map

$$\text{Spd}\, \mathcal{O}_E \to \text{Gr}_{\mathcal{T},\text{Spd}\, \mathcal{O}_E}$$

which is the closure of the map $\text{Spd}\, E \cong \text{Gr}_{\mathcal{T},\text{Spd}\, E,\mu}$.

Let $\underline{T(\mathbf{Q}_p)/\mathcal{T}(\mathbf{Z}_p)}$ denote the disjoint union of copies of $\text{Spd}\, \breve{\mathcal{O}}$ over all elements of $T(\mathbf{Q}_p)/\mathcal{T}(\mathbf{Z}_p)$.

Proposition 25.2.1. If $b = 1$ and $\mu = 0$, there is a natural isomorphism

$$\mathcal{M}_{(\mathcal{T},1,0)}^{\text{int}} \cong \underline{T(\mathbf{Q}_p)/\mathcal{T}(\mathbf{Z}_p)}$$

of v-sheaves over $\text{Spd}\, \breve{\mathbf{Z}}_p$, where both sides are naturally groups over $\text{Spd}\, \breve{\mathbf{Z}}_p$. In general,

$$\mathcal{M}_{(\mathcal{T},b,\mu)}^{\text{int}} / \text{Spd}\, \breve{\mathcal{O}}$$

is a torsor under $\mathcal{M}_{(\mathcal{T},1,0)}^{\text{int}}$.

Proof. The space $\mathcal{M}_{(\mathcal{T},1,0)}^{\text{int}}$ parametrizes shtukas with no legs. In particular, $\mathcal{T}$-torsors over $S \mathbin{\dot{\times}} \text{Spa}\, \mathbf{Z}_p$ with an isomorphism with their Frobenius pullback are equivalent to exact $\otimes$-functors from $\text{Rep}_{\mathbf{Z}_p} \mathcal{T}$ to $\mathbf{Z}_p$-local systems on S, by the Tannakian formalism and the relative version of Proposition 12.3.5. By Proposition 22.6.1, this is equivalent to a pro-étale $\mathcal{T}(\mathbf{Z}_p)$-torsor on S. The trivialization over $\mathcal{Y}_{[r,\infty)}(S)$ is equivalent to a trivialization of the corresponding $T(\mathbf{Q}_p)$-torsor. Thus, $\mathcal{M}_{(\mathcal{T},1,0)}^{\text{int}}$ is given by $\underline{T(\mathbf{Q}_p)/\mathcal{T}(\mathbf{Z}_p)}$.

To see that $\mathcal{M}^{\mathrm{int}}_{(\mathcal{T},b,\mu)} \to \operatorname{Spd} \breve{\mathcal{O}}$ is a torsor under $\mathcal{M}^{\mathrm{int}}_{(\mathcal{T},1,0)}$, note first that it is a quasitorsor, as given to points of $(\mathcal{P}_1,\dots)$, $(\mathcal{P}_2,\dots)$ of $\mathcal{M}^{\mathrm{int}}_{(\mathcal{T},b,\mu)}$, the torsor of isomorphisms between $\mathcal{P}_1$ and $\mathcal{P}_2$ defines a point of $\mathcal{M}^{\mathrm{int}}_{(\mathcal{T},1,0)}$. Moreover, $\mathcal{M}^{\mathrm{int}}_{(\mathcal{T},b,\mu)} \to \operatorname{Spd} \breve{\mathcal{O}}$ is a v-cover as one can find a $\operatorname{Spd} \mathbf{C}_p$-point and any such extends to a $\operatorname{Spa} \mathcal{O}_{\mathbf{C}_p}$-point, as any $\mathcal{T}$-torsor over the punctured $\operatorname{Spec} A_{\mathrm{inf}}$ extends to $\operatorname{Spec} A_{\mathrm{inf}}$ by a theorem of Anschütz, [Ans18]. $\qquad\square$

25.3 NON-PARAHORIC GROUPS

As the conjecture of Kudla-Rapoport-Zink involves non-parahoric level structures, we need to understand what happens when we only assume that $\mathcal{G}^\circ$ is parahoric. The moduli problem $\mathcal{M}^{\mathrm{int}}_{(\mathcal{G},b,\mu)}$ can be defined as before. An essential problem is that in general it is not an integral model of $\mathcal{M}_{G,b,\mu,\mathcal{G}(\mathbf{Z}_p)}$. Namely, Proposition 22.6.1 fails when $\mathcal{G}$ is not parahoric. For this reason, we make the following assumption on $\mathcal{G}$.

Assumption 25.3.1. The map $H^1_{\text{ét}}(\operatorname{Spec} \mathbf{Z}_p, \mathcal{G}) \to H^1_{\text{ét}}(\operatorname{Spec} \mathbf{Q}_p, G)$ is injective.

Note that $H^1_{\text{ét}}(\operatorname{Spec} \mathbf{Z}_p, \mathcal{G}) = H^1_{\text{ét}}(\operatorname{Spec} \mathbf{Z}_p, \mathcal{G}/\mathcal{G}^\circ)$ and that under the Kottwitz map

$$\mathcal{G}(\breve{\mathbf{Z}}_p)/\mathcal{G}^\circ(\breve{\mathbf{Z}}_p) \hookrightarrow \pi_1(G_{\overline{\mathbf{Q}}_p})_I .$$

Taking the coinvariants under Frobenius, the left-hand side becomes $H^1_{\text{ét}}(\operatorname{Spec} \mathbf{Z}_p, \mathcal{G})$, and the right-hand side becomes $\pi_1(G_{\overline{\mathbf{Q}}_p})_\Gamma = B(G)_{\mathrm{basic}}$, which contains $H^1_{\text{ét}}(\operatorname{Spec} \mathbf{Q}_p, G)$ as a subgroup. Thus, the assumption is equivalent to the condition that the displayed map stays injective after taking coinvariants under Frobenius.

Proposition 25.3.2. Under Assumption 25.3.1, the generic fiber of $\mathcal{M}^{\mathrm{int}}_{(\mathcal{G},b,\mu)}$ is given by $\mathcal{M}_{G,b,\mu,\mathcal{G}(\mathbf{Z}_p)}$.

Proof. The essential point is to show that the exact $\otimes$-functor from $\operatorname{Rep}_{\mathbf{Z}_p} \mathcal{G}$ to $\mathbf{Z}_p$-local systems on the generic fiber is pro-étale locally trivial. By the argument of Proposition 22.6.1, it is enough to check that it is trivial at all geometric points. By Assumption 25.3.1, it is in turn enough to check that the corresponding exact $\otimes$-functor to $\mathbf{Q}_p$-vector spaces is trivial. But this corresponds to a semistable G-torsor on the Fargues-Fontaine curve that is a modification of type μ of the G-torsor corresponding to b. As $b \in B(G, \mu^{-1})$, it follows that it is trivial by using Proposition 24.1.2; indeed, otherwise it would be some nontrivial element $b' \in B(G)_{\mathrm{basic}}$, and applying Proposition 24.1.2 to $J_{b'}$ and the $J_{b'}$-torsor corresponding to b which is then a modification of type μ of the trivial $J_{b'}$-torsor and has κ-invariant $\kappa(b) - \kappa(b')$, we see $\kappa(b) - \kappa(b') = -\mu^\natural$. But as $b \in B(G, \mu^{-1})$, we know that $\kappa(b) = -\mu^\natural$, so that $\kappa(b') = 0$, and so b' is trivial.   □

Finally, we assemble everything and prove the conjectures of Rapoport-Zink and Kudla-Rapoport-Zink, [RZ17], [KRZ].

25.4   THE EL CASE

In this section, we consider the EL data of [RZ17] as recalled in the Appendix to Lecture 21. We fix the flat local model considered there. There is a unique element $b \in B(G, \mu^{-1})$, which we fix.

Again, there is the simplest case where $\mu_\varphi = (0, \ldots, 0)$ for all $\varphi \neq \varphi_0$. In this case, the moduli problem

$$\mathcal{M}_{\mathcal{D}} \cong \mathcal{M}_{\mathrm{Dr}}$$

is known as the Drinfeld formal scheme; cf. [RZ96, Sections 3.54–3.77]. It is a regular p-adic formal scheme over $\mathrm{Spf}\,\breve{\mathscr{O}}_F$ whose generic fiber is given by $\mathbf{Z}$ copies of the Drinfeld upper-half space

$$\Omega^n_{\breve{F}} = \mathbb{P}^{n-1}_{\breve{F}} \setminus \bigcup_{H \subset \mathbb{P}^{n-1}_{\breve{F}}} H \,,$$

where H runs over all F-rational hyperplanes. One can single out one copy by fixing the height of the quasiisogeny ρ to be 0. Following [RZ17], we pass to these open and closed subfunctors, which we denote by a superscript ^{0}; this reduces the $J_b(\mathbf{Q}_p)$-action to a $J_b(\mathbf{Q}_p)_1$-action, where $J_b(\mathbf{Q}_p)_1 \subset J_b(\mathbf{Q}_p)$ is the subgroup of all elements whose reduced norm is a p-adic unit.

Theorem 25.4.1. For any $\mathcal{D}$ as in [RZ17], there is a natural $J_b(\mathbf{Q}_p)_1$-equivariant isomorphism of formal schemes

$$\mathcal{M}^0_{\mathcal{D}} \cong \mathcal{M}^0_{\mathrm{Dr}} \times_{\mathrm{Spf}\,\breve{\mathscr{O}}_F} \mathrm{Spf}\,\breve{\mathscr{O}}_E \,.$$

Proof. By [RZ17], the formal scheme $\mathcal{M}^0_{\mathcal{D}}$ is p-adic and flat over $\breve{\mathscr{O}}_E$. By Corollary 25.1.3 and Proposition 18.4.1, it suffices to prove that for any $(\mathcal{G}, b, \mu)$ as in the theorem with associated adjoint data $(\mathcal{G}'_{\mathrm{ad}}, \bar{b}, \bar{\mu})$ (which are the same in all cases), the map

$$(\mathcal{M}^{\mathrm{int}}_{(\mathcal{G}, b, \mu)})^0 \to \mathcal{M}^{\mathrm{int}}_{(\mathcal{G}'_{\mathrm{ad}}, \bar{b}, \bar{\mu})} \times_{\mathrm{Spf}\,\breve{\mathscr{O}}_F} \mathrm{Spf}\,\breve{\mathscr{O}}_E$$

is a closed immersion which is an isomorphism on the generic fiber.[3] Indeed, then in all cases the left-hand side is the closure of the generic fiber of the

---

[3]It is probably an isomorphism.

right-hand side. Here, $\mathcal{G}'_{\mathrm{ad}}$ is the smooth model of G_{ad} whose $\mathbf{Z}_p$-points are the maximal compact subgroup of G_{ad}; this is not parahoric, but it satisfies Assumption 25.3.1, as in fact the Kottwitz map $\mathcal{G}'_{\mathrm{ad}}(\breve{\mathbf{Z}}_p) \to \pi_1(G_{\mathrm{ad},\overline{\mathbf{Q}}_p})_I$ is an isomorphism.

For this, one first checks that the map is quasicompact, which can be done by the same technique of choosing an infinite set of points; we omit this. As the map is partially proper, it remains to prove that it is bijective on $(C, \mathcal{O}_C)$-valued points when C is of characteristic 0, and injective on the special fiber. In characteristic 0, this follows by analysis of the period map: In both cases, the target is the projective space $\mathbb{P}^{n-1}(C)$, the image is Drinfeld's upper half-plane by the admissibility condition, and the fibers are points by the choice of level structures.

It remains to handle points of characteristic p. For this, note that the left-hand side is p-adic, so it is enough to prove that if k is a discrete algebraically closed field of characteristic p, then

$$(\mathcal{M}^{\mathrm{int}}_{(\mathcal{G},b,\mu)})^0(k) \hookrightarrow \mathcal{M}^{\mathrm{int}}_{(\mathcal{G}'_{\mathrm{ad}},\overline{b},\overline{\mu})}(k) \ .$$

Here, the left-hand side parametrizes $\mathcal{G}$-torsors $\mathcal{P}$ over $W(k)$ together with an isomorphism to the $\mathcal{G}$-torsor over $W(k)[\frac{1}{p}]$ given by b, such that the relative position of $\mathcal{P}$ and its image under Frobenius is bounded by μ. This is a closed subset of the affine flag variety, called an affine Deligne-Lusztig variety, which is entirely contained in one connected component. Similarly, the target can be embedded into an affine flag variety for $\mathcal{G}'_{\mathrm{ad}}$. As the induced map of affine flag varieties is injective on connected components (cf. proof of Proposition 21.5.1), we get the result. $\qquad\square$

25.5   THE PEL CASE

Finally, we consider the PEL data of [KRZ]. Again, we take the flat local model considered in the Appendix to Lecture 21, and there is a unique element $b \in B(G, \mu^{-1})$, which we fix.

Lemma 25.5.1. Assumption 25.3.1 is satisfied, i.e., the map

$$H^1_{\mathrm{\acute{e}t}}(\operatorname{Spec} \mathbf{Z}_p, \mathcal{G}) \to H^1_{\mathrm{\acute{e}t}}(\operatorname{Spec} \mathbf{Q}_p, G)$$

is injective.

Proof. If F is unramified over F_0, this is clear as then the special fiber of $\mathcal{G}$ is connected, so assume that F is ramified over F_0. Recall that there is an exact sequence

$$1 \to \operatorname{Res}_{\mathcal{O}_{F_0}/\mathbf{Z}_p} \mathcal{U} \to \mathcal{G} \to \mathbf{G}_{\mathrm{m}} \to 1$$

and $G_{\mathrm{ad}} = \mathrm{Res}_{F_0/\mathbf{Z}_p} U_{\mathrm{ad}}$. It suffices to see that the composite map

$$
\begin{aligned}
H^1_{\text{ét}}(\mathrm{Spec}\,\mathcal{O}_{F_0}, \mathcal{U}) &\to H^1_{\text{ét}}(\mathrm{Spec}\,\mathbf{Z}_p, \mathcal{G}) \to H^1_{\text{ét}}(\mathrm{Spec}\,\mathbf{Q}_p, G) \\
&\to H^1_{\text{ét}}(\mathrm{Spec}\,\mathbf{Q}_p, G_{\mathrm{ad}}) = H^1_{\text{ét}}(F_0, U_{\mathrm{ad}})
\end{aligned}
$$

is injective. But now the Kottwitz map

$$
\mathcal{U}(\breve{F}_0)/\mathcal{U}^\circ(\breve{F}_0) \to \pi_1(U_{\overline{\mathbf{Q}}_p})_{I_{F_0}} = \pi_1(U_{\mathrm{ad},\overline{\mathbf{Q}}_p})
$$

is an isomorphism, where I_{F_0} is the inertia group of F_0; in fact, everything is given by $\mathbf{Z}/2\mathbf{Z}$. Thus, taking coinvariants under Frobenius, the result follows. $\square$

Theorem 25.5.2. For any $\mathcal{D}$ as in [KRZ], there is a natural isomorphism of formal schemes

$$
\mathcal{M}_{\mathcal{D}}^0 \cong \mathcal{M}_{\mathrm{Dr}}^0 \times_{\mathrm{Spf}\,\breve{\mathcal{O}}_{F_0}} \mathrm{Spf}\,\breve{\mathcal{O}}_E \;,
$$

where the superscript ^{0} refers to the open and closed subspaces where the quasi-isogeny ρ has height 0, and the Drinfeld moduli space is the one for $n = 2$ and the field F_0.

Proof. Again, $\mathcal{M}_{\mathcal{D}}^0$ is p-adic and flat over $\breve{\mathcal{O}}_E$ by [KRZ]. We use Corollary 25.1.3 and Proposition 18.4.1 and observe that the corresponding adjoint data $(\mathcal{G}'_{\mathrm{ad}}, \bar{b}, \overline{\mu})$ agree with those in the EL case for $n = 2$ and the field F_0. Thus, it is enough to prove that the map

$$
(\mathcal{M}_{(\mathcal{G},b,\mu)}^{\mathrm{int}})^0 \to \mathcal{M}_{(\mathcal{G}'_{\mathrm{ad}},\bar{b},\overline{\mu})}^{\mathrm{int}} \times_{\mathrm{Spf}\,\breve{\mathcal{O}}_{F_0}} \mathrm{Spf}\,\breve{\mathcal{O}}_E
$$

is a closed immersion which is an isomorphism on the generic fiber; for this, one proceeds as in the EL case. $\square$

Bibliography

[AIP18]   F. Andreatta, A. Iovita, and V. Pilloni, *Le halo spectral*, Ann. Sci. Éc. Norm. Supér. (4) **51** (2018), no. 3, 603–655.

[Ans18]   J. Anschütz, *Extending torsors on the punctured* $\mathrm{Spec}(A_{\mathrm{inf}})$, arXiv:1804.06356.

[BD96]   A. A. Beilinson and V. G. Drinfeld, *Quantization of Hitchin's fibration and Langlands' program*, Algebraic and geometric methods in mathematical physics (Kaciveli, 1993), Math. Phys. Stud., vol. 19, Kluwer Acad. Publ., Dordrecht, 1996, pp. 3–7.

[Ber]   P. Berthelot, *Cohomologie rigide et cohomologie rigide à supports propres*, http://perso.univ-rennes1.fr/pierre.berthelot/ publis/Cohomologie_Rigide_I.pdf.

[Ber80]   ———, *Théorie de Dieudonné sur un anneau de valuation parfait*, Ann. Sci. École Norm. Sup. (4) **13** (1980), no. 2, 225–268.

[Ber93]   V. G. Berkovich, *Étale cohomology for non-Archimedean analytic spaces*, Inst. Hautes Études Sci. Publ. Math. (1993), no. 78, 5–161 (1994).

[Ber08]   L. Berger, *Construction de (ϕ, Γ)-modules: représentations p-adiques et B-paires*, Algebra Number Theory **2** (2008), no. 1, 91–120.

[BGR84]   S. Bosch, U. Güntzer, and R. Remmert, *Non-Archimedean analysis*, Grundlehren der Mathematischen Wissenschaften [Fundamental Principles of Mathematical Sciences], vol. 261, Springer-Verlag, Berlin, 1984, A systematic approach to rigid analytic geometry.

[BL95]   A. Beauville and Y. Laszlo, *Un lemme de descente*, C. R. Acad. Sci. Paris Sér. I Math. **320** (1995), no. 3, 335–340.

[BLR95]   S. Bosch, W. Lütkebohmert, and M. Raynaud, *Formal and rigid geometry. IV. The reduced fibre theorem*, Invent. Math. **119** (1995), no. 2, 361–398.

[BMS18]   B. Bhatt, M. Morrow, and P. Scholze, *Integral p-adic Hodge theory*, Publ. Math. Inst. Hautes Études Sci. **128** (2018), 219–397.

[Bor98]   M. Borovoi, *Abelian Galois cohomology of reductive groups*, Mem. Amer. Math. Soc. **132** (1998), no. 626, viii+50.

[Bro13]   M. Broshi, *G-torsors over a Dedekind scheme*, J. Pure Appl. Algebra **217** (2013), no. 1, 11–19.

[BS15]   B. Bhatt and P. Scholze, *The pro-étale topology for schemes*, Astérisque (2015), no. 369, 99–201.

[BS17]   ———, *Projectivity of the Witt vector affine Grassmannian*, Invent. Math. **209** (2017), no. 2, 329–423.

[BS19]   ———, *Prisms and prismatic cohomology*, arXiv:1905.08229.

[BV18]   K. Buzzard and A. Verberkmoes, *Stably uniform affinoids are sheafy*, J. Reine Angew. Math. **740** (2018), 25–39.

[Cer76]   I. V. Cerednik, *Uniformization of algebraic curves by discrete arithmetic subgroups of* $\mathrm{PGL}_2(k_w)$ *with compact quotient spaces*, Mat. Sb. (N.S.) **100(142)** (1976), no. 1, 59–88, 165.

[Col02]   P. Colmez, *Espaces de Banach de dimension finie*, J. Inst. Math. Jussieu **1** (2002), no. 3, 331–439.

[Con08]   Brian Conrad, *Several approaches to non-Archimedean geometry*, p-adic geometry, Univ. Lecture Ser., vol. 45, Amer. Math. Soc., Providence, RI, 2008, pp. 9–63.

[CS17]   A. Caraiani and P. Scholze, *On the generic part of the cohomology of compact unitary Shimura varieties*, Ann. of Math. (2) **186** (2017), no. 3, 649–766.

[CTS79]   J.-L. Colliot-Thélène and J.-J. Sansuc, *Fibrés quadratiques et composantes connexes réelles*, Math. Ann. **244** (1979), no. 2, 105–134.

[Dem74]   M. Demazure, *Désingularisation des variétés de Schubert généralisées*, Ann. Sci. École Norm. Sup. (4) **7** (1974), 53–88, Collection of articles dedicated to Henri Cartan on the occasion of his 70th birthday, I.

[dJ95]   A. J. de Jong, *Crystalline Dieudonné module theory via formal and rigid geometry*, Inst. Hautes Études Sci. Publ. Math. (1995), no. 82, 5–96 (1996).

[dJvdP96]   J. de Jong and M. van der Put, *Étale cohomology of rigid analytic spaces*, Doc. Math. **1** (1996), No. 01, 1–56.

[Dri80]   V. G. Drinfeld, *Langlands' conjecture for* GL(2) *over functional fields*, Proceedings of the International Congress of Mathematicians

(Helsinki, 1978), Acad. Sci. Fennica, Helsinki, 1980, pp. 565–574.

[Elk73] R. Elkik, *Solutions d'équations à coefficients dans un anneau hensélien*, Ann. Sci. École Norm. Sup. (4) **6** (1973), 553–603 (1974).

[Fal99] G. Faltings, *Integral crystalline cohomology over very ramified valuation rings*, J. Amer. Math. Soc. **12** (1999), no. 1, 117–144.

[Fal02a] _____, *Almost étale extensions*, Astérisque (2002), no. 279, 185–270, Cohomologies p-adiques et applications arithmétiques, II.

[Fal02b] _____, *A relation between two moduli spaces studied by V. G. Drinfeld*, Algebraic number theory and algebraic geometry, Contemp. Math., vol. 300, Amer. Math. Soc., Providence, RI, 2002, pp. 115–129.

[Fal10] _____, *Coverings of p-adic period domains*, J. Reine Angew. Math. **643** (2010), 111–139.

[Far] L. Fargues, *L'isomorphisme entre les tours de Lubin-Tate et de Drinfeld et applications cohomologiques*, L'isomorphisme entre les tours de Lubin-Tate et de Drinfeld, Progr. Math., vol. 262, pp. 1–325.

[Far16] _____, *Geometrization of the local Langlands correspondence: An overview*, arXiv:1602.00999.

[Far17] _____, *G-torseurs en théorie de Hodge p-adique*, to appear in Comp. Math.

[FF18] L. Fargues and J.-M. Fontaine, *Courbes et fibrés vectoriels en théorie de Hodge p-adique*, no. 406, 2018, With a preface by Pierre Colmez.

[FGL08] L. Fargues, A. Genestier, and V. Lafforgue, *L'isomorphisme entre les tours de Lubin-Tate et de Drinfeld*, Progress in Mathematics, vol. 262, Birkhäuser Verlag, Basel, 2008.

[Fon82] J.-M. Fontaine, *Sur certains types de représentations p-adiques du groupe de Galois d'un corps local; construction d'un anneau de Barsotti-Tate*, Ann. of Math. (2) **115** (1982), no. 3, 529–577.

[Fon94] _____, *Le corps des périodes p-adiques*, Astérisque (1994), no. 223, 59–111, With an appendix by Pierre Colmez, Périodes p-adiques (Bures-sur-Yvette, 1988).

[Fon13] _____, *Perfectoïdes, presque pureté et monodromie-poids (d'après Peter Scholze)*, Astérisque (2013), no. 352, Exp. No. 1057, x, 509–534, Séminaire Bourbaki. Vol. 2011/2012. Exposés 1043–1058.

[FS] L. Fargues and P. Scholze, *Geometrization of the local Langlands correspondence*, in preparation.

[FW79]   J.-M. Fontaine and J.-P. Wintenberger, *Extensions algébrique et corps des normes des extensions APF des corps locaux*, C. R. Acad. Sci. Paris Sér. A-B **288** (1979), no. 8, A441–A444.

[Gör01]   U. Görtz, *On the flatness of models of certain Shimura varieties of PEL-type*, Math. Ann. **321** (2001), no. 3, 689–727.

[GR03]   O. Gabber and L. Ramero, *Almost ring theory*, Lecture Notes in Mathematics, vol. 1800, Springer-Verlag, Berlin, 2003.

[GR16]   ———, *Foundations of almost ring theory*, 2016, arXiv:math/0409584v11.

[Hed14]   M. H. Hedayatzadeh, *Exterior powers of π-divisible modules over fields*, J. Number Theory **138** (2014), 119–174.

[Hoc69]   M. Hochster, *Prime ideal structure in commutative rings*, Trans. Amer. Math. Soc. **142** (1969), 43–60.

[HP04]   U. Hartl and R. Pink, *Vector bundles with a Frobenius structure on the punctured unit disc*, Compos. Math. **140** (2004), no. 3, 689–716.

[Hub93]   R. Huber, *Continuous valuations*, Math. Z. **212** (1993), no. 3, 455–477.

[Hub94]   ———, *A generalization of formal schemes and rigid analytic varieties*, Math. Z. **217** (1994), no. 4, 513–551.

[Hub96]   ———, *Étale cohomology of rigid analytic varieties and adic spaces*, Aspects of Mathematics, E30, Friedr. Vieweg & Sohn, Braunschweig, 1996.

[HV11]   U. Hartl and E. Viehmann, *The Newton stratification on deformations of local G-shtukas*, J. Reine Angew. Math. **656** (2011), 87–129.

[JN16]   C. Johansson and J. Newton, *Extended eigenvarieties for overconvergent cohomology*, arXiv:1604.07739.

[Ked04]   K. S. Kedlaya, *A p-adic local monodromy theorem*, Ann. of Math. (2) **160** (2004), no. 1, 93–184.

[Ked16]   ———, *Noetherian properties of Fargues-Fontaine curves*, Int. Math. Res. Not. IMRN (2016), no. 8, 2544–2567.

[Ked18]   ———, *On commutative nonarchimedean Banach fields*, Doc. Math. **23** (2018), 171–188.

[Ked19a]   ———, *Sheaves, stacks, and shtukas*, Perfectoid Spaces: Lectures from the 2017 Arizona Winter School (Bryden Cais, ed.), Mathematical Surveys and Monographs, vol. 242, Amer. Math. Soc., Providence, RI, 2019, pp. 58–205.

[Ked19b]   _____, *Some ring-theoretic properties of* A_{inf}, arXiv:1602.09016.

[Kir18]   D. Kirch, *Construction of a Rapoport-Zink space for* GU(1,1) *in the ramified 2-adic case*, Pacific J. Math. **293** (2018), no. 2, 341–389.

[Kis06]   M. Kisin, *Crystalline representations and F-crystals*, Algebraic geometry and number theory, Progr. Math., vol. 253, Birkhäuser Boston, Boston, MA, 2006, pp. 459–496.

[KL15]   K. S. Kedlaya and R. Liu, *Relative p-adic Hodge theory: Foundations*, Astérisque (2015), no. 371, 239.

[KL16]   _____, *Relative p-adic Hodge theory: Imperfect period rings*, arXiv:1602.06899.

[Kot85]   R. Kottwitz, *Isocrystals with additional structure*, Compositio Math. **56** (1985), no. 2, 201–220.

[KP18]   M. Kisin and G. Pappas, *Integral models of Shimura varieties with parahoric level structure*, Publ. Math. Inst. Hautes Études Sci. **128** (2018), 121–218.

[KRZ]   S. Kudla, M. Rapoport, and T. Zink, *On Čerednik uniformization*, in preparation.

[Laf97]   L. Lafforgue, *Chtoucas de Drinfeld et conjecture de Ramanujan-Petersson*, Astérisque (1997), no. 243, ii+329.

[Laf02]   _____, *Chtoucas de Drinfeld et correspondance de Langlands*, Invent. Math. **147** (2002), no. 1, 1–241.

[Laf18]   V. Lafforgue, *Chtoucas pour les groupes réductifs et paramétrisation de Langlands globale*, J. Amer. Math. Soc. **31** (2018), no. 3, 719–891.

[Lau]   E. Lau, *On generalized D-shtukas*, Thesis, 2004.

[Lau13]   _____, *Smoothness of the truncated display functor*, J. Amer. Math. Soc. **26** (2013), no. 1, 129–165.

[Lau18]   _____, *Dieudonné theory over semiperfect rings and perfectoid rings*, Compos. Math. **154** (2018), no. 9, 1974–2004.

[LB18]   A.-C. Le Bras, *Espaces de Banach-Colmez et faisceaux cohérents sur la courbe de Fargues-Fontaine*, Duke Math. J. **167** (2018), no. 18, 3455–3532.

[Liu13]   T. Liu, *The correspondence between Barsotti-Tate groups and Kisin modules when $p = 2$*, J. Théor. Nombres Bordeaux **25** (2013), no. 3, 661–676.

[Lou17]   J. Lourenço, *The Riemannian Hebbarkeitssätze for pseudorigid spaces*, arXiv:1711.06903.

[Mat80]   H. Matsumura, *Commutative algebra*, second ed., Mathematics Lecture Note Series, vol. 56, Benjamin/Cummings Publishing Co., Inc., Reading, Mass., 1980.

[Mih16]   T. Mihara, *On Tate's acyclicity and uniformity of Berkovich spectra and adic spectra*, Israel J. Math. **216** (2016), no. 1, 61–105.

[Mil76]   J. S. Milne, *Duality in the flat cohomology of a surface*, Ann. Sci. École Norm. Sup. (4) **9** (1976), no. 2, 171–201.

[Oor01]   F. Oort, *Newton polygon strata in the moduli space of abelian varieties*, Moduli of abelian varieties (Texel Island, 1999), Progr. Math., vol. 195, Birkhäuser, Basel, 2001, pp. 417–440.

[PR03]   G. Pappas and M. Rapoport, *Local models in the ramified case. I. The EL-case*, J. Algebraic Geom. **12** (2003), no. 1, 107–145.

[PR05]   ———, *Local models in the ramified case. II. Splitting models*, Duke Math. J. **127** (2005), no. 2, 193–250.

[PR08]   ———, *Twisted loop groups and their affine flag varieties*, Adv. Math. **219** (2008), no. 1, 118–198, With an appendix by T. Haines and Rapoport.

[PR09]   ———, *Local models in the ramified case. III. Unitary groups*, J. Inst. Math. Jussieu **8** (2009), no. 3, 507–564.

[PZ13]   G. Pappas and X. Zhu, *Local models of Shimura varieties and a conjecture of Kottwitz*, Invent. Math. **194** (2013), no. 1, 147–254.

[Rap90]   M. Rapoport, *On the bad reduction of Shimura varieties*, Automorphic forms, Shimura varieties, and L-functions, Vol. II (Ann Arbor, MI, 1988), Perspect. Math., vol. 11, Academic Press, Boston, MA, 1990, pp. 253–321.

[Rap05]   ———, *A guide to the reduction modulo p of Shimura varieties*, Astérisque (2005), no. 298, 271–318, Automorphic forms. I.

[Rap18]   ———, *Accessible and weakly accessible period domains*, Appendix to "On the p-adic cohomology of the Lubin-Tate tower" by P. Scholze, Annales de l'ÉNS (4) 51 (2018), no. 4, 811–863.

[RG71]   M. Raynaud and L. Gruson, *Critères de platitude et de projectivité. Techniques de "platification" d'un module*, Invent. Math. **13** (1971), 1–89.

[RR96]   M. Rapoport and M. Richartz, *On the classification and specialization of F-isocrystals with additional structure*, Compositio Math. **103** (1996), no. 2, 153–181.

[RV14]   M. Rapoport and E. Viehmann, *Towards a theory of local Shimura varieties*, Münster J. Math. **7** (2014), no. 1, 273–326.

[RZ96]   M. Rapoport and T. Zink, *Period spaces for p-divisible groups*, Annals of Mathematics Studies, vol. 141, Princeton University Press, Princeton, NJ, 1996.

[RZ10]   L. Ribes and P. Zalesskii, *Profinite groups*, second ed., Ergebnisse der Mathematik und ihrer Grenzgebiete. 3. Folge. A Series of Modern Surveys in Mathematics [Results in Mathematics and Related Areas. 3rd Series. A Series of Modern Surveys in Mathematics], vol. 40, Springer-Verlag, Berlin, 2010.

[RZ17]   M. Rapoport and T. Zink, *On the Drinfeld moduli problem of p-divisible groups*, Camb. J. Math. **5** (2017), no. 2, 229–279.

[Sch12]   P. Scholze, *Perfectoid spaces*, Publ. Math. Inst. Hautes Études Sci. **116** (2012), 245–313.

[Sch13]   ———, *p-adic Hodge theory for rigid-analytic varieties*, Forum Math. Pi **1** (2013), e1, 77.

[Sch15]   ———, *On torsion in the cohomology of locally symmetric varieties*, Ann. of Math. (2) **182** (2015), no. 3, 945–1066.

[Sch17]   ———, *Étale cohomology of diamonds*, arXiv:1709.07343.

[Sch18]   ———, *On the p-adic cohomology of the Lubin-Tate tower*, Ann. Sci. Éc. Norm. Supér. (4) **51** (2018), no. 4, 811–863, With an appendix by Michael Rapoport.

[SR72]   N. Saavedra Rivano, *Catégories Tannakiennes*, Lecture Notes in Mathematics, Vol. 265, Springer-Verlag, Berlin-New York, 1972.

[Sta]   The Stacks Project Authors, *Stacks Project*, http://stacks.math.columbia.edu.

[Ste65]   Robert Steinberg, *Regular elements of semisimple algebraic groups*, Inst. Hautes Études Sci. Publ. Math. (1965), no. 25, 49–80.

[SW13]   P. Scholze and J. Weinstein, *Moduli of p-divisible groups*, Camb. J.

Math. **1** (2013), no. 2, 145–237.

[Tat67]   J. Tate, *p-divisible groups*, Proc. Conf. Local Fields (Driebergen, 1966), Springer, Berlin, 1967, pp. 158–183.

[Tsu99]   T. Tsuji, *p-adic étale cohomology and crystalline cohomology in the semi-stable reduction case*, Invent. Math. **137** (1999), no. 2, 233–411.

[Var04]   Y. Varshavsky, *Moduli spaces of principal F-bundles*, Selecta Math. (N.S.) **10** (2004), no. 1, 131–166.

[Wed14]   T. Wedhorn, *Introduction to Adic Spaces*, http://www3.mathematik.tu-darmstadt.de/hp/algebra/wedhorn-torsten/lehre.html, 2014.

[Wei17]   J. Weinstein, $\mathrm{Gal}(\overline{\mathbf{Q}}_p/\mathbf{Q}_p)$ *as a geometric fundamental group*, Int. Math. Res. Not. IMRN (2017), no. 10, 2964–2997.

[Zhu17a]   X. Zhu, *Affine Grassmannians and the geometric Satake in mixed characteristic*, Ann. of Math. (2) **185** (2017), no. 2, 403–492.

[Zhu17b]   _____, *An introduction to affine Grassmannians and the geometric Satake equivalence*, Geometry of moduli spaces and representation theory, IAS/Park City Math. Ser., vol. 24, Amer. Math. Soc., Providence, RI, 2017, pp. 59–154.

Index